IET TRANSPORTATION SERIES 29

Mobility as a Service

Other related titles:

Volume 1	**Clean Mobility and Intelligent Transport Systems** M. Fiorini and J-C. Lin (Editors)
Volume 2	**Energy Systems for Electric and Hybrid Vehicles** K.T. Chau (Editor)
Volume 5	**Sliding Mode Control of Vehicle Dynamics** A. Ferrara (Editor)
Volume 6	**Low Carbon Mobility for Future Cities: Principles and applications** H. Dia (Editor)
Volume 7	**Evaluation of Intelligent Road Transportation Systems: Methods and results** M. Lu (Editor)
Volume 8	**Road Pricing: Technologies, economics and acceptability** J. Walker (Editor)
Volume 9	**Autonomous Decentralized Systems and Their Applications in Transport and Infrastructure** K. Mori (Editor)
Volume 11	**Navigation and Control of Autonomous Marine Vehicles** S. Sharma and B. Subudhi (Editors)
Volume 12	**EMC and Functional Safety of Automotive Electronics** K. Borgeest
Volume 15	**Cybersecurity in Transport Systems** M. Hawley
Volume 16	**ICT for Electric Vehicle Integration with the Smart Grid** N. Kishor and J. Fraile-Ardanuy (Editors)
Volume 17	**Smart Sensing for Traffic Monitoring** Nobuyuki Ozaki (Editor)
Volume 18	**Collection and Delivery of Traffic and Travel Information** P. Burton and A. Stevens (Editors)
Volume 20	**Shared Mobility and Automated Vehicles: Responding to socio-technical changes and pandemics** Ata Khan and Susan Shaheen
Volume 23	**Behavioural Modelling and Simulation of Bicycle Traffic** L. Huang
Volume 24	**Driving Simulators for the Evaluation of Human–Machine Interfaces in Assisted and Automated Vehicles** T. Ito and T. Hirose (Editors)
Volume 25	**Cooperative Intelligent Transport Systems: Towards high-level automated driving** M. Lu (Editor)
Volume 26	**Traffic Information and Control** Ruimin Li and Zhengbing He (Editors)
Volume 30	**ICT Solutions and Digitalisation in Ports and Shipping** M. Fiorini and N. Gupta
Volume 32	**Cable Based and Wireless Charging Systems for Electric Vehicles: Technology and control, management and grid integration** R. Singh, S. Padmanaban, S. Dwivedi, M. Molinas and F. Blaabjerg (Editors)
Volume 34	**ITS for Freight Logistics** H. Kawashima (Editor)
Volume 36	**Vehicular Ad Hoc Networks and Emerging Technologies for Road Vehicle Automation** A.K. Tyagi and S. Malik
Volume 38	**The Electric Car** M.H. Westbrook
Volume 39	**Electric Vehicle Components and Charging Technologies Design, Modeling, Simulation and Control** S. Singh, S. Gairola and S. Dwivedi (Editors)
Volume 43	**AI Techniques in EV Motor and Inverter Fault Detection and Diagnosis** Y. Hu, X. Zhang and W. Lang
Volume 45	**Propulsion Systems for Hybrid Vehicles** J. Miller
Volume 79	**Vehicle-to-Grid: Linking electric vehicles to the smart grid** J. Lu and J. Hossain (Editors)

Mobility as a Service

Its development, deployment, and future

Beate Kubitz and James Gleave

The Institution of Engineering and Technology

Published by The Institution of Engineering and Technology, London, United Kingdom

The Institution of Engineering and Technology is registered as a Charity in England & Wales (no. 211014) and Scotland (no. SC038698).

First published 2024

The Institution of Engineering and Technology
Futures Place
Kings Way, Stevenage
Hertfordshire, SG1 2UA, United Kingdom

www.theiet.org

British Library Cataloguing in Publication Data
A catalogue record for this product is available from the British Library

ISBN 978-1-83953-143-9 (hardback)
ISBN 978-1-83953-144-6 (PDF)

Typeset in India by MPS Limited

Cover Image: Oscar Wong/Moment via Getty Images

Contents

About the authors

Beate Kubitz is a transport consultant and the director of Beate Kubitz Associates Ltd, UK. Beate has extensive experience in future mobility, including the role it plays in carbon reduction through research, innovation, and policy development. She received a fellowship from the Foundation for Integrated Transport in 2020 and is a visiting lecturer on the MSc Transport Planning and a research fellow in the Smart Mobility Unit at the University of Hertfordshire.

James Gleave is a transport consultant and the founder and director of Mobility Lab, UK. James has delivered and assisted in the delivery of many transport and infrastructure strategies and policies. He has supported the delivery of numerous innovative technology, autonomous vehicles, smart infrastructure, and mobility-as-a service projects. He is also a board member of the Transport Planning Society.

Acknowledgements

The authors would like to thank the following people for their expertise, time, and words of support in writing this book.

Liliane Abdul-Reda
Richard Armitage
John Austin
Giles Bailey
Chas Ball
Ross Bassnet
Guillermo Campoamor
David Carnero
Beth Cocker
Jeremy Dalton
James Datson
Paula Dauncey
Liz Davidson
Crissy Ditmore
Antony De Heveningham
Stephen Joseph OBE
Piia Karjalainen
Keith Kelly
Jenny Milne
Lukas Neckermann
John Paddington
Giles Perkins
Julian Scriven
Christopher Slade
Nick Tasker
Sandra Witzel
Lucy Yu

We authors would also like to thank those closest to us who have been part of our writing experience, both the good and the bad, and whose support has made this possible.

Karen
Chipps Chippendale
and
Belle, Jasper, and Jess

Chapter 1
Introduction

The world of transport has been subject to a significant degree of technological change over the last 20 years. Whilst the development and increasing adoption of new technologies has driven this change [1], wider demographic changes, advances in data analytics, changing attitudes to technology, and the presence of ongoing significant grand challenges such as climate change have created fertile ground for the deployment of new transport technologies.

Within this context, Mobility-as-a-Service (or MaaS) has emerged as one of the most talked about concepts in the world of transport, especially among practitioners. We will later discuss how to define MaaS (there have been several attempts) and various approaches to the area from commercial and public service perspectives. But for now, it is sufficient to consider MaaS as shifting transport from primarily an ownership method of consumption (you own a car, therefore you must take more car trips to get the best value) towards a service method of consumption (you only pay for transport when you use it) enabled by digital platforms.

Such a concept is not new in transport – arguably the use of public transport and various forms of hire and rental have established the service method of consumption in transport for many decades (if not over a century) – but MaaS is a radically different take on this. It has presented policymakers, entrepreneurs, local government, users, and practitioners with several significant challenges in defining the concept, testing it, and understanding its impact.

The purpose of this book is to be a 'stocktake' of the development of MaaS. This is a field of transport that is emerging and still evolving, though at this point several key lessons can already be learned from the work undertaken to date.

Reference

[1] Wong Y, Hensher D, and Mulley C., Institute of Transport and Logistics Studies. Emerging transport technologies and the modal efficiency framework: A case for mobility as a service (MaaS) [Internet]. 2018 [cited 2023 August 11]. Available from: https://ses.library.usyd.edu.au/bitstream/handle/2123/19100/ITLS-WP-18-04.pdf

Chapter 2
The history of Mobility-as-a-Service

The history of Mobility-as-a-Service is scattered with theoretical papers produced by various national and intergovernmental organisations and by universities that conceptualise and theorise about MaaS or its features, that have not been practically implemented. At the other end of the scale, technology start-ups and city authorities, regions, and even community organisations lay claim to practical iterations of the concept. This maelstrom of 'top-down' conceptualising and 'bottom-up' (or, in some cases, 'investment capital up') trialling, piloting, and development make the history both confusing and colourful.

Several of the underpinning principles behind Mobility-as-a-Service predate the concept by decades (although by how long depends on the definition of Mobility-as-a-Service). Core concepts such as the ability to pay for multiple services using a single payment mechanism, and journey planning across multiple modes of transport (including both public and shared transport modes) have been standard practice within the transport sector for around 30 years. Shared transport has a similar long-standing history. Bike share can trace its antecedents as far back as the 1960s, whilst car sharing had early instances in the 1980s. Optimising car use has historical roots in hitch-hiking, but more formalised systems were launched around the start of the 21st century (see Chapter 12 on shared transport).

Within public transport, the concept of paying a single fare or capped fare for travel across multiple modes of public transport has become an established practice across many major cities. For example, in 1984 the city of London introduced the One Day Travelcard for off-peak travel, allowing unlimited travel on buses, the London Underground, and from 1989 railway services and the Docklands Light Railway. For this to be introduced it was necessary to simplify the delivery of combined payment mechanisms, and also streamline the previous 'point-to-point' fare system to a zonal fare system. This work on fares formed the basis of the successor to the Travelcard, the Oystercard. Such has been the success of Oyster that by 2023, the Mayor of London was able to propose the abolition of the One Day Travelcard add-on for National Rail tickets, though at time of writing this plan has been rescinded.

Similar zonal fare systems have become a common practice among cities across the world. The public transport network of Île-de-France Mobilités (which covers Paris) is divided into five fare zones, although single tickets are still sold throughout the network. Public Transport Victoria, meanwhile, operates two zonal fare systems – the Metropolitan Zones (3) covering the city of Melbourne, whilst the Regional Fare Zones (13) cover the rest of Victoria.

Integrated journey planning has been feasible, if somewhat tricky, since the introduction of the paper timetable. Regional and city bus and train timetables have made it possible to plan journeys over multiple modes, but the introduction of information and communication technologies to ticketing systems enabled integrated journey planning to become simplified and understandable to regular passengers.

In 2000 in the United Kingdom, Traveline was established, becoming the first national integrated public transport journey planner in the United Kingdom. A partnership between local transport authorities and public transport operators, this service was originally offered via telephone and with a limited online presence. Now, the dataset that underpins Traveline, the Traveline National Dataset (TNDS), is widely used by modern journey planners including Google Maps and Apple Maps.

In parallel with developments in the public transport sector, in 2007 José Viegas precipitated what became known as 'The Lisbon Study' by writing a paper on the potential of shared taxis. Several years after his initial observations that many taxis and private hire vehicles were only carrying one passenger – whilst often travelling along the same routes between similar destinations – International Transport Forum (ITF) researchers were able to collect sufficient data and build a highly detailed model of mobility.

They focused on mobility data and modeling the city of Lisbon, Portugal. They discovered that if they removed all private car trips and replaced them with trips in shared vehicles they could reduce the number of vehicles needed to fulfil all the journey requirements of the time. They tested different configurations including self-driving shared vehicles, electric vehicles, 6-seater Shared Taxis and 8- and 16-seater 'Taxi-Buses'. The prima facie results were stunning: in the shared mobility city, only 10% or less of the number of vehicles were needed to get citizens where they wanted when they wanted. Congestion disappeared, CO_2 emissions fell by one-third and on-street parking space was no longer needed. Whilst only some versions showed these extreme results, the long-lasting impact of the paper was to create a new ideal. It showed that private vehicles could be replaced by shared transport with positive benefits and cities have recognised it among the possible versions of better urban living.

Meanwhile, various practical attempts to develop services that could be seen as proto-MaaS were underway. One early application was developed in Hannover in Germany, combining public transport and car sharing in 2004 (with limited success), whilst the later trial, by the city of Vienna, which attempted to combine travel planning and payments in 2012, gained more traction.

The modern concept of Mobility-as-a-Service was first defined in an academic sense by Heikkilä [1]. Examining industry transformations in four industries (telecommunications, energy, airlines, and railways), the author concluded that the passenger transport industry is starting to experience the same disruptive trends that affected these industries. These key trends were identified as:

- **Social change**, notably the attitudes and consumption patterns of millennials that were shifting away from ownership-based to more the consumption of services. With particular reference to changing concepts of freedom and higher access to mobile phones.
- **Urbanisation and globalisation**, with cities forecast to become larger and host a higher percentage of the global population over the next 30 years (including Helsinki), and an increasing geographical scale of economic, political, social, and cultural interactions.
- **Servicing and sharing**, with the emergence of the sharing economy at the time, with people willing to forsake maintenance, storage, and responsibility for items. Whilst the technological element was important, and the emergence of sharing platforms such as TaskRabbit was noted, the cultural and environmental aspects of the change were also noted.
- **Scarcity of resources** in a multi-faceted fashion. The study identified three particularly important aspects of this. The first was the increasing scarcity of public sector resources due to funding cuts. The second was the impact of ageing and the changing needs of older people, particularly as older people become a greater proportion of the population. The third was the environmental angle, with environmental resources being increasingly exploited and the need to reduce environmental impacts of transport being especially acute.
- **Technological advances**, of which the digital transformation of society and the economy enabled by the internet was most notable, as well as the ubiquity and usability of smartphones. Meanwhile, rapid advances in transport technology such as intelligent transport systems and the onset of autonomous vehicles.
- **Paradigm change** in the transport sector, reflecting a desire to define transport issues differently, and set a new vision for how transport systems should be planned and built, signifying a shift away from the reductionist approaches of previous transport planning.

This challenge, particularly of digitisation and the provision of the offer as a service, was one that the public transport operations in Helsinki were not able to develop on their own, at the time.

The results of this study were to describe '*a way to reorganize the passenger transport sector so that it would promote the concept of "Mobility as a Service" (MaaS), that is, convenient provision of a versatility of attractive mobility services. This study provides a suggestion of a transformed mobility sector'*. From the user's perspective, this was a simple concept (Figure 2.1).

Since this point, interest in Mobility-as-a-Service as a concept (definitions of MaaS are covered further in this introduction) and a service offering has increased significantly and become a common part of the transport lexicon.

In 2015, the first independent organisation dedicated to the development and promotion of MaaS was established – the MaaS Alliance. A public-private partnership based in Brussels, Belgium, the purpose of the group has been to establish the common frameworks needed for MaaS to scale. This has included thought-

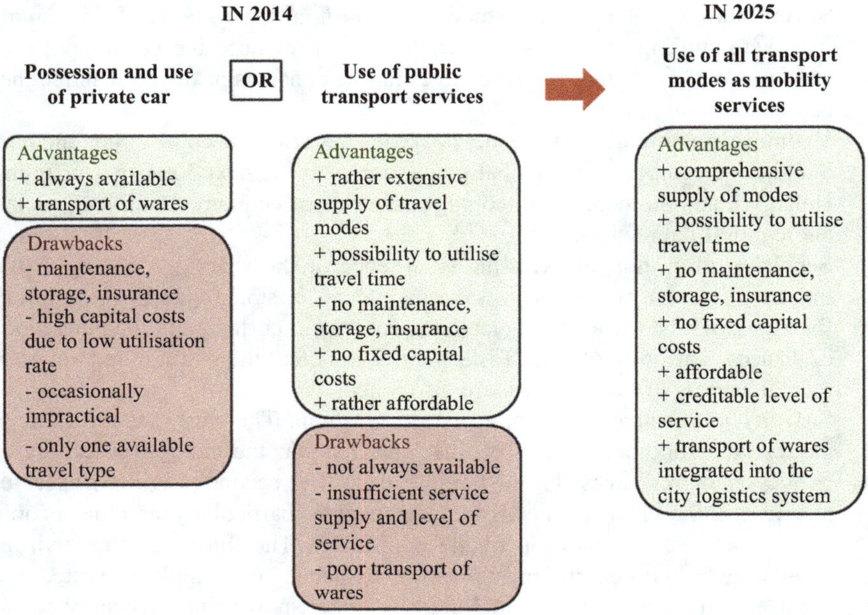

Figure 2.1 The transformation of mobility to Mobility-as-a-Service [1]

leadership on matters such as establishing a market, passenger rights, data, and interoperability.

In this book, we will look at the development of MaaS, in its widest sense, through the lenses of the business models of MaaS, the geographies of MaaS and the technologies of MaaS. We will examine how regulatory frameworks have shaped MaaS throughout its development. Over a decade since the first MaaS experiments, we will look at which models are maturing, and where they are most mature.

The business models of MaaS include some of the social elements of MaaS, and broader questions such as 'who benefits from MaaS' and 'who pays for MaaS'. They also cover the complexity of the relationships MaaS creates. The interplay between different types of transport, operators and the platform generates questions that are both contractual and affect the utility of MaaS for the passenger: for instance, where the guarantees in the system lie for both travelers and mobility providers and which element of the system covers each. We will explore these questions including 'how should the platform itself be funded?' and 'how can it derive revenue?' These questions will be explored in the Business Models chapter and to some extent in the Regulation chapter.

In addition, we will look at the technology of MaaS. The developments that have facilitated MaaS are largely connected to smartphone technology and include:

- Connectivity: The Internet of Things (IOT) has achieved vast leaps in the last decade from simple e-scooters to V2V and V2X communication.

- Positioning: a wide variety of means to locate connected objects have become highly accessible – from GPS, mobile data, narrow band IOT and other sources including small localised networks, WiFi, and mobile mast triangulation – enabling everything to be located, from mobile phones to bikes and increasingly in remote areas.
- Mapping: online mapping has become increasingly detailed (although with notable gaps particularly in the high resolution required for informative active travel maps).
- Smart payments and ticketing: NFC and the technology of fare capping have become widespread, integrations with RFID and wearables are all possible and management of data required for account-based ticketing has become increasingly able.

Technology providers and transport operators have developed suites of services that support the provision of MaaS, from APIs to plug-ins and Software Development Kits and Software as a Service. Business models are, in some cases, derived from these services.

MaaS platforms are often defined by geography. Whilst the potential for travellers to travel anywhere using seamless links between transport modes is clearly attractive, the reality is that the geographies of MaaS are quite often defined by the operators or authorities which build the platform – and their business models. There is increasing geographical coverage by MaaS platforms but it is neither ubiquitous nor homogenous. For instance, MaaS platforms built by long-distance rail providers cover much wider areas than city-based MaaS systems. These latter generally cover dense urban areas served by city authorities. Sometimes a common platform provider will offer to use MaaS for multiple areas (even if they are not otherwise connected). Equally where MaaS is solving a geographic problem – low-density neighbourhoods, rural areas, or connecting an area severed from the main network – the conditions on the ground will shape the form and content of the system.

There are national and regional variations. In some cities and countries (for instance Vienna and Japan) a version of MaaS is fundamental to the way transport is organised, with passengers expecting seamless multi-modal travel and transport authorities facilitating it. In others (notably the United Kingdom and the United States) it could be argued that MaaS has entered what Gartner refers to as the 'trough of disillusionment'. This is where, after significant hype of a concept, the reality of the idea sets in, and where organisations realise the value, and issues, of MaaS as a commercial proposition and service offering. This does not mean that the idea has been found to be a poor idea, but it means that the practical realities of delivering on the promise are being understood, ahead of future steady growth and adoption.

This book will summarise the current state of Mobility-as-a-Service and the constituent service offerings and technologies involved. But within the wider context, MaaS is an idea in transition. Many of the initial trials of the idea have been undertaken, and some of the initial commercial ventures into MaaS have also become established, and the relative strengths and weaknesses of the concept have become clearer.

Whilst many of the technological issues are solved, delivery now depends more on the social and regulatory context than it does on the technology.

Reference

[1] Heikkilä S. Mobility as a Service – A Proposal for Action for the Public Administration, Case Helsinki. Aaltofi [Internet]. 2014. Available from: https://aaltodoc.aalto.fi/handle/123456789/13133

Chapter 3
Defining Mobility-as-a-Service

3.1 Introduction

Over the years, there have been several attempts to make a definition of what constitutes MaaS, taken from a variety of different perspectives. Many definitions have sought to provide what is, in effect, a 1990 'elevator pitch' or a one-paragraph sound bite. Others have been more technical, driven by technology or institutional factors. All with a different purpose and different perspective on the idea.

The intention of this chapter is not to provide another definition, nor to be precise as to what such a definition should consist of. It is to introduce the reader to the variety of definitions of MaaS that have been put forward by numerous sources over the years. In practice, the definition of MaaS is still subject to some discussion, even if many practitioners could identify many of the common components that most could identify as being part of 'MaaS'.

3.2 Conceptual definitions

Digitisation and platformisation have been a constant theme for most of the last two decades. Our lives have become increasingly centred on smartphones and apps that manage our communications, relationships, our access to news, our finances, and our time.

Whilst transport has seemingly come later to the party, it is notable that precursors to the multi-modal digital transport platforms were emerging quietly well over a decade ago. Bike-share company Nextbike has been in existence since 2004, Uber was founded in 2009, and multi-modal journey planner Citymapper in 2011.

Mobility-as-a-Service (MaaS) has, to a certain extent, become synonymous with the digitalisation of journey planning – a concept with over 20 years of history. In reality, the ability to book and pay for the planned transport across multiple modes is more recent.

Numerous attempts have been made over the years to define what MaaS is more precisely. These multiple attempts have generated multiple versions of 'the definition of MaaS', making a definitive version somewhat difficult to pin down. For instance, Deloitte defines MaaS as:

> At its core, MaaS relies on a digital platform that integrates end-to-end trip planning, booking, electronic ticketing, and payment services across all

> modes of transportation, public or private. It's a marked departure from where most cities are today, and from how mobility has been delivered until now [1].

One of the earliest definitions of MaaS was offered by Sampo Hietenan, who describes MaaS as a mobility distribution model that delivers users' transport needs through a single interface of a service provider [2]. The ITF adopted a similar definition for MaaS, defining it as:

> based on the idea of accessing via a single medium – a smartphone – a variety of mobility services, including public transport (PT) and shared mobility services. The concept underlying MaaS is that, in the near future, urban mobility has to become fully multimodal in order to reduce the external costs of urban mobility [3].

Meanwhile, the UK's House of Commons Transport Committee stated, in 2018, that:

> MaaS is the term for the digital platforms (often smartphone apps) through which people can access a range of public, shared and private transport, using a system that integrates the planning, booking and paying for travel [4].

When undertaking a critical review of MaaS definitions, it is challenging to define specifically what MaaS is, as whilst there may be some degree of consensus as to the core components of MaaS, their relative importance is subject to some discussion. Kamargianni *et al.* [5] summarises that at the core of each MaaS solution are three key components that, in conjunction, provide users with seamless multi-modal journeys:

- *Ticket & Payment integration*: when one smart card or ticket can be used to access all the modes taking part in the service and one account is charged for the use of those services.
- *Mobility package*: when customers can pre-pay for a specific amount (in time or distance) for a combination of mobility services.
- *ICT integration*: when there is a single application or online interface that can be used to access information about the modes.

In addition, there have been numerous academic studies that seek to define MaaS. Based on an extensive review of the literature, nine core characteristics of MaaS can be defined (Table 3.1).

However, as this book will explore, MaaS is an emerging term with multiple definitions that largely fall within a catch-all set of new digital, demand-driven, transport, innovative services, and multi-modal integrations. Examples that fall into this set include smart transport solutions aimed at driving down car use, innovative (and simplified) transport purchasing models, new services that are multi-modal at their outset, and those that start with a single mode and add additional modes to their offer.

Early MaaS developments and forerunners are guides to the aspirations for MaaS, and possibly its definition.

Table 3.1 Core characteristics of MaaS [6]

Core characteristics of MaaS	Description
Integration of transport modes	A goal of MaaS schemes is to encourage the use of public transport services, by bringing together multi-modal transportation and allowing the users to choose and facilitating them in their intermodal trips. The following transport modes may be included: public transport, taxi, car-sharing, ride-sharing, bike-sharing, car rental, and on-demand bus services. Envisioning a service beyond the urban boundaries, it will also embrace long-distance buses and trains, flights, and ferries.
Tariff option	MaaS platform offers users two types of tariffs in accessing its mobility services: 'mobility package' and 'pay-as-you-go'. The package offers bundles of various transport modes and includes a certain amount of km/minutes/points that can be utilised in exchange for a monthly payment. The pay-as-you-go charges users according to the effective use of the service.
One platform	MaaS relies on a digital platform (mobile app or web page) through which the end-users can access all the necessary services for their trips: trip planning, booking, ticketing, payment, and real-time information. Users might also access other useful services, such as weather forecasting, synchronisation with personal activity calendars, travel history reports, invoicing, and feedback.
Multiple actors	MaaS ecosystem is built on interactions between different groups of actors through a digital platform: demanders of mobility (e.g., private customer or business customer), a supplier of transport services (e.g., public or private), and platform owners (e.g., third party, PT provider, authority). Other actors can also cooperate to enable the functioning of the service and improve its efficiency: local authorities, payment clearing, telecommunication, and data management companies.
Use of technologies	Different technologies are combined to enable MaaS: devices, such as mobile computers and smartphones; a reliable mobile internet network (WiFi, 3G, 4G, LTE); GPS; e-ticketing and e-payment system; database management system and integrated infrastructure of technologies (i.e., IoT).
Demand orientation	MaaS is a user-centric paradigm. It seeks to offer a transport solution that is best from the customer's perspective to be made via multi-modal trip planning feature and inclusion of demand-responsive services, such as taxi.
Registration requirement	The end-user is required to join the platform to access available services. An account can be valid for a single individual or, in certain cases, an entire household. The subscription not only facilitates the use of the services but also enables service personalisation.
Personalisation	Personalisation ensures end users' requirements and expectations are met more effectively and efficiently by considering the uniqueness of each customer. The system provides the end-user with specific recommendations and tailor-made solutions based on her/his profile, expressed preferences, and past behaviours (e.g., travel history). Additionally, they may connect their social network profiles with their MaaS account.

(Continues)

Table 3.1 (*Continued*)

Core characteristics of MaaS	Description
Customisation	Customisation enables end users to modify the offered service option according to their preferences. This can increase MaaS' attractiveness among travellers and its customers' satisfaction and loyalty. They may freely compose a specified chained trip or build their mobility package with a different volume of usage of certain transport modes to better achieve their preferred travel experiences.

In Vienna, the Smile project was a precursor of MaaS. A 3-year project to test the use of an integrated mobility service was launched in 2012. A thousand people were selected to test the service which offered a single account for all means of transport, access to multi-modal information, and the ability to book, pay, and use the various mobility offers via smartphone. However, the user had separate contracts with each operator and separate tickets for each element of the journey (although they appeared integrated in the application). The app provided a summary of usage – although it did not provide for a single invoice and payment system [7].

Besides this more comprehensive offer, several app-based journey-planning services were launched in the early 2010s. Among them Transit, a Canadian journey-planning app that later incorporated mobile ticketing payments to become a MaaS platform [8], and Skedgo which moved from its launch as a free trip-planning app (established in 2011) to a customised travel service for companies and local areas by 2015. Skedgo too now has a full MaaS offer [9].

The concept of Mobility-as-a-Service as a direct replacement for a private car saw its first practical iteration in 2013. A Swedish visionary and pioneer, Hans Arby, launched a pilot project MaaS in Gothenburg for 6 months, which was called Ubigo. Participants put at least one car in storage whilst they took part in the trial – and many found that they were happy with the Ubigo model of car replacement with multiple transport options. The main takeaway from the pilot was that people adapted their mobility behaviour, were happy with their new travel options, and did not regret giving up their private car [10].

Similar – although area-wide – modelling took place in 2015 when the Lisbon Study calculated that, if all private cars were to be removed from the city's streets, the transport needs of the population could be satisfied by a limited number of shared vehicles [11]. This demonstrated that in theory, all the trip requests could be met by a shared fleet consisting of less than 10% of the vehicles on the city's roads – providing it was dispatched by a single 'operator' and private vehicles were banned. Whilst the parameters were extreme, the modelling provided a glimpse of a different possible future, with fewer vehicles, less congestion, and better air quality.

The modelling tools used in the Lisbon Study fused transport data with demand data and created optimal journey patterns from which the necessary fleets and infrastructure could be calculated and iterated for different KPIs from

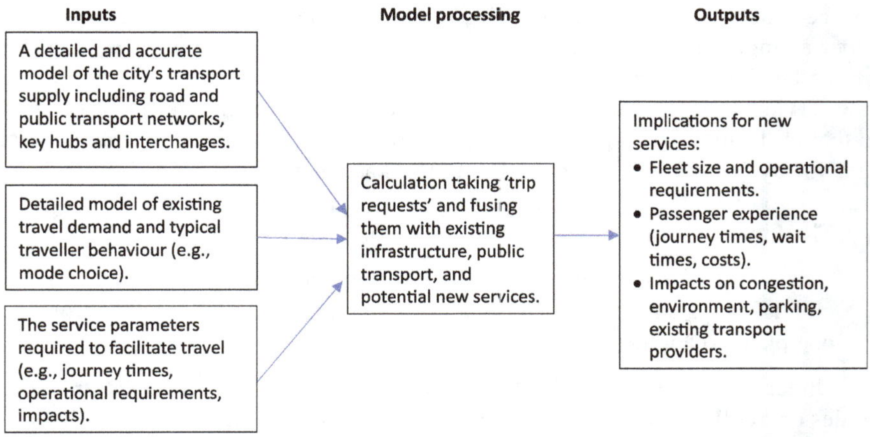

Figure 3.1 The inputs into the Lisbon study [11]

operational efficiency to social impacts. Effectively the requirements in terms of rules and vehicles could be calculated for different possible traffic scenarios (Figure 3.1).

It is important to note that only the most extreme versions of the Lisbon Model experiment which included banning private vehicles, were beneficial. When the ITF's model was run to achieve a 50% replacement of private vehicles overall road traffic doubled [12].

This highlights interesting regulatory issues around both MaaS and the environment it is embedded in. The requirement for governance to both facilitate MaaS and regulate it is covered in Chapter 17.

There are some particularly radical examples of regulation enabling MaaS, and perhaps the most comprehensive example is the case of Finland. However, the radical Finnish transport law was not spontaneous. The groundwork for Finnish transport and digitalisation regulation was laid by companies trialling digital innovations. In Turku, Finland, the public transport company Föli, began to base its ticketing architecture on personal identification from 2014, enabling tickets to be sold to individuals and linked to their accounts. At the same time, it opened its APIs to third parties who could then sell its tickets (although also signing an agreement is part of the process). In 2016 a local start-up, Tuup, created an app to provide information about transport options via a route planner that compared routes on foot, by bike, by public transport, by car or by taxi plus information on real-time public transport departures. It enabled the purchase of public transport tickets if that mode was chosen.

The Finnish government spotted the potential and enacted a new, radical, transport law, which itself has become part of the development of MaaS in Finland. The 2017 Finnish national transport law, the Finnish Transport Code, introduced to digitalise, automate and robotise transport became a key enabler for services such as Tuup. The Finnish Transport Code brings all transport modes under a single law, eliminating all specific laws referring to means of transportation. The key element supporting MaaS is the requirement that public and private service providers provide an open Application Programming Interface (API) so that all forms of transport can be integrated into one seamless travel chain, all paid for by one mobile system, essentially that all transport modes can be integrated into one holistic system.

The Finnish Transport Minister at the time, Anne Berner [13]:

> This will create a possibility for all and everyone to attach themselves to one ticket system which includes all different opportunities to move from one place to another.

Under this vision, future infrastructure will connect different transport modes and collect data, which can be applied to provide further possibilities for all kinds of other services, helping companies in the sector to come up with better solutions for users.

Whilst the new laws did not necessarily smooth an uninterrupted path for MaaS in Finland – for instance, Tuup ended its app in 2018 – the abundance of pilots and the surrounding legal framework provided the foundations for many future developments that have furthered and further defined the concept of MaaS.

The advent of the Finnish Transport code enabled another MaaS visionary, Sampo Heitanen, to launch his company, MaaS Global, and trial their first product in Finland: Whim. Heitanen had been conceptualising Mobility-as-a-Service since 2006 and initially presented his vision to the Finnish Science Centre Heureka in 2014. Two years later, the first trip using the Whim app was made [14].

Although these Nordic projects can claim to have crystallised a version of MaaS in which multi-modal transport apps and platforms are specifically targeted (and priced) to replace privately owned vehicles, other projects and business models emerged that map many of the features and benefits of these initial versions.

Other visions of MaaS emerged in different regions and presented alternative conceptual frameworks. This is particularly true of the type of MaaS that intersects with the automotive industry. Rapid development and adoption of autonomous driving was assumed to be imminent in the late 2010s and the use of MaaS to plan journeys, summon vehicles, and pay for trips was routinely envisioned by industry. Huge investments were made to bring this version of MaaS into being. In 2015 Uber set up a unit devoted to developing the self-driving vehicle and invested over $1 billion in the technology over 5 years [15]. By 2020, it became apparent that autonomous vehicle development will take longer and cost more than anticipated in those early days. One consequence was that Uber sold off its self-driving unit in 2020. Although this vision has not entirely faded, there is an acceptance that its application is likely to be more limited, and potentially much more focused on shared or even mass transit vehicles. Those trials in existence by 2023 conform either to shared or public transport or 'robotaxi' models.

In between these two extremes, other concepts and ideas of MaaS have emerged. Their core feature has been a journey planner and ticketing or payment for travel on a digital platform (mainly but not exclusively served via app). For the purposes of this book, we will consider those services that include more than one mode within the broader envelope of MaaS.

Teasing out the themes that enable the development of MaaS shows that the interplay between governance, geography, technology, and other factors leads to specific versions of MaaS.

The Finnish Transport Act, which mandated open transport data was a major impetus in Finland. Sparse populations spread across rural geographies in Scotland have motivated the Scottish government to fund pilots of rural MaaS. Technological advances that enable the ingestion of APIs, detailed mapping, and rapid calculation have contributed to the ambition of MaaS. Meanwhile, the details in the regulations governing transport in some parts of the world have shaped MaaS towards private or public transport operators. We will look at some of these relationships in subsequent chapters.

3.3 Technical definitions

Regardless of the conceptual definition used, a common element of MaaS definitions is the requirement for a digital platform to play the role of integrator, aggregator, journey planner, and payment mechanism (or any mixture of these). Consequently, a significant amount of work has been undertaken on defining MaaS in a technological sense.

This does not mean that having a digital system is an essential prerequisite of a solution being considered as MaaS. But with the convergence of digital connectivity and transport services, as well as a rapidly developing data analytics capability across much of the transport industry, many organisations consider MaaS to be a 'digital first' solution.

An early technical framework or architecture suggested by the UK's Transport Systems Catapult in 2016 set out a layer-based structure, with a central API gateway to ingest APIs from public and shared transport.

This work reflects thinking that MaaS should not be developed as a specific technological architecture in a closed technological system, but instead based on the development of an ecosystem of services and service offerings, with interconnectivity facilitated through existing open technological protocols. MaaS can also be defined as a business ecosystem (Figure 3.2). Of particular importance within this ecosystem are several actors, most notably:

1. transport operators (this term also includes mobility service providers, i.e., parking space providers);
2. data providers;
3. technology and platform providers (technical back-end providers);
4. ICT infrastructure;
5. insurance companies;
6. regulatory organisations;
7. universities and research institutions.

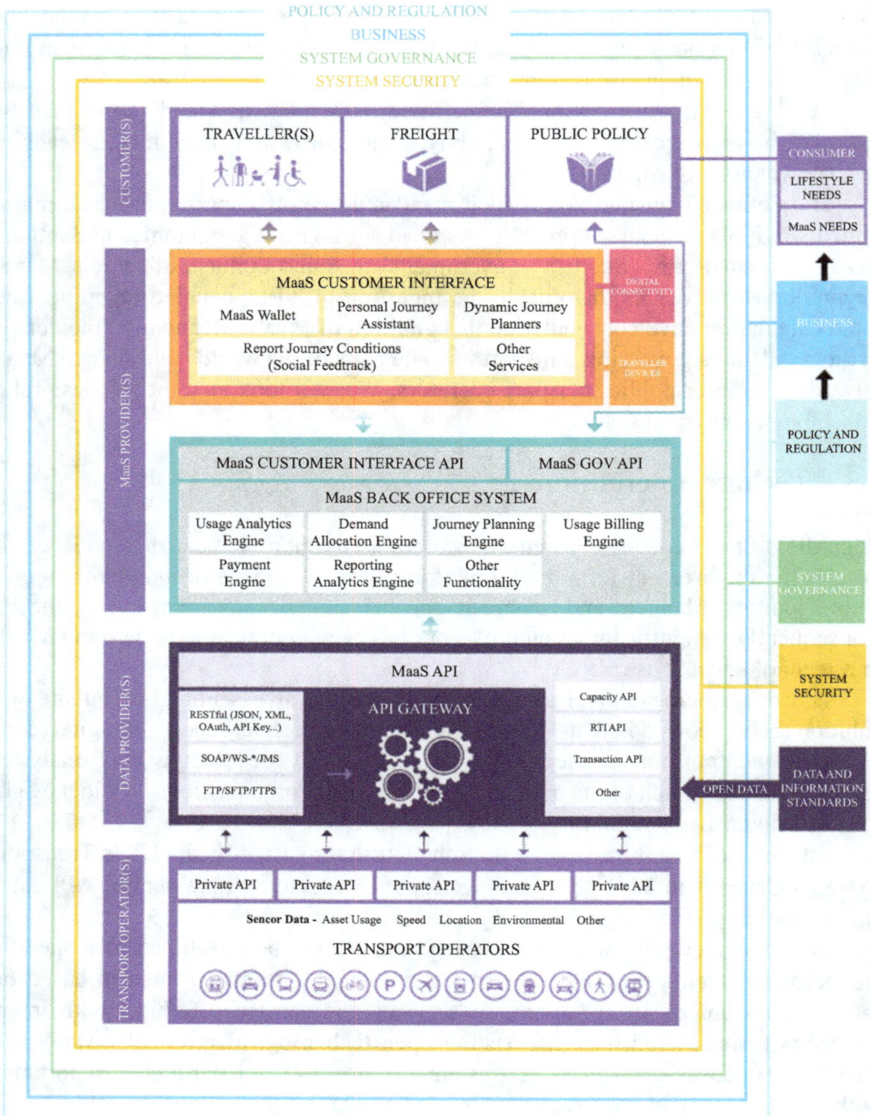

Figure 3.2 A Mobility-as-a-Service reference architecture [16]

In its work on interoperability for data and data models in MaaS, the MaaS Alliance sought to simplify the technological definition of MaaS by seeking to standardise the way by which information and data are exchanged between different stakeholders within the MaaS ecosystem (Figure 3.3). The conclusion from their work, however, was that [18]:

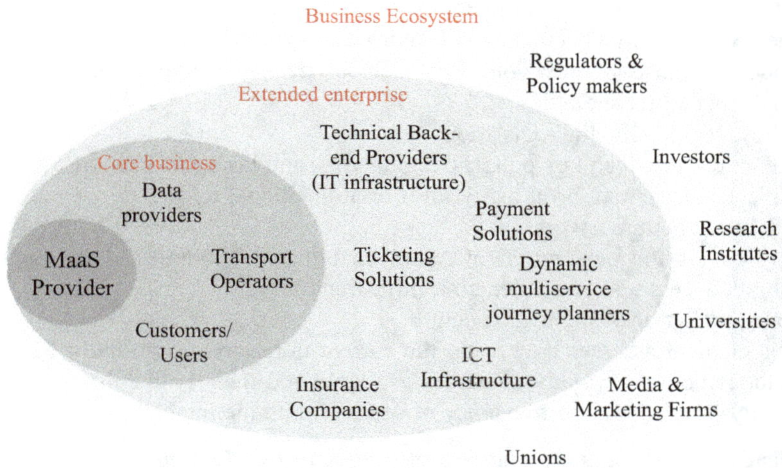

Figure 3.3 The business ecosystem of MaaS [17]

- There is no need to create yet another standard for MaaS but there is a crucial need for better, more fruitful conversation between all standardisation stakeholders to build consensus together, realigning models from a relationship point of view.
- Pivotal Points of Interoperability, Consensus Framework, and Minimum Interoperability Mechanism (MIM) are approaches that help to build interoperability.
- Mapping data models at the concept level is difficult. It requires a lot of time and detailed knowledge about each of them. This can only be achieved as a continuous and collective effort.

However, this is not universally accepted. For instance, the French government has suggested that standardisation for MaaS would be more effective, and is investigating a unified MaaS API.

3.4 MaaS readiness

Besides the technical architecture, there have also been efforts to standardise the understanding of how ready a city or area is for the introduction of MaaS. In 'The Mobility as a Service Maturity Index: Preparing the Cities of the Mobility as a Service Era' [19], the authors set out five dimensions for analysis.

1. Transport operators' data sharing and openness
   The extent to which transport operators share data and make APIs available to third parties. This includes whether data and APIs are made 'open' (i.e., freely available to use, redistribute and alter).

2. Citizen familiarity and willingness
   The extent to which citizens' lifestyles and behaviour aligns with a MaaS model of transport provision. This includes travel behaviour and the use of MaaS-related technologies.
3. Policy, regulation, and legislation
   The extent to which key policies, regulations, and laws that support MaaS are in place. These may be at a city level or at a national level.
4. ICT infrastructure
   This covers the penetration of MaaS enabling technologies. These include internet access and smart ticketing infrastructure.
5. Transport services and infrastructure
   This element assesses how ready the current transport system is for MaaS. It includes both the variety of modes available and their availability – i.e., the density of services, the frequency of services, and the integration of services.

The MaaS readiness index provides a broader set of evaluation criteria than previous indices and covers the users, operators, and the statutory environment. In this, it highlights that behavioural elements are an important part of MaaS adoption, besides the availability and digitisation of transport infrastructure.

The illustration below shows the complexity of the index, with sublevels nested below the headings (Figure 3.4).

Whilst the paper questions whether areas with public or privately operated public transport systems are better able to support MaaS, the evidence in the paper itself – which evaluates London and the West Midlands' readiness – appears to indicate that the more comprehensively transport network designed and provided by the public operator in London gives the city the advantage.

One important point that the authors are at pains to point out is that the transport requirement for MaaS must be both comprehensive and not concentrated in the centre of the area.

More specifically MaaS cannot be successful with only public transport. One reason for this is that the appeal of MaaS is its ability to offer 'door to door' travel. Therefore other 'demand responsive' modes such as bike sharing, ride sharing, car sharing, taxi and ride-hailing are needed to provide 'last mile' solutions (i.e., to connect people with public transport stations/stops and their final destination) (Lund, Date unknown). Offering a wide variety of transport modes gives MaaS providers greater choice meaning they are more likely to be able to meet the needs and preferences of their customers.

This connects with their firm assertion that MaaS cannot be successful with only public transport.

Whilst the authors use these and other experiences to weigh their index, the guiding principles have been used in various other city readiness assessments. Examples of the deployment of this model include Turin, Rome, and Genoa in Italy [20], as well as Dumfries in Scotland [21].

There is a degree of circularity in the assessment of MaaS readiness. The observed pattern of public transport density tending to map to population density

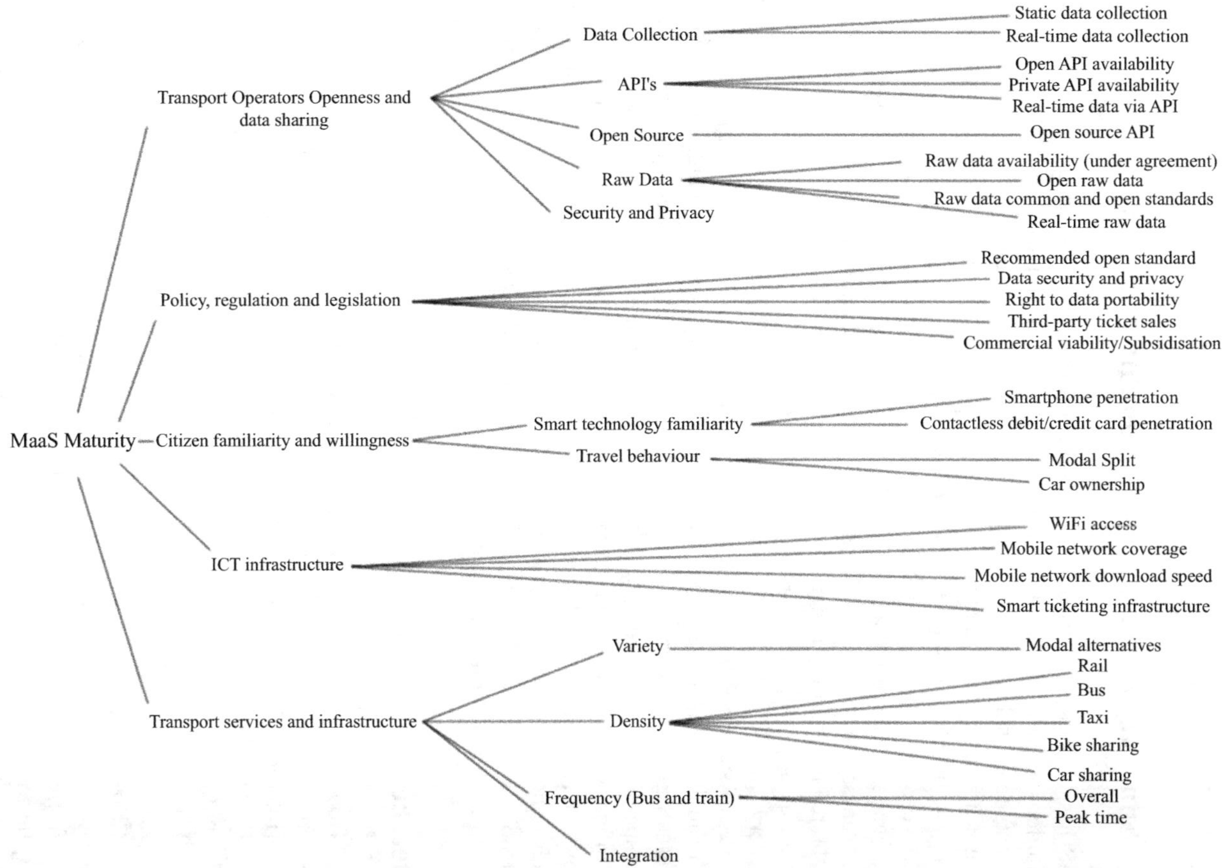

Figure 3.4 Development of a MaaS maturity index [19]

(i.e., more dense in densely populated urban areas) could potentially be compensated by last-mile transport made available through MaaS platforms. The absence of such transport reduces the degree of MaaS maturity – however, the advent of MaaS could improve access, user base, and provision of last-mile services.

3.5 Summary of defining what Mobility-as-a-Service is

MaaS has been defined by a very wide cohort of organisations, from academic institutions to national and international transport bodies, local and national governments to consultancies, start-ups, pioneers, and visionaries.

It is also been defined by its core components, its functionality, its technology, and by the benefits anticipated to flow from it.

A very loose, broad definition is the ability to plan journeys, book and pay for travel across multiple public and shared transport modes via a website or application. Many of the aspirations for MaaS include the reduction of private vehicle use and the reduction of the negative impacts of private car use (including congestion, poor air quality, and CO_2 emissions).

Very specific definitions of the architecture have been drawn up, and broader schemes of standards that enable the platforms to carry out these functionalities.

In addition, the degree to which areas (cities, regions, and rural areas) are ready for MaaS points to a practical definition of MaaS. The requirements for MaaS encompass the physical components (the availability of public and shared transport and infrastructure), the digital elements (both open data and good connectivity to enable data exchange), and then the public framework in terms both of public policies and individual preparedness to use digital resources. The levels of these indicators give a 'MaaS readiness' score which indicates the potential for the creation of MaaS and its uptake in any given area.

References

[1] Goodall W, Fishman T, Bornstein J, and Bonthron B. Issue 20 | 2017 The Rise of Mobility as a Service Reshaping How Urbanites Get Around [Internet]. 2017. Available from: https://www2.deloitte.com/content/dam/Deloitte/nl/Documents/consumer-business/deloitte-nl-cb-ths-rise-of-mobility-as-a-service.pdf

[2] Hietanen S. "Mobility as a Service" – the New Transport Model? *Eurotransport*. 12(2):2–4.

[3] Crozet Y. Mobility as a Service a New Ambition for Public Transport Authorities Discussion Paper [Internet]. [cited 16 Aug 2023]. Available from: https://www.itf-oecd.org/sites/default/files/docs/maas-ambitions-public-transport-authorities.pdf

[4] Mobility as a Service – Eighth Report of Session 2017-19 [Internet]. House of Commons Transport Committee; Dec 2018 [cited 16 Aug 2023].

Available from: https://publications.parliament.uk/pa/cm201719/cmselect/cmtrans/590/full-report.html

[5] Kamargianni M, Li W, Matyas M, and Schäfer A. A Critical Review of New Mobility Services for Urban Transport. *Transportation Research Procedia*. 2016;14:3294–303.

[6] Jittrapirom P, Caiati V, Feneri AM, Ebrahimigharehbaghi S, González MJA, and Narayan J. Mobility as a Service: A Critical Review of Definitions, Assessments of Schemes, and Key Challenges. *Urban Planning*. 2017;2(2):13.

[7] MaaS in Europe: Lessons from the Helsinki, Vienna and Hanover Experiments [Internet]. Available from: https://www.cerema.fr/system/files/documents/2020/04/cerema_parangonnage_maas_rapport_complet_eng.pdf

[8] Transit – About [Internet]. transitapp.com. Available from: https://transitapp.com/about

[9] Witzel, Skedgo S. 2022.

[10] UbiGo: Swedish Mobility-as-a-Service App Brings Smart Mobility [Internet]. Fluidtime. [cited 16 Aug 2023]. Available from: https://www.fluidtime.com/en/ubigo/

[11] Wilson S. Urban Mobility System Upgrade [Internet]. ITF. 2015 [cited 16 Aug 2023]. Available from: https://www.itf-oecd.org/node/14579

[12] Docherty I. New Governance Challenges in the Era of "Smart" Mobility. In: Marsden G, Reardon L, (eds.) *Governance of the Smart Mobility Transition*. Bingley: Emerald Insights; 2018. p. 19–32.

[13] ERTICO. Finland's Transport Code Focuses on Digitalisation of Transport [Internet]. ERTICO Newsroom. 2017 [cited 16 Aug 2023]. Available from: https://erticonetwork.com/finlands-transport-code-focuses-digitalisation-transport/

[14] A Brief History of MaaS Global, The Company Behind the Whim App [Internet]. Whim Helsinki. 2019. Available from: https://whimapp.com/helsinki/en/history-of-maas-global/

[15] Marshall A. Uber Gives Up on the Self-Driving Dream [Internet]. Wired. [cited 16 Aug 2023]. Available from: https://www.wired.com/story/uber-gives-up-self-driving-dream

[16] Yu L, and Datson J, Mobility as a Service Exploring the Opportunity for Mobility as a Service in the UK [Internet]. 2016. Available from: https://cp.catapult.org.uk/wp-content/uploads/2021/07/Exploring_the_Opportunity_for_Mobility.pdf

[17] Matyas M. The Business Ecosystem of Mobility-as-a-Service [Internet]. 2017. Available from: https://discovery.ucl.ac.uk/id/eprint/10037890/1/a2135d_445259f704474f0f8116ccb625bdf7f8.pdf

[18] Interoperability for Mobility, Data Models, and API [Internet]. [cited 16 Aug 2023]. Available from: https://maas-alliance.eu/wp-content/uploads/2021/11/20211120-Def-Version-Interoperaability-for-Mobility.-Data-Models-and-API-_-FINAL.pdf

[19] Goulding R, and Kamargianni M. The Mobility as a Service Maturity Index: Preparing the Cities for the Mobility as a Service Era [Internet]. 2018 [cited

16 Aug 2023]. Available from: https://discovery.ucl.ac.uk/id/eprint/10063087/1/Contribution_10902_fullpaper.pdf%C2%A0

[20] Caballini C, Corazza MV, Costa V, Delponte I, and Olivari E. Assessing the Feasibility of MaaS: A Contribution from Three Italian Case Studies. *Sustainability*. 2022;14(24):16743.

[21] Arrieta Solís J. MaaS as a Driver of Sustainable Planning: An Evaluation of Dumfries' Readiness [Internet]. www.theseus.fi. 2022 [cited 16 August 2023]. Available from: https://www.theseus.fi/handle/10024/781578

Chapter 4
How technology has enabled MaaS

4.1 Introduction

The rise of smart technology over the latter part of the twentieth century has been fundamental to the development of MaaS. Without developments in transport ticketing, mobile phone technology, and geographical information systems as well as increasingly advanced computing, MaaS would not be possible.

Much has been written about the 'smart city' concept, generally focusing on the potential for technology – particularly data and connectivity – to be used to shape services and interactions between the city and the people living there. The much-hyped and subsequently abandoned Google Sidewalk Labs plans for redeveloping a section of Toronto waterfront into a newly built sensor-clad 'wired town' was probably the epitome of the technology-focused idea of the Smart City. However, a broader definition possibly more useful – and more appropriate to the 'smart' evolution of those cities established centuries ago – such as that suggested by Lukas Neckermann in Smart Cities, Smart Mobility:

> We propose that a smart city is one that combines its data, its resources, its infrastructure and its people to continually focus on improving liveability. A smart city is an aggregation of power and creativity, but also a body of data and live analysis. It has a soul; it sets goals and shares its passion. And if a smart city can so be equated with a human, smart mobility is a city's circulatory system.

To work, this circulatory system does depend on technology. However, the essential feature of these technologies (from ticketing to communications systems) is that they are able to work together and for the benefit of the city inhabitants and visitors. Whilst early developments in simplifying fare collection in large cities like London and Paris predate the emergence of the 'smart city' description (covered in the Introduction and Chapter 6), the benefits of simple fares and payment means were recognised and the systems rapidly and permanently adopted both by transport users and city authorities.

4.2 Emergence of intelligent transport systems

Much energy – at governmental and inter-governmental levels has been devoted to ensuring the development of these technologies. In particular, this has required that standards are developed and interoperability encouraged. Smart transport technology, also known as Intelligent Transport Systems (ITS), is now integrated with national, regional, and local infrastructures across much of the world. Within the realms of ticketing specifically, standards such as the Integrated Transport Smartcard Organisation (ITSO), as adopted by public transport co-ordinating bodies and operators, have also helped to facilitate the adoption of integrated ticketing to some degree.

Internationally, the rise of ITS is a key element that has enabled the development of MaaS. It includes fundamental technologies on which MaaS is based including account-based ticketing on public transport, Geographic Information Systems (GIS), digitised information 'feeds' (usually APIs), and the algorithms that create routes and itineraries.

ITS has been promoted internationally, often through intergovernmental organisations like the International Transport Forum (ITF) which is a think tank for transport policy of the OECD and also by the European Union.

These initiatives are, in many ways, the precursors – and enablers – of MaaS. The ITF was established in 2006. The EU's activities aimed at adopting and improving technologies that improve urban mobility are of a similar vintage. In 2010, ITS Directive (2010/40/EU) gave the necessary legal framework to the EU member states to accelerate the implementation of smart technologies in the transport sector, whilst also giving the freedom to each country to decide their priorities [1].

It has updated advice to member states and promoted standards across the sector over the subsequent decades, citing the potential of ITS in its key communication on 'resource-efficient urban mobility' in 2013:

> More specifically, ITS help to optimise the use of existing infrastructure through a variety of means, such as traffic signals, journey planners, smart ticketing or cooperative systems (including vehicle-to-vehicle and vehicle-to-infrastructure communication systems). They foster the coordinated management of road and public transport networks and ease the implementation of urban logistics and access restriction schemes. To ensure greater coherence and interoperability of ITS solutions across Europe, the Urban ITS Expert Group20 developed a set of Guidelines for deployment of key applications of urban ITS21. Best practice, as well as standardisation needs for urban ITS, have been identified in this context.
>
> Combined efforts are needed to make urban areas benefit from full potential of ITS solutions, while ensuring cost-effective, continuous and seamless ITS deployment across the EU [2].

This iterative process between technology developers and standard setters is often replicated or furthered at the national level. In the United Kingdom, Smartex, an umbrella group for organisations working in the smart technology sector, set up the Smart Transport Forum in 1997, at the request of the UK's Department for Transport (DfT). It is a membership-based association open to all who are involved professionally with the transport application of smart cards and other 'smart customer media' within the United Kingdom. This covers smart ticketing, smart payment, journey management, and reporting elements – plus the underlying technologies supporting these applications. Some key smart ticketing technologies were used in London's Oyster card from 2003 – loading value onto a plastic contactless smart card that was used to tap in and tap out of the Transport for London Underground or tap in (only) on buses. The card uses an RFID chip to identify itself on electronic readers and to update its balance. The back-office software manages the record of the balance and reconciles these taps every night. Oyster – and other UK smart transport cards in the United Kingdom – are developed using ITSO, the UK's smart transport standard specification. This specification ensures that operators' fare-charging systems speak the same language [3].

4.3 Consumer technologies

Whilst the ITS elements of MaaS are fundamental, it is also enabled by consumer technology – namely the ubiquity of smartphones in the hands of users. Enabling travellers to access mapping, information, and payment directly empowers them, greatly streamlining their journeys and boosting their trust in the travel experience. Always accessible real-time location information, in particular, increases people's confidence and feeling of having choice and control over their journey.

Whilst some providers of MaaS (particularly those that have social equity responsibilities – for instance, city authorities) also use backward compatible technology to ensure that people without smartphones or even bank accounts can access services, typically, these methods represent more traditional means of connecting to the central MaaS platform. The attractiveness of MaaS for most travellers is the information and control it provides over journeys wherever they are on their personal devices.

The technologies that have evolved to facilitate MaaS are a blend of consumer-oriented technologies, infrastructure, and computing platforms.

Smartphones. Essential to (most) versions of MaaS is the smartphone. The smartphone provides means to plan journeys, buy tickets, hold tickets or identify the passenger, and communicate in various ways. The specific elements of smartphone technology that are used in MaaS applications are:

- Location
    - GPS, mobile, and WiFi connections all enable location identification.
    - Sensors, e.g., accelerometer, gyroscope, and compass are used to detect motion, direction of travel, etc., and are used in relation to locating the user and relating their position to maps within journey planners.
- Ticketing and payment
    - NFC chips in smartphones are used in ticketing, payments, and wallets (see below).

  - 2FA systems – 2-factor authentication uses combinations of authentication (e.g., password and SMS) for payments and logging into sensitive systems.
  - Cameras are used for scanning (e.g., QR codes) and also for ID verification – including biometric data (for instance face ID).
- Bluetooth connectivity
  - Bluetooth is sometimes used to detect and connect with shared mobility and micromobility vehicles.
  - It is also used for digital car keys and to confirm location proximity to other Bluetooth devices (for instance in ride-sharing scenarios or in car parks).
- App ecosystem
  - The apps available for smartphones determine the availability of MaaS apps and sometimes the functions that can be used in MaaS systems (e.g., authentication apps or payment modes).
- Social media
  - Although not primarily required for MaaS, can be a way of validating ID and social networks for ride-sharing, and also a way of sharing feedback, routes, and information.
- Audio
  - Microphones are used for speech commands.
  - Loudspeakers are used for spoken instructions and alerts (sometimes including using the vibration function as an alert).
  - These functions can help improve accessibility to transport systems for people with vision impairments.

Of course, smartphone functionality depends on connection to the mobile network, with more data-heavy applications requiring more recent 'generations' of mobile networks – the 'G' in 3G, 4G, and 5G. Each new generation permits a higher rate of data transfer. The mobile network has moved on from 2G to 4G and 5G mobile operations, although 2G and 3G technologies are still in use. Coverage of the higher Gs – 4G and 5G – is less comprehensive than that of 2G and 3G. In some rural areas, there are what is known as 'not spots' where no coverage by any operator is available (partial not spots are areas where some carriers are not available). Many governments recognise that this is an issue and have plans to improve connectivity in all areas. The UK plans are illustrated below (Figure 4.1) [4].

Whilst universal coverage is an aspiration, 'not spots' can be an issue for MaaS (and indeed any connected service or vehicle) in mountainous and very rural areas. Making services robust in areas of low coverage is a challenge that some platforms particularly have engaged with – enabling as much offline functionality as possible.

4.3.1 Connectivity and location technology

Connectivity and location technology are primary to MaaS. It is at the core of the provision of real-time information. It enables passengers to be located – and

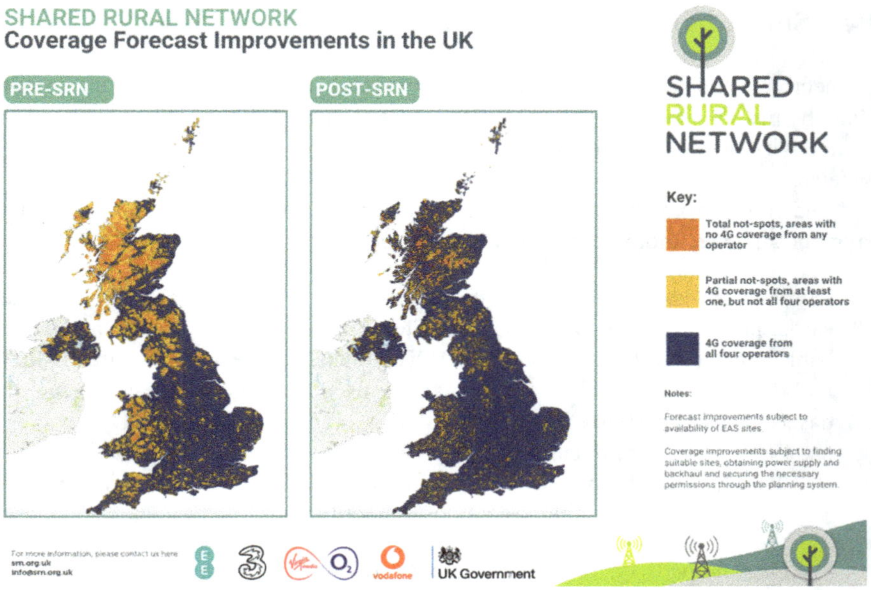

Figure 4.1 Shared rural network – coverage forecast improvements in the United Kingdom [5]

to locate themselves with respect to services, quite literally, to see where they are on maps (using their smartphone GPS) or where vehicles are in real-time.

- GPS is the Global Positioning System. It depends on networks of satellites orbiting the Earth. By triangulating between three satellites the exact position of the GPS receiver (for MaaS, the smartphone) can be identified. Triangulation with a fourth satellite provides altitude. GPS is used to locate the passenger in journey planners and also to locate vehicles in shared and micromobility. Some public transport vehicles use GPS – for instance, recent bus ticketing machines include GPS functionality enabling the bus to be located both for operational reasons and – if desired – to be communicated in journey planners and real-time information.
- IoT is the Internet of Things. Narrowband IoT is a protocol for wirelessly connection over a low-power wide area network or LPWAN [6]. Narrowband-IoT is one of four basic technologies deployed in the LPWAN sector. The others are long range (LoRa), long-term evolution for machines (LTE-M), and Sigfox [7]. Narrowband-IoT is the most utilised of the four technologies. It is useful for connecting vehicles used in shared mobility and micromobility so that they can be located.

4.4 Smart payment and ticketing

Connectivity and location technology play crucial roles in both the utilisation of MaaS by passengers and its provision, there are additional forms of vehicle-specific connectivity deployed in relation to modes, as well as smart payment and ticketing systems.

The move from cash to electronic payment is sometimes typified as 'smart payment'. The technologies that enable this include:

- Payment cards: Payment cards in transport can be plastic transport payment cards, pre-loaded with value or connected to automated payment mandates or bank cards (credit and debit cards). Their use has become integral to new mobility and MaaS. They are sometimes used in conjunction with account-based ticketing systems for 'tap on tap off' systems like London's Oyster. They use EMV chip and pin cards which rely on:
    - RFID: Radio Frequency Identification (generally smart card systems)
    - NFC: Near field communication (generally bank card systems)

  RFID and NFC systems use an embedded integrated circuit chip and antenna that exchanges data with the validation or payment system. The cards are not powered – there is sufficient energy from the signal in the card reader to generate a response.

  The card readers are generally linked to a back office system that reconciles transactions. As technology has developed these have become smaller and lighter. What used to require static gates with inbuilt readers linked to huge servers by fixed wire connections can now be achieved by lightweight card readers on ticket machines with mobile connections. Whilst the computing power to reconcile transactions is still required these more agile systems are more accessible to smaller operations in smaller towns, cities, and areas.
- Online banking and banking apps: online banking developed at the end of the 20th century. Adoption has increased as banking services have also become available through phone apps. Whilst online banking is not core to smart payment systems, it was a development that facilitated increased familiarity with carrying out financial transactions and management through smartphones. It also facilitated the development of mobile wallets on smartphones.
- Mobile wallets: Mobile wallets store payment card information on smartphones. This means that phones can be used to pay for ticketing (particularly where it is being purchased via an online transaction or app). The market leaders in mobile payment systems are Apple Pay and Google Pay. Mobile wallets can be used to store value (e.g., vouchers and tokens) and also tickets.
- In addition, traditional direct debit mandates (also known as continuous payment authorities) can be used to pay for subscriptions. The regular automated monthly collection of subscriptions (either regular or variable amounts) is key to some models of MaaS and some shared transport modes.

'Smart ticketing' is a catch-all term for new technologies that can improve overall efficiency and service levels of public transport. It includes electronic ticketing systems (like London's Oyster card) and is recognised that simple integrated tickets enable and promote multi-modal travel.

As technologies have converged (using the same NFC bank card as both ticket and payment), smart ticketing and smart payment have become increasingly intertwined.

Technologies used in ticketing:

- Ticketing tokens
  - For tap-on tap-off systems with account-based ticketing, the bank cards used to tap on and tap off are also the travel tokens, conferring permission to travel. Where bank cards are stored on smartphones in digital wallets, they can be used in the same way, accessing the phone's NFC chip.
  - Other systems use the unique patterns encoded in QR or Astec codes as ticketing tokens. These can be read from either paper or mobile phone screens making them versatile.
- Account-based ticketing systems
  - Simple account-based ticketing is used for tap-on tap-off trips. The bank card is both a ticket and payment method. Only limited data is held, just the card details so that at the end of the day the number of taps on and off can be reconciled and the card charged the correct final amount. Most such ticketing systems also allow travellers to set up an account with more details and register their regularly used payment cards. This can help safeguard against fraud, and in some cases permits better functionality and choice in the types of tickets (e.g., for season tickets).
  - It is possible that biometrics will be integrated into account-based systems in the future. Just as fingerprints and facial recognition have been incorporated in smartphone security to enable the phone to be unlocked or as part of app-based banking processes, this technology can be linked to account-based systems and biometric readers could be activated at the start and finish of journeys to record trips and charge accordingly.

ITSO, the UK smart card standard provides a useful diagram outlining the relationship between cards/NFC chips and readers, showing where the ITSO standard secure electronic data processing modules (ISAMs) are used to check bank cards, authenticate and validate electronic tickets and compile journey data. ISAMs are the size of a mobile phone SIM card and are inserted in railway ticketing gates and other ticketing machines (Figure 4.2).

The weakness of NFC-based systems is that they are only slowly being adapted for iOS. As of 2023, ITSO on mobile (provided by Yotra) was only supported by Google [9].

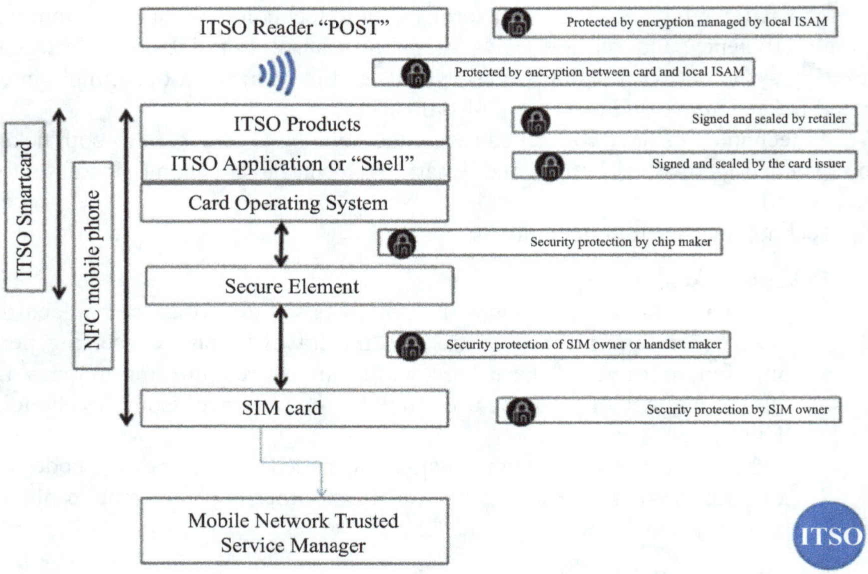

Figure 4.2 The technology behind ITSO [8]

4.5 Connected and automated vehicles

The potential for connected and autonomous vehicles to change the transport landscape has been discussed for a decade or so. Some concepts of MaaS theorised flotillas of driverless cars summoned at the touch of a button. Whilst the proximity of the horizon for truly driverless vehicles seems to have been overstated for some time, there are numerous technologies in use and development that are incrementally bringing it closer.

Several technologies enable vehicles to connect to and interact with infrastructure and/or with each other. Limited use cases have been used for specific areas and vehicles and are well established (for instance in the bus priority traffic signal case above). However, broader adoption in the wider vehicle fleet depends on the development of use cases and standards enabling universal communications.

For instance, in a report published by the UK's Centre for Connected and Autonomous Vehicles, it was suggested that information such as vehicle location and speed of travel could be used by MaaS journey platforms to inform journey planning and booking car parking or tickets for onward travel via public transport. However, although 71% of new vehicles registered in the United Kingdom were connected in 2019, according to a report published by the SMMT [10], the harnessing of the data they produce has yet to show wider benefits. At present, however, connected vehicle data is generally used directly by vehicle operators, fleet managers, and insurers to inform their operations rather than for wider purposes.

Connected private vehicles produce rich data streams with the potential to enable new services with benefits to road stakeholders and beyond. In theory, this data could be incorporated into navigation systems or integrated with Universal Traffic Management Control (UTMC) systems, however, there are no trials at the scale of this, nor is adoption mandated in any territory.

The development of autonomous vehicles is similarly still at the trial stage. Whilst there are many technologies that automate some aspects of driving, there are few trials on open roads and even fewer that do not employ a 'safety driver' intended to take control of the vehicle if the automated driving systems appear not to be working. Fully automated driving requires an estimated billion lines of code, with cars processing huge amounts of external and internal data including input from external sensors including LiDAR and cameras, mapping data and GPS as well as the functional state of the vehicle. Driving requires synthesising this with the rules of physics as well as the rules of the road. Because this is vastly complex, six levels of automation have been identified (Figure 4.3).

As a rule, vehicles in shared use have been used at higher levels of automation. Some production cars are currently sold with limited automation (levels 2–3) and can drive with supervision in certain environments. Automation at level 4 has been

Levels 0-2: Human is monitoring the environment:

Level 0	No driving automation, the human is in control at all times.
Level 1	Driver assistance – e.g., adaptive cruise control – the human is in control of vehicle systems and is monitoring the driving environment.
Level 2	Partial driving automation – includes advanced driver assistance (ADAS) including automated lane keeping – whilst the vehicle can detect some elements in its surroundings and steer and control the speed, the human is required to be ready to take control at all times.

Levels 3-5: Automated system monitors the environment:

Level 3	Conditional driving automation – the vehicle detects the environment and can perform most driving tasks, but the human is required to be ready to take control at any time.
Level 4	High driving automation – these vehicles can drive automatically in limited situations, for instance in a limited area (for instance with high level of mapping), on certain types of roads or within a strict speed restriction. It is possible for a human driver to override the computer and there are controls with which to drive.
Level 5	Full driving automation – such vehicles will have no controls for humans and will be entirely self-driving.*

*https://www.sae.org/news/2019/01/sae-updates-j3016-automated-driving-graphic

Figure 4.3 Levels of driving automation [11]

trialled in on-demand transport and taxi systems that are or could be connected with MaaS (see Section 5.9).

4.6 Behind the platforms

Key building blocks of platforms include APIs, algorithms, and mapping. As the software requirements have become more complex for elements of transport (including ticketing and payment systems as well as 'full MaaS'), building blocks have been created separately and different ways of incorporating them invented.

Mapping:
GIS: Geographic Information Systems enable digital mapping. GIS is designed to store, retrieve, manage, display, and analyse geographic and spatial data. This is key to providing the maps that are the foundation of journey planning. Besides showing roads and rails, they also encode the rules of the road, timetables, and other information. They are increasingly complex as more and more detailed information can be encoded in them. Having the correct information coded into a GIS is also crucial. Whilst information about traffic rules and regulations for cars (such things as one-way streets, low-emission zones, and speed limits) has become ubiquitous and sound across most territories, good information for people on foot, bike or in wheelchairs is less so. The presence and width of pavements or the presence of obstacles that may make a path unnavigable for some users (e.g., a chicane or steps that would stop a wheelchair user or cyclist) is often not recorded or the data is fragmented. GIS is increasingly used to manage databases of city traffic regulations (e.g., the layout of parking, loading, and other, sometimes time-bound, rules). It is also foundational to connected and autonomous vehicle technologies.

Several terms are used to describe the basic processes and building blocks of platforms:

Algorithms:
Journey planning software uses algorithms to create routes, basically following a set of rules to solve the problem of A to B. In journey planning algorithms, the algorithm will evaluate options against the constraints set by the programmers – which usually include factors like the time of arrival, total journey time, and total cost. Optimising algorithms generally involves additional processes to consider all the possible opportunities and identify the limitations of each approach. The key to effective algorithms is to constrain them sufficiently to produce timely results (considering huge sets of options across numerous sources can be time-consuming – timely information is essential for travellers) but not so much as to exclude good options or yield useful results.

APIs – application programming interfaces:
APIs are a common way of providing data about transport services. They provide structured data at an accessible point so that it can be queried – e.g., a timetable API

provides a set of timetable data that can find a particular address, structured in such a way that a journey planning algorithm can query it. The format of the data itself could conform to different standards – typically JSON or REST.

Software as a Service (SaaS):
Rather than installing and maintaining software, the buyer accesses applications over the internet. Examples of SaaS in transport include DDRT platforms (where the platform is maintained by the provider and the operator runs buses using it to pick up the people directed by the platform along the routes it provides) and ticketing solutions (for instance mobile ticketing solutions work like this, with the unique tokens and validations managed by a remote platform).

SDKs Software development kits:
SDKs enable service providers to add functionality to their app or platform through a simplified process. This tends to be an easier way for platforms to add information or services. For instance, Splyt can provide access to its network by offering an SDK that enables an existing app to search the extended network and incorporate trips facilitated by the providers on that network.

White label:
In addition, the concept of the 'white label app' has been used for developers to create a set of functions and architecture without specific branding which can then be made specific to an operator incorporating – potentially area-specific – information, APIs, and rules. For instance, bike share schemes might be commissioned by a local authority. The bike share provider could offer a 'white label' app that has all the functionality that enables people to find, hire, and return bikes (potentially with a journey planner as well) which would then be branded and made specific to the scheme commissioned. Some MaaS providers work as 'white label' MaaS app providers – e.g., Skedgo and Trafi (Sections 5.4 and 5.5).

4.7 The transport environment

In addition, the transport environment is increasingly digitised, with roads, motorways, and urban areas often connected. Rail and tram have incorporated automated elements for some time, although the systems tend to work differently. Some of these key technologies are summarised here.

4.7.1 Road technologies and urban traffic control (UTC) systems

Traffic management has become increasingly sophisticated, and data-driven. Highway authorities have access to technologies including sensors, mobile phone-based traffic congestion information, and cameras to enable real-time information about the status of the road network.

Urban traffic control (UTC) systems coordinate traffic signals over wide areas to manage traffic flows on the road network. Traffic signals are linked together and adjusted to achieve specific objectives (such as minimising stops or shortening travel time over particular routes). Objectives and signal patterns can be adjusted, as can digital road signs (including flexible lane technology) and vehicle-to-infrastructure connected systems.

One relevant example is that of bus priority traffic signals. In this system the aim is to reduce the time buses wait at traffic lights. The signal's green light periods are determined by approaching buses and can use rules such as 'green light extensions' for approaching buses or reset the cycle to ensure that the signal will be green on the approach of the bus. There are multiple methods to identify and locate buses with respect to the traffic lights including on-board GPS, transponders, and induction loops embedded in the road surface at the required distance before the traffic light to effect the desired phase change to light [12].

Innovations in traffic and road management have the potential to interact with MaaS applications. Only a few trials have suggested this at present (see Amsterdam case study in Section 5.5) however, it has strong potential as a way of fostering behaviour change.

Two systems in particular could have a role in changing travel patterns; charging for use of particular roads and parking management.

4.7.2 Electronic fee collection and congestion charging

Automated payment systems have been developed to facilitate the collection of clean air zones, congestion charges or tolls for using infrastructure. These systems work using automated number plate recognition (ANPR). Road users are advised to set up accounts and register their bank account details to facilitate automatic payment by direct debit. They may also download, register, and connect their banking details with apps.

Systems linked to vehicles can be incorporated into MaaS which incorporates personal or shared vehicles (for instance including 'park and ride' schemes).

4.7.3 Smart parking

Various ways of identifying either the vehicle or the driver are used to digitise and automate parking and payment, however, they are generally based on one of two methods:

- ANPR
  Payment is linked to the number plate of the vehicle, either manually via a ticket machine into which the user types their number plate or by using an app. In the latter case, the app registers the number plate and submits payment via connected (pre-registered) bank details.
- Mobile phone and Bluetooth sensor
  This system is based on sensors in parking bays that connect to the driver's smartphone over Bluetooth. The connection is validated by an app and payment is managed in the app through connected account details.

The increasingly digitised nature of parking enables parking to be incorporated in MaaS. Examples include lift-sharing and parking integrations (see Section 12.2) and park and ride integrations.

Some MaaS platforms include their own vehicle options in the trip planner. For instance, the Skedgo trip planner modes include train, bus, taxi, subway, metro, cab, tram, and the traveller's own or shared car or bike, motorcycle or ride share. Users can set their preferences within the possibilities, use concession pricing and adjust the minimum transport times, and walking/roll/cycling speed or enable wheelchair mode. This enables the user to compare their own car (including its costs) with other options or incorporate it into their journey [13].

As vehicles become increasingly connected (even though full 'autonomous driving' may be some way off), communication with infrastructure has the potential to shape where they go, how fast, where they park and for how long, as well as associated costs. There is also the potential for more dynamic management of roads, kerbs, and parking, as well as prioritisation of public and shared transport over private or single occupancy vehicles so that space and utilisation of cities and highway networks can be optimised.

4.8 Summary of how technology has enabled MaaS

MaaS has been enabled not by a specific technology, but through the development of a suite of technologies over many decades of investment, development, and refinement. All of these technologies pre-date the MaaS concept, and many have been widely deployed across the transport sector for many decades. This reinforces the idea that MaaS is a service offering and not a technological one.

Individual technologies have enabled MaaS in specific ways, but the following technologies and groups of technologies have been particularly important in the development of MaaS, and even in defining the concept itself:

- The **smartphone** has been important in terms of providing the primary interface with the consumer. MaaS operators are able to utilise the technological capabilities of the smartphone, notably locational technologies like GPS and payment technologies like contactless payment.
- **The internet and wider digital connectivity** provide the data streams and connectivity necessary for MaaS operators to offer services, although access to high-speed internet is not ubiquitous between and within countries.
- **Smart and integrated payments** provide the technological capability not only for customers to pay for services, but also for operators and MaaS providers to provide a variety of different prices and tickets. Note that this also provides the opportunity to validate trips.
- **Supporting Smart City technologies** that provide an ecosystem of service offerings and data that could complement MaaS offerings. This includes experimentation in technologies that have yet to become commercially viable, such as autonomous vehicles.

The technological aspects of MaaS have been extensively discussed in academic papers and industry thought leadership. However, the technology is merely the means to provide the MaaS, and much of the technology enabling MaaS is not new. To define MaaS purely by its technological capabilities is, therefore, a limited framing by which to view the MaaS concept.

References

[1] EUR-Lex – 32010L0040 – EN – EUR-Lex [Internet]. Europa.eu. 2018 [cited 15 Aug 2023]. Available from: https://eur-lex.europa.eu/legal-content/EN/ALL/?uri=CELEX%3A32010L0040

[2] EN EN [Internet]. Available from: https://eur-lex.europa.eu/resource.html?uri=cellar:82155e82-67ca-11e3-a7e4-01aa75ed71a1.0011.02/DOC_3&format=PDF

[3] Homepage: ITSO [Internet]. www.itso.org.uk. [cited 15 Aug 2023]. Available from: https://www.itso.org.uk

[4] The Shared Rural Network (SRN) Campaign | Mobile UK [Internet]. www.mobileuk.org. [cited 16 Aug 2023]. Available from: https://www.mobileuk.org/shared-rural-network

[5] Forecast Coverage Improvements [Internet]. Shared Rural Network. Available from: https://srn.org.uk/forecast-coverage-improvements/

[6] NB-IoT Explained: A Complete Guide to Narrowband-IoT [Internet]. i-SCOOP. Available from: https://www.i-scoop.eu/internet-of-things-iot/lpwan/nb-iot-narrowband-iot/

[7] 5 Things to Know about the LPWAN Market in 2021 [Internet]. IoT Analytics. 2021. Available from: https://iot-analytics.com/5-things-to-know-lpwan-market/

[8] What is Smart Ticketing? ITSO [Internet]. www.itso.org.uk. [cited 16 Aug 2023]. Available from: https://www.itso.org.uk/about-us/what-is-smart-ticketing/

[9] General 2 [Internet]. yotra. [cited 16 Aug 2023]. Available from: https://www.yotra.co.uk/itso-on-mobile

[10] Connected and Autonomous Vehicles 2019 Report/Winning the Global Race to Market [Internet]. SMMT; 2019 [cited 19 Aug 2023]. Available from: https://www.smmt.co.uk/wp-content/uploads/sites/2/SMMT-CONNECTED-REPORT-2019.pdf

[11] SAE International. SAE J3016 Automated-Driving Graphic [Internet]. sae.org. 2019. Available from: https://www.sae.org/news/2019/01/sae-updates-j3016-automated-driving-graphic

[12] Selective Vehicle Priority Methods [Internet]. Leeds.ac.uk. 2023 [cited 19 Aug 2023]. Available from: https://www.its.leeds.ac.uk/projects/spruce/utmc1rev.html#_Toc428247153

[13] Journey Planner Widget [Internet]. SkedGo. [cited 19 Aug 2023]. Available from: https://skedgo.com/journey-planner-widget/

Chapter 5
MaaS business models

5.1 Introduction

One of the major issues discussed, trialled, and continually critiqued is the concept of the 'Mobility as a Service Business Model'. The basic and simple idea that people should be able to replace a privately owned vehicle with easy, digitally mediated, access to transport that demonstrates value for money when compared to car ownership contains a number of hidden complexities.

The concept of MaaS is demonstrably beneficial to society and the environment by reducing the resources required for journeys (see Chapter 3). Making it beneficial to travellers is key to uptake – through the well-tried means of ensuring cost, convenience, and a degree of control over their journeys. But the final part of the jigsaw, how the business model captures the contracts between the platform, the traveller, and the transport operators in an attractive way – and how the revenue it derives is apportioned – is not settled. There are multiple different MaaS business models, with different MaaS eco-systems attached.

5.2 Classification of business models

In some ways, these different business models have developed from their national or regional case studies or examples. The free market, start-up propagating, soil of the United States has grown individual businesses trying to expand their reach to become MaaS operators (generally more anchored in the automotive industry or based on ride-hailing). The more public service-centred systems in Europe have favoured MaaS growing out of public transport systems and radical open data. Other models, with various flavours of in-between, have also evolved.

However, whatever the seed and seedbed of MaaS, the questions that circle any discussion of the business model of MaaS are:

- Who is the 'customer' of MaaS? Cities, individual travellers, the transport operators, government authorities, or others?
- Who owns the platform that enables MaaS?
- What are the relationships between the platform and the transport operators, and how does the platform mediate the relationship between transport operators and customers?

5.2.1 Who is the 'customer' of MaaS? Cities, individual travellers, or the transport operators?

This question could be reframed as 'Who benefits from MaaS?'. As there are externalised costs to society, the economy and the environment for car travel, so there are external benefits whenever car travel is reduced. These benefits accrue to cities and those who live in them through better air quality, less congestion, safer streets, and a lower carbon footprint.

This car-replacement element in particular is easily understood by cities as an environmental and social good. It can also be linked to the concept of 'mobility budgets' or 'mobility credits' enabling citizens with poor access to transport to travel more, potentially accessing employment, education, and services, thus enabling social and economic equity. In some cities, mobility budgets have been used as levers to encourage people to trade in their cars in exchange for credit to their mobility budget. One trial in the UK city of Coventry that commenced in 2020 offered volunteers were offered 'mobility credits' totalling up to £3,000 a year to give up their cars. Motorists taking part were required to scrap their cars and in return were given between £1,500 and £3,000 worth of mobility credits to spend, using a smartphone app or Swift travel card (the regional public transport ticket), on public transport, taxis, bike share or a car club [1]. This scheme was funded as part of the UK's Future Transport fund.

Beneficiaries of MaaS could also include transport operators who benefit from increased numbers of passengers (and therefore revenue). Potentially they could also benefit from better journey data to help tailor their operations better – with insights over when people want to travel and the journeys they want to make potentially available. However, the transparency of this data will depend on the operators' agreements around their participation on a MaaS platform. This is, however, a complex area requiring operators to be more open about their own data in order to be able to see the broader picture of trip demand. It is particularly fraught where operators are required to be in competition with each other (as happens in the UK bus services where bus operators running commercial services are required to comply with competition law and are specifically prohibited from cooperating).

Individual travellers can also benefit from better information so that they can plan and execute their journey. A MaaS platform may include more options within a single interface (combining public transport with on-demand or shared modes) and more real-time information and alternatives. MaaS could also benefit them financially through capped prices, best-price guarantees, or mobility subscriptions. In addition, it offers a way of managing a mobility budget if such is available via their employer or state incentives.

5.2.2 Who owns the platform that enables MaaS?

Over time, several different owners of MaaS platforms have emerged. Some MaaS providers launched with confidence that the platform itself is monetisable (e.g., MaaS Global) through the concept of mobility subscriptions equivalent to the costs of car ownership. Other platforms were commissioned by cities or companies that own the platform and maintain the relationship with the traveller. In some cases, access to multi-modal transport is offered by transport operators or journey planners, effectively turning their own single mode or service platform into a MaaS platform.

The ownership model of MaaS is important as it helps to determine the key objectives that the MaaS platform should seek to achieve. Whilst there are variations in practice globally, the ownership of the MaaS ecosystem by the public sector is generally seen to be seeking to achieve broader public sector and 'social good' outcomes such as tackling climate change or tackling social exclusion. Whereas ownership in the private sector – whilst wider social benefit may be part of corporate goals – is focussed on generating ridership with consequent monetisation – ultimately for profit.

5.2.3 What are the relationships between the platform, its users, and the transport operators?

The details of these relationships include data obligations, financial transactions, legal agreements about which party bears which duties towards the traveller, and how the traveller can show their permission to travel (ticket or equivalent). These relationships can be very complex because they cover relationships between several or all of local (and wider) government authorities; transport operators; entrepreneurs; and other relevant providers (e.g., payment and ticketing system providers). This effectively creates a complex interplay between monetisation, ticketing, passenger assurance, and customer care.

The questions that arise, for any multi-modal app, include:

- Who guarantees arrival if one of the modes is delayed, cancelled or unavailable?
- Where does responsibility for refunds lie?
- How are fares apportioned between operators if there are multiple route possibilities for the same journey?
- Where do transactions occur (within the platform or off-platform via operators)
- Can (or should) some take place on the transport operators' servers?
- Who bears the costs of doing business (for instance card processing costs and other layers such as fraud prevention and ticket provision)?

The reality has been multiple combinations of answers to these questions, conceptualising MaaS in a variety of ways. Where and how the MaaS platform developed often depends on the answers and how the relationship between the platform, operators, and travellers is defined.

The biggest question that emerges, and on which the existence of each service is predicated, is the ultimate question: 'How should the platform itself be funded?'. Put another way: 'How can a MaaS platform derive revenue?'.

In the course of the emergence of MaaS, this question has been answered in different ways, which has led to a diversity of underpinning business models and a diversity of perspectives on it. Rather than there being one version of MaaS, we have identified seven different models, each with its own implications. These models are as so:

- Single Operator Business Model
- Single Function Business Model
- City Authority Business Model
- Resold to Cities Business Model
- Resold to Companies Business Model
- Resold to Destinations Model
- Autonomous Vehicle Model

5.3 Single-mode operator business model

This is a platform developed by a transport provider which is then expanded to add additional providers (either through ownership or partnership). For instance, Uber created a highly used app connecting travellers to private hire services. It then bought Jump bikes (in the United States) and added access to them in the app. Uber derived revenue from both modes (however this integration has now ceased). This is not solely a private-sector model. Spanish national rail provider Renfe has developed its own platform (see below). Other operators have started to show multi-modal journey planning in their apps. For instance, First Bus in the United Kingdom includes possible rail connections, however, this is not currently linked to ticketing.

This model represents an evolution of an existing transport service into Mobility-as-a Service. There are examples of private-sector transport operators and public-sector transport operators developing into multi-modal services.

For each, the reasons are subtly different. Whilst both aim to increase revenue, the public sector providers have social, environmental, and social equity targets to meet. These goals mean that public sector providers need to increase access to their services and the appeal of these services to meet their goals.

However, the common theme is that expanding the digital transport offer means adding revenue streams to the core offer. In addition, the additional functionality can make the core operator's services more attractive increasing the use of that app (and reducing the likelihood of customers/travellers looking elsewhere or using multiple apps).

5.3.1 Case study: Uber

The story of Uber is interesting in this regard. Founded in 2009, Uber has evolved from its initial function as a ride-hailing platform. Its app is widely downloaded (it was the most widely downloaded travel app in the United States in 2018 and in several other markets). However, it has struggled to reach profitability despite rapid diversification and evolutions in services.

One of its key strategies intended to enable it to attain profitability has been to add functions – increasing the number of associated income streams. Whilst food deliveries was one such diversification, of more relevance to MaaS was its acquisition, in 2017, of JUMP bikes. JUMP was a bike share operator which, at the time, managed schemes in over 40 different markets. This acquisition allowed bike share to be visible and bookable through the Uber app in the United States. Uber users could, as a result, see how long their trip would be by private hire vehicle (and how long it would take for it to arrive) and where hire bikes were located with how long the trip would take by bike.

Based in New York, JUMP started in 2010 as Social Bicycles and had grown before its acquisition by Uber.

The original Social Bicycles was branded as a 'service by JUMP Bikes that provides the hardware and software behind the most loved bike share systems in the world'. Social Bicycles operated 15,000 bikes in over 40 different markets – including in Brighton in the United Kingdom. These bikes were generally publicly procured docked systems in which the local authority collaborated on the design and delivery of the network whilst also subsidising it.

As a dockless brand, JUMP was able to operate where it saw an opportunity as a private sector provider – mainly in big cities where revenue would provide a return on the cost of the bikes and operations. It rolled out a new flagship high-profile JUMP branded dockless electric bike service in San Francisco (with 250 bikes), and Washington, DC, in 2018.

Uber's initial plan was to retain the standalone JUMP app and increase the reach of the JUMP bikes through integration with the Uber app.

Early reports from Uber showed that at congested times people were selecting JUMP bikes to beat the traffic (covered in Section 5.10).

However, this is not the only 2017 project that Uber undertook to expand its multi-modal offer. It added a pilot project in San Francisco that let users rent a vehicle from within the app via Uber's partnership with the peer-to-peer car-sharing startup Getaround.

In addition, it started selling integrated tickets for public transit systems in cities around the world within the year, using Masabi's JustRide platform that enables mobile payment options for public transport systems around the world [2].

At the time, Uber Chief Executive Dara Khosrowshahi claimed that the deal furthered Uber's goal of offering 'the fastest or most affordable way to get where you're going, whether that's in an Uber, on a bike, on the subway, or more' [3].

The ingestion of modes did not stop there. In 2018 Uber joined forces with electric scooter rental company Lime in a partnership deal that would see the ride-hailing giant invest $335 million in Lime and promote it on its app. Lime was founded in 2017 in the United States, originally as a bike share company. They developed e-bike share in 2018 and also added e-scooters that year before the deal with Uber.

However, growth has not been guaranteed. Both Uber and Lime suffered a decline in business during the coronavirus pandemic, so a certain amount of streamlining has taken place. In 2020 Uber transferred its own electric bike and scooter division (JUMP) to Lime, with the companies working to further integrate their apps [4]. By 2022 Lime was in 226 cities [5].

The actual offer that Uber can provide depends very much on location. Whilst the app is almost universally available to customers, the core ride-hailing offer depends first on independent private hire drivers signing up for the app. Drivers are generally offered city locations to drive in if they sign up – meaning customers are much less likely to find rides beyond the urban core. Additional services like e-scooters and hire bikes also depend on their local distribution, which is predicated on a combination of potential profitability (most free-floating schemes will look for an appropriate level of urban density and other transport to launch in a location) and whether local bye-laws permit on-street free-floating vehicles. The public transport ticketing availability via Uber for any location will depend on whether the area uses the Masabi JustRide platform.

The overall impact is that some cities – and particularly their urban cores – will have excellent provision of transport through the Uber app. However, this choice will not extend to rural areas or even to less profitable neighbourhoods.

Over time, it is clear that Uber's business model has shifted towards providing services to the government, a change discussed in Section 5.10.

Uber is probably the most successful example of private sector MaaS by agglomeration. Other app-based transport services have added modes on a much smaller scale. In the United Kingdom, a community car club, Co-Cars (car-sharing) added e-bikes to their app and augmented their services in the southern city of Exeter [6]. Meanwhile, e-scooter-sharing brands Tier, Dott, and Voi have added e-bikes to their offer. In 2022, Tier completed the acquisition of Nextbike, augmenting its free-floating e-mobility services with Nextbike's enormous network of bike-share operations, encompassing schemes tendered by cities as well as those operated independently.

Whilst this trend for increasing the offer of private sector digital transport operators is likely to continue, more recently it is being mirrored by other modes – in the public transport sector as well.

A contrast to Uber is the development of a MaaS app by Renfe, the Spanish railway operator. In some ways, the development of Uber and Renfe mirror each other. Ride-hailing is largely a local 'last mile' business. Adding public transport ticketing enables Uber to add long-distance travel. Conversely, Renfe is a long-distance operator adding the last mile through its MaaS development.

5.3.2 Case study: Renfe as a service

Spanish railway operator Renfe has placed MaaS at the core of its strategy to develop a sustainable railway network that sits within the Spanish future transport system. Renfe is placed to be a key instrument of government policy as it is owned by the Spanish government [7].

The MaaS platform, Dōcō, is key to a new multi-modal approach to the railway. It will integrate different modes of private and public transport through a single booking and payment process on a smartphone app. The aim is to improve connectivity to Renfe's rail services by offering door-to-door and on-demand services.

There is not one, universal, national rollout, however. The complexity of MaaS means that the app will be rolled out on a location-by-location basis, initially with 14 Spanish cities (autumn 2022). It will then be expanded to cover 27 locations in 2023 [8].

The business models differ, however, where the ownership of the technology is concerned. Whilst Uber is its own technology provider, Renfe has procured technology from a third party – Siemens in this case. This reflects several elements of the relationship between transport operators and MaaS. Public transport operators have long and established histories before the concept of MaaS. Whilst some have early involvement with the development of Maas (for instance in Vienna), this is the exception rather than the rule. The later development of a MaaS platform for Renfe comes at a point when mature third-party MaaS technology providers exist. Some of those providers sprung out of MaaS-focused start-ups (for instance Trafi, Skedgo) and others out of existing technology providers (for instance Mentz, Fluidtime) as discussed in Sections 5.4 and 5.5.

There is another model positioned between services like Uber and Renfe, known as the German MaaS platform 'Mobility Inside'. Mobility Inside aspires to be a MaaS solution for the whole of Germany. It is the creation of public transport operators in Berlin, Münich, and the Rhein-Main area (which covers Frankfurt and several smaller cities in the area) and an attempt to grow the city and area-based MaaS apps into a national scheme. Their vision is to create national 'networked mobility'; a Germany-wide mobile app that includes both local and long-distance public transport, as well as all the sharing operators across all the regions, in one app. The development has been supported by the Germany government at the national level through BMDV (Federal Ministry of Transport and Digital Infrastructure) funding of €10 million.

5.3.3 Case study: Mobility Inside

Mobility Inside is marketing itself to other public transport operators as a comprehensive service that will 'network' their existing offer so that it is part of the national app – as well as create their own regional white-label MaaS apps.

The founding partners, Mobility SVM (Munich) RMV (Rhein-Main) and Berlin BVG claim that the app will cover 40% of the German population (in 2022), which they aim to increase to 70% by the end of 2025. They offer prospective partners a simplified integrated business model with 'strong technology partners' and contracts without procurement processes.

What they described as their 'Networked' offer is the delivery of Germany-wide through-ticketed journeys, with a neutral choice of sharing operators, that can be booked from one app, so that customers can cross regional boundaries. It bundles services from operators into one platform and makes local offers available across regions. In Germany, regional ticketing could be quite complex before the inception of the single €49 Germany Ticket (which was launched in 2023 and covers all local and regional transport). The idea behind the Mobility Inside offer is that (providing all the regional transport operators join forces) it can provide a single app for all operators across Germany.

The intention is that a single app with increased buying power reduces costs through communal purchasing power, programming and development of customers' interfaces and by using standards for data including real-time information, fare data, and ticketing.

Like other MaaS platforms, Mobility Inside offers its partners the choice of white label app or API connection and supports digital development through the digitalisation of tariffs, development of e-Ticket validation, and readying of RTI systems. Recognising that there are varying degrees of development among its prospective partners – the German public transport operators – there are three levels of offer:

- Networking model – for those who have their own platform
- Support model – for operators that have a digital channel but without a regional platform
- Full-Service Model – creates the local and regional offers including:
    - digitalisation of tariffs
    - Real-time data
    - Enabling e-ticket validation
    - Any required technical infrastructure (that does not already exist) required for MaaS
    - Includes working with national MaaS groups and using the national PT body (VDV) standards

Effectively it is offering a type of 'Mobility Platform as a Service' model. However, the business model depends on both sufficient willing operators and ensuring that joining together multiple local transport providers is done smartly to ensure that the app does not consume vast amounts of data or work too slowly to be useful. To get around the sheer volume of detailed information about transport in each city, many apps which cover multiple locations require a manual shift between cities, to localise searches to a specific area. How this is managed in a country-wide through-ticketing app that includes all the available public and shared transport will be interesting.

Other issues with Mobility Inside are common to MaaS apps, as the Mobility Inside offer is effectively another layer of cost for operators (many of which are already running their regional/city systems). As with other MaaS apps, it is questionable whether it can be funded from ticket sales. The other issue is adoption – where people are already adopting one MaaS app, the question is, will they adopt another? For instance, the Deutsche Bahn train ticket app already provides a very user-friendly journey planner that sells multi-modal public transport tickets (although not yet shared transport) and there is potential inertia to move from one to the other. The development of Mobility Inside will be interesting to watch.

5.4 Single function business model

In this case, a single-function app (which could be journey planning, banking or payment) gains a customer base by providing a good, trustworthy user experience. It then needs to monetise that experience either through booking fees, advertising revenue or by creating subscription services. This can effectively become a MaaS by bundling together multiple modes that are generally pay-as-you-go and deriving income from them.

MaaS developing from journey planning has occurred in a few different situations. Citymapper has adopted this model, with its highly respected journey planner also offering a travel pass providing access to public transport, cycle hire, and taxis in London. Despite its excellent reputation as a journey planner and city simplifier, Citymapper has spent considerable time and resources searching for a business model, or at least a way of monetising its services fully.

5.4.1 *Case study: Citymapper*

Citymapper was founded in 2011, enabled by the provision of real-time open transport data by Transport for London. It prided itself on providing a choice of detailed step-by-step itineraries so that users could choose the cheapest, the most active, the fastest route and take it with confidence. It expanded the cities it covered as access to open data grew, adding New York, Paris, and Berlin in 2013 – and continuing to grow worldwide with around 400 cities in 2023. It was regularly hailed as being the best source of information for getting around complex cities and prided itself on both its data and its ability to work closely with cities and use local skills, knowledge, and information to get the best quality routes.

Users are issued with a contactless card linked to their app account which gives them access to unlimited London Underground travel in zones one and two, and citywide TfL bus travel. The upgrade 'Super Duper' version includes unlimited 30-min cycle trips on Santander bikes – bikes are unlocked using the contactless card. In addition, the Super Duper pass covers £10 of taxis booked via the app or via a linked account on Kapten. In 2019 basic pass cost £31 a week, cheaper than the £35 charged by TfL, with the Super Duper pass priced at £39 per week [2022 prices] [9].

In 2017 it sought to use search data to identify gaps in the network that it could fill with transport provision and launched a shared taxi service and a night bus service. The aim was to revolutionise transport and demonstrate how data could form the foundation for better public transport services.

Whilst the idea was good, and the aspirations bold – to create a bus service that adapted to demand and changed route accordingly – the reality was that such a service would not fulfil bus registration requirements in London. Citymapper was given approval to run buses at night on a set route in east London – the CM2.

This bus was a minibus format that proved possible and 'right sized' for the service. With its cool features and filling a service gap it proved pretty popular. However, Citymapper discovered some of the frustrations of running a bus in London (and indeed the United Kingdom) the hard way. Within a weekend of launching the service, Citymapper had additional data to support a change in the route – but a statutory delay is built into changes to bus timetables and routes so any tweaks took weeks to be permitted.

It also created a shared taxi service in 2018 (which is simpler to register than a bus within the regulatory framework in London).

The CMX1 pop-up route ran on a circular route over Waterloo and Blackfriars' bridges in central London. The route was fixed and the bus ran between bus stops.

People were able to hop on and hop off. Bus information was integrated both with the Citymapper app and with departure boards at the bus stops and was provided as open data to other apps. The buses themselves were smaller, with charging points and onboard information. The idea was to run the pop-up route to enable them to 'rethink' how buses and routes operate. The integration with Citymapper was designed so that the route would show up in A to B routing whenever the algorithm decides – based on their viability and frequency. The idea was to provide a multi-modal ready transport element [10].

Citymapper then created Citymapper Pass, which it describes as 'a subscription to all mobility'. The pass combines Citymapper's powerful journey-planning capabilities with a travel card that pays for tickets (and can be loaded onto smartphones). Users get a 'complete mobility experience' this way. A single weekly subscription covering either the 'basic' package of unlimited public transport or the upgraded 'super duper' package combining unlimited public transport with a £10 weekly credit for bike share, e-scooters, and ride-hailing from a number of suppliers.

TfL's account-based ticketing system – which works with both NFC bank cards and its own Oyster RFID cards – made it possible for Citymapper to launch its own travel card. This card was basically a bank card that could be used to buy travel tickets and shared transport and would draw down a monthly direct debit from the user's main account to cover the costs.

However, Citymapper added advertising to enable it to derive income from its app in 2023. In a post, it explained that it had decided to include advertising to keep the app free to users. Alternatively, users could subscribe to a premium service, 'CLUB' to remove ads and 'access additional features'.

Shortly after this development, Citymapper was bought by Via, an interesting consolidation covered in Section 5.4.

The development and business model of Citymapper is in contrast to the development of Transport for London's own routing and transport data provision. TfL launched its travel planner in 2002 [11]. This was an innovation at the time – providing real-time information and routing across buses, rail, and the underground system. TfL contracted Mentz GmbH, a Germany mobility software developer, to create this travel planner. This followed a more traditional software development model, in which the authority pays for the development plus annual fees (for instance a licence and service fee) that cover maintaining the software and the server power required for the number of information requests per day. In some instances the charges can be related to licensing plus fees scaled to the number of requests per day – however, the essence is the same. The third-party technology provider is reimbursed for their work and the transport operator bears the costs of providing detailed information. Where ticketing is integrated (in this model) the operator receives all the fare boxes. This is in contrast to some other businesses where the MaaS app derives revenue from either a percentage of fares, per-ticket-sold fees or unused ticket/travel allowances or some combination of the above.

Numerous other journey-planning apps have become MaaS apps, however, they have done so by switching the focus of their business model from B2C to B2B or B2G. There are several instances of the 'clever journey planner in search of a business model', which have made this pivot. And one of the most notable is the company Skedgo with its TripGo app.

5.4.2 Case study: Skedgo and TripGo

Skedgo was founded in 2009 and launched its TripGo app in 2012. TripGo is a free-to-use travel planner that combines multiple modes – including public transport, taxis, shared, and personal transport. This latter is one of its key differentiators – it claims to be the only app that integrates multi-modal travel with users' own car, bike or motorcycle. It also includes parking (where data exists). It offers a 'connect to calendar' feature to provide automatic trip planning from and between events, including automatic trip and leave alerts with real-time information. Supported locations include many Australasian and European cities, some North American and a scattering of others across South America, Asia, and the Middle East.

Skedgo launched its TripGo API in 2016. Over time its business has focused on providing services for clients, from large business organisations (Pon, Conduent (previously Xerox)) to transport authorities and cities (Transport for Greater Manchester, City of Darwin) as well as other providers within the transport industry (Ticketer). It has developed from providing a multi-modal journey planner to being able to integrate booking and payments across public and shared transport. It sets out its store as a modular provider of MaaS, based around the TripGo API.

The TripGo API provides transport information to enable developers to provide seamless and personalised door-to-door trip planning using any public, private or commercial mode of transport. The existing API covers thousands of transport providers and (as of 2023) claimed to cover over 400 cities and regions, including real-time data as well as integrated bookings and payments.

Skedgo has transparent pricing models and offers its API for a monthly fee based on the number of transactions per month (with a free option for low transaction rates, handy for developers). This enables Skedgo to offer 'white label' apps – branded for clients and in some cases with integrated custom services. These are useful for companies requiring trip-planning apps for staff. Its pricing strategy involves a tiered approach, starting with 'off-the-shelf' plans that provide basic API access at an introductory rate. These are published on its website. More advanced plans, which include additional support, integrations, expanded scopes of work, and special conditions, are available at higher rates.

The business model is based on providing the API plus software development kits (SDK) to integrate the trip planning into existing front ends. Alternatively (and additionally where new operators or service providers require integration) Skedgo sets out the elements of MaaS it can provide in detail. Skedgo is one of the few MaaS providers to provide very transparent, modular, pricing. Its pricing formulae follow the formats of software as a service provider in other industries (for instance business processes like accounting, internal communications, or marketing channels) [12].

The pricing for each element of MaaS is set out to include:

- The front end – web-based journey planners, apps, and all the elements that could be included in them (e.g., integration with personal calendar, subscriptions, rewards, booking and payment, feedback, and analytics)

- Public transport integrations (for operators not included in the 3,200 transport service providers integrated)
- Shared transport integrations (including car share, bike share, taxi type, and parking) with and without booking
- Active travel, accessibility, and other integrations – including elevation and a terrain data for cyclists, wheelchair users, pedestrians with prams, etc., accessibility data, real-time traffic integration, turn-by-turn navigation for driving and cycling using HERE or TOMTOM, weather, other data integration, e.g., air quality, scenic info, and points of interest
- Region – review city or region and fare calculation set-ups

This generally includes a set-up fee and monthly charges at different levels (basic, enterprise, silver, gold, and platinum).

The automotive sector has also not been immune from the development of a single-function service. The most notable of which is Waze. Waze is an automotive-focused platform that went from routing to transport provider, although it eventually reverted to providing routing only.

5.4.3 Case study: Waze

Waze started with mapping, growing out of an online community mapping project, FreeMap, in Tel Aviv, Israel in 2006. This basically involved people driving different routes, and as they did, recording them on their phones and then later uploading their routes to the map. Users would then connect 'their' roads to the map. This project has evolved through several iterations (and via another name – Linqmap) to become Waze. However, the essential model built around people adding information to the map has remained the same.

Waze developed from a static map to a source of real-time information as access to better (and cheaper) mobile connectivity became available. People could add information about hazards and others could see live updates of those hazards on their routes.

Waze is a prime example of the utility of crowd-sourcing and the network effect. As it has grown, it has become more useful – with more people using it both to navigate and share hazard information it generates more data in more detail over a wider area enabling more accurate journey time predictions and better routing. Waze was acquired by Google in 2013.

By 2022, Waze estimated 150 million monthly active users worldwide.

The insights from users also created enough data to show that people were looking at making similar trips at similar times. The Waze carpooling feature grew out of this, and Waze added this shared transport offer to its mapping and routing tools in 2016 [13]. The Carpool feature was only active in some areas – launching in California. It matched drivers with passengers and collected a small payment from the passengers which was transferred to the drivers. The carpool feature was axed in 2022 [14]. Parent company Google cited changes in transport patterns and behaviour in the wake of the COVID-19 pandemic [15]. The Waze business model continues to be based on selling advertising.

5.5 City authority business model

This service is developed by a city or area authority, often based on its multi-modal public transport card. It generally includes a journey planner that offers people trips including shared mobility (bike-share, e-scooters, car clubs) and other pay-as-you-go services (private hire vehicles or taxis). The degree of integration depends on local circumstances. Many cities can easily integrate public transport through pre-existing relationships and contracts. For other, privately owned services, where the booking and transactions happen can vary between the MaaS platform and the operators' platforms via deep links or partial integrations.

The development of MaaS was accelerated by the city of Vienna and its interest in digitising its transport system. This early adopter, its proto-Maas in development since 2010, also highlights some of the possible ownership models of MaaS platforms. Its early trial project led to the development of two separate MaaS platforms with different ownership models. This is different from Helsinki, where the first mover in the creation of MaaS was the technology platform provider.

In parallel, Üstra Hannover began a journey that led to the development of its MaaS platform in 2004. It launched a multi-modal bundle, HANNOVERmobil, which augmented the annual public transport pass with car-sharing (car club) membership, a 25% reduction on German rail (the BahnCard 25) and other discounts on mobility and related services (from bike shops to taxis).

Whilst HANNOVERmobil had a big impact on those who used it in terms of modal shift and reducing car ownership (around 50% either disposed of their car or decided not to buy one), only 1,300 people actually bought the original HANNOVERmobil bundle.

However, the initiative laid the groundwork for future developments. By 2013, a mobility platform was in development with VW and a local car-sharing platform that integrated taxis and car sharing. Although the elements VW developed were discontinued, Üstra took the platform idea forward.

By 2016, Üstra had developed and launched a Mobility Shop 'selling' public transport, car-sharing, bike share, and taxis. Üstra has the overall contract with the user, registering them with individual shared transport services and billing them for usage at the end of the month. In addition, Üstra disbursed payments to the operators. Users can buy different tariff bundles; however, Üstra regards the platform as a way to test offers and new modes. There are no subscription or registration fees for the basic tariff so that customers can try services – those who would like to use them more often then upgrade their tariff-level (the HANNOVERmobil option still exists for those who would like a bundled multi-modal subscription) [16].

Other cities have taken slightly different approaches. Their adoption of MaaS has taken first mover experiences into account, often steering a course between the Vienna model where ownership of the platform is almost wholly with the city and the Helsinki model where the platform provider is the 'first mover' in the creation of MaaS (see Section 5.6).

Whilst Vienna is developing MaaS in line with the strategic role of public transport, Amsterdam began its MaaS journey in 2018, as a result of pressures created by major disruptions anticipated by huge road rebuilding over the next 10 years.

For the Amsterdam Zuidas district, MaaS is being tested as an instrument of behaviour change to relieve the pressure on commuters during long-term traffic disruption. The technology is subordinate to the concept of a mobility budget used to shape behaviour.

In Berlin, the Jelbi app was the first public-facing MaaS platform that provided multi-modal journey planning and ticketing using the Trafi platform. However, this partnership approach was augmented by the public transport agency, Berliner Verkehrsbetriebe (BVG) commissioning an additional trip planner app, Fahrinfo. This multiple approach highlights the nervousness of some areas to rely on partnerships and insist on keeping ownership of at least some technology within their control.

As MaaS has become more widespread, procurement has played a major role. Cities have become involved in processes to acquire MaaS platforms or work with platform providers. In some instances, they facilitate a commercial platform setting up shop in their city (without subsidy) whilst in others they procure existing technology to create a data platform and MaaS solution for their city. This process reflects the maturing of the market and of the technology with multiple providers of MaaS solutions.

The final case studies look at the procurement and development processes for MaaS platforms in the United Kingdom in 2021.

5.5.1 Case study: Vienna and Wien Mobil

Upstream Mobility was thus created in 2016, with five employees. By 2019 the organisation had grown to 60 people. As projected, Upstream Mobility required subsidies from the City of Vienna for its first 2 years. It secured income from revenue streams associated with support for areas developing MaaS (including consulting, IT development, and supply of the digital platform that integrates data and manages interfaces with mobility operators, the MaaS operator, and the user) and is now financially self-sustaining.

The WienMobil app is operated by Wiener Linien, and developed by Upstream Mobility which is owned by Wiener Stadtwerke (itself a public company) which owns Wiener Linien.

The WienMobil app provides information on all modes of transport in Vienna (including electric scooters and mopeds) and a route search for public transport, bicycles, bike-sharing, car-sharing, private cars, and taxis. It also offers multi-modal trips. The parameters used for route calculation can be changed by users. The user can authorise the use of the phone's GPS data collected via the application and made anonymous. About 50% of people authorise this.

Vienna, or Wien as it is known in German, is the capital city of Austria. The population of the metropolitan area, an area of 414.9 km^2, was just short of 2 million people by 1 January 2021 [17].

Public transport is prized in Vienna and plays a key role in the city's development and ambitions. The public transport system, Wiener Linien is wholly owned by public services and utility provider Wiener Stadtwerke (which is also publicly owned). It is an extensive system comprising region express lines (the S-Bahn), metro lines, tramways, and buses.

Among its successful initiatives, the annual ticket to the city's public transport system, Wiener Linien, the 'Jahreskarte' is priced at an affordable €365 (as of 2022). Effectively a euro a day. A whopping 852,300 were sold in 2019 (equivalent to 44% of the city's population). Whilst year on year, cars within Vienna have grown slowly over the last decade, the number of annual pass holders more than doubled [18]. Around 38% of the modal share is public transport, with 7% cycling and 30% walking, leaving 25% modal share for private cars [19].

Given the longstanding emphasis Vienna has placed on public transport and sustainable mobility, it is perhaps not surprising that it has also had a longstanding interest in providing digital access to its services. In some respects, it was also fortunate to have a technology company on its doorstep able to help fulfil those ambitions. In 2010–11, local technology company, Fluidtime designed Numo, a journey planner, which laid the foundations for multi-modal services with travel information and payment.

In 2012, a partnership of the Vienna transport operator (Wiener Linien) and the national railway operator (ÖBB) answered a European call for projects. Their successful bid was the Smile project. This 3-year project aimed to develop an innovative integrated mobility service. In interviews with researchers commissioned by Cerema (the French national agency for Climate and Territories) and project leaders themselves admitted at the time, this seemed utopian. Nevertheless, the project was a resounding success.

The service offered the highest level of integration possible at the time, with a single account for all means of transport, access to multi-modal information, and the ability to book, pay, and use the various mobility offers with a smartphone. However, the user had separate contracts with each operator, and separate tickets for each modal journey, all of which were integrated into the application. A summary of usage was provided to the user, but this was not a real bill. The technology was provided by Fluidtime (as for the Numo project).

Initially, it was hard to convince the first mobility operators to join, however, once the process had begun, it was easier to persuade others to follow. A thousand people were selected to test the system, and the project concluded that it was successful in terms of technology, governance, and usage. At the end of the Smile project, the success motivated both project partners to develop a MaaS system of their own, however, each had a radically different approach.

The ÖBB sought a private investor to develop the WegFinder application. By 2022, the app provided an Austria-wide route planning, booking, and payment service for public transport, taxis, rental cars, bikes, and e-scooters (depending on the availability of the services) [20].

For their part, Wiener Linien chose to internalise the skills. Having received €500,000 per year for the Smile project, financed as a European project, the City of Vienna then financed this continuation with additional funding. Wiener Linien continued its relationship with Fluidtime to develop the BeamBeta application.

This 'single global invoice' based service was not popular and further development was deemed too risky on a large scale. The application was renamed WienMobilLab and the single global invoice model abandoned returning to the more popular single account function. Wiener Linien then set up a subsidiary dedicated to MaaS, convinced that the authorities themselves should take responsibility for their own developments so as not to be dependent on the commercial logic of private entities.

For each proposed route, the application provides information about the availability of services (number of bikes shared in the station, type of vehicle for car-sharing, etc.). However, it only directly facilitates the purchase of public transport tickets and links to partner apps for registration and booking of other modes. For taxis, it provides a link to manage the phone call.

The history of the Wien Mobil app is an indication of how the level of integration with privately owned shared mobility can shift. Whilst the platform aims to work as closely as possible with shared mobility operators, direct booking and payment through the app has proved harder to achieve. For instance, there was a proposal for direct booking for car-sharing with DriveNow, however, subsequently Car2go and DriveNow merged to become ReachNow. The strategy of the new operators, Daimler and BMW, was to develop a dedicated app for ReachNow, rather than integrating the service into the Wien Mobil application. By 2022, however, they changed strategy again and sold the SHARE NOW venture (including ReachNow) to Stellantis (which owns Free2Move) [21].

In the opposite direction, the expiry of the JCDecaux contract with the city for bike-sharing in 2021 presented an opportunity for the city to specify better integration with the WienMobil app as part of a subsequent contract. This was eventually granted to Nextbike, with greater integration specified in the procurement.

Meanwhile, it is important to note that the development of MaaS is set within national frameworks that collect, consolidate, and open up data as a matter of national development. As part of this, Austrian public authorities collaborate on a shared reference system for transport infrastructure data: GIP [22].

This provides fare system APIs set up at the regional level and route calculation APIs developed at the national level. It contains all roads with profile information and rapid updates, public transport timetables, cycling infrastructure, bike-sharing stations, and car club locations (car-sharing). Each federal state populates the GIP for its area. All official decisions from road closures, changes to intersections, and speed limits are produced in a standard form that allows automatic integration into the GIP.

In addition, every Austrian town centralises housing permits, so has a comprehensive and detailed list of inhabited addresses. Towns connect these addresses to the nearest road, cycle, and pedestrian network in the GIP's geographic information system. The data is harmonised and enhanced by ITS Vienna region. The system was initiated primarily to improve emergency services access, however, it provides an extraordinarily detailed base map for the GIP.

A national real-time traffic information system – EVIS – is also in development which would also improve journey planning within the MaaS apps. This is part of a broader effort to collect, categorise, and organise transport data both at infrastructure and movements level [23].

The development of MaaS has been embedded in Vienna and in Austria in several ways. Whilst initial funding of the Smile project came from Europe, additional funding enabling the 'spin off' of Upstream Mobility was provided by the city authorities. A major national ecosystem supports MaaS development with AustriaTech acting as a centre of research and expertise on ITS (financed by the Austrian Ministry of Transport and self-raised European funding).

It should also be noted that, in Austria, there is financial cooperation between the various urban transport authorities, which enables them to purchase services jointly and share the costs according to the size of the area. This model was followed for the WienMobil application, where initial investment was borne by Wiener Linien but which was then extended to other Austrian cities (Graz, Salzburg, and Linz) at a lower cost. In addition, Upstream Mobility built the platform for the City of Hamburg.

The model is one of reducing costs to cities to the minimum – however, the assumption is that cities will still ultimately bear the running costs of the MaaS platform. There does not appear to be any suggestion other than that the platform be provided at the city's expense (through publicly owned companies) as a tool for improving the city's sustainability. There appear to be no plans to monetise it in any other way (e.g., charging shared mobility operators for inclusion).

In their 2019 interviews, the delegates from Cerema were not able to find a huge amount of economic data, however, they concluded that, in general, to develop a MaaS application using this model, a local authority would need to commit to:

- IT costs that can range from several hundred thousand euros to a million euros. These costs cover:
  - adapting products developed by companies such as Fluidtime or Upstream Mobility to local specificities;
  - deployment and installation of these products locally;
  - the development of new features according to the desired level of integration (information, routes, booking, ticket, payment) and the number of operators involved.
- Legal costs for support from lawyers on the agreements to be signed between the various entities involved (mobility service providers, MaaS application providers, etc.).

Their research indicated that in the case of WienMobil, the development of the application did not lead to specific payment between the City of Vienna and Wiener Linien. Most of the costs were covered by the Smile project, and the subsequent adaptation costs are covered as part of Wiener Linien's Research and Development budget, which includes it in its contractual assignments. Depending on future developments, additional funding requests may be made to the local authority.

In principle, for each mobility service operator, a double contract must be envisaged: between Upstream Mobility and the mobility operator (for technical integration subjects) and between Wiener Linien (as a MaaS operator) and the mobility operator.

WienMobil coexists with other MaaS platforms. Whim was released in Vienna in September 2019. After entering into a discussion with Upstream Mobility, MaaS Global launched Whim using Upstream Mobility's open platform [24].

5.5.2 Case study: Amsterdam

In 2018 the Zuidas district of Amsterdam, an important business district with many multinational company headquarters, realised it was facing extreme disruption as the road network was being reconfigured over the next 10 years. The local authorities and businesses realised that there would need to be a reduction in car travel into the area to keep congestion under control.

Whilst alternative options were available, they also recognised the huge hassle for commuters to find them and decide which were appropriate and could help in the circumstances. They put together both a long-term plan – the MaaSterplan – which had a high level of commitment and also came up with a short-term, 'bottom up' intervention based on user engagement.

The willingness of employers to embrace MaaS was key to this. The area also had both mobility demand and supply in the form of multiple options. The circumstances – a pressing need – helped ensure that people were keen to organise solutions.

The delivery of MaaS in Amsterdam was led by the City of Amsterdam. Within the city, two trials were run. The first targeted employees who drove into the area at least 3 times a week, and who had a company car plus available parking. They were offered the opportunity to hand in their company car keys for a month, in return for a €1,000 Mobility Budget. At the end of the trial, the average budget spent was €606. Of the participants, 50% said they would prefer a monthly budget with MaaS over a company car.

The second targeted employees who travelled to the area by various means of transportation. Their offer depended on their current travel situation, and all were offered between €250 and €1,000 per month as a Mobility Budget in return for driving less (or not at all). Similarly, this group found they spent less than half the budget (46% on average) and all participants were positive about the way they have learned to travel differently and the alternative modes.

The trials concluded that to successfully change behaviour, both 'sticks' and 'carrots' were needed. Sticks – pressure to change, came in the form of 'pain points' – and that major construction works could create this kind of pressure. As for carrots, to encourage uptake, incentives are needed. MaaS would need to include rewards for good behaviour. In addition, the role of employers in encouraging and organising uptake was key.

These trials were carried out with very lightweight technology, to inform the procurement process for MaaS [25].

5.5.3 Case study: Berlin

The Jelbi MaaS platform was launched in June 2019. It was launched as a cooperation between the Berlin public transport provider, Berliner Vehrkehrsbetriebe (BVG) and Lithuanian technology provider Trafi in June 2019.

Berlin rolled out its first MaaS platform at a rapid pace. Rather than being wholly owned by the Berlin authorities or its arm's length public transport provider BVG, Jelbi is the result of an agreement between BVG and Trafi reached in January 2019. From that point, roll-out proceeded at a rapid pace. The recruitment of mobility providers began in the following month and a 'soft launch' went live with bus, rail plus one provider for each bike share, e-scooter share, and car club/car-sharing service in June of that year [26].

From its start, the Jelbi service in Berlin offered multi-modal route planning with the capability for users to select, book, and pay for their trip without leaving the app. Trafi built, developed, and maintained the Jelbi platform whilst BVG marketed it and communicated with end users.

Rather than a financially motivated business case, the case for Jelbi was about collecting data for the BVG. In an interview with Jill Petzinger, Germany Correspondent, Yahoo Finance UK, Henrik Haenecke, BVG Transport CFO said that as a public company, BVG is not aiming to make money from Jelbi, and breaking even cost-wise would be fine for them down the line.

Like many, he anticipates this will happen: *'[there is] definitely a business there, because Uber and Daimler and BMW are doing exactly the same thing: They are combining their services with public transit, because they want to make money'.*

In the first instance, Jelbi is essentially a research project that is able to collect data that has not previously existed on how Berliners get from A to B.

BVG will share trip data from the app with the companies on the platform. *'We don't need personalised data—we don't want to know how you move around in the city—but we want to understand how does Berlin move around',* Haenecke said.

This 'business case' thinking is underlined by the fact that by 2022 several German providers, like Sixt Share and Share Now (at that time BMW and Daimlers' joint car share operations including Car2Go and Drive Now) had not joined the Jelbi app. This was viewed by Haenecke as their efforts to retain ownership of their customers and brand without wider participation.

> ... at the moment they have not expressed interest to be part of this. We said, okay we're going to build this up with the partners that are interested. Maybe they will join us at a later point in time [27].

In fact, by mid-2023, Sixtshare had joined the platform whilst the Share Now joint venture had been sold to Stellantis [28].

One important integration that could be specified by BVG is that of BerlKönig. Jelbi became one of the ways that people could book and pay for this on-demand bus service available in East Berlin. This service, launched in 2018, was a new development for BVG, enabling it to offer affordable shared minibuses serving areas where existing fixed-route transport options are limited. The on-demand platform is run by Via, known for its app-based service management (see also Section 5.10 and Chapter 13), however, people can also book and pay for rides at some BVG kiosks using cash.

Jelbi branding also draws together 'real world' options. There are dozens of Jelbi stations across Berlin. They are mobility hubs strategically located near bus and train stops with a high amount of foot traffic every day from commuters. The Jelbi Stations are hubs for the vehicles on offer from mobility providers that partner with the app. There are two types – the fully-fledged stations with bikes, e-scooters, mopeds, car-sharing, and ride-sharing – and Jelbi 'Points' where only two-wheeled modes are on offer.

As of 2021 [29]:

- 8% of Berliners use Jelbi
- 45,000 vehicles and 12 mobility offerings are integrated into the platform
- Jelbi users' average trip duration: 36 min (all modes together)
- Type of vehicles integrated: E-Scooters, E-Mopeds, Taxis, Ridehailing, Carsharing (station-based, free-floating), Bike sharing
- Providers integrated in the application: BerlKönig, Emmy, Lime, Nextbike, Miles, Mobileeee, Nextbike, Taxi Berlin, Tier, Voi, VBB [29]

The Jelbi app was a significant milestone for Trafi. The company started out with a business-to-consumer app – similar to that of London-based Citymapper (acquired by Via in 2023) or Israel's Moovit (which was acquired by Intel in May 2020). It launched in Vilnius in 2013, and by 2014 was live in the Baltics, Indonesia, Brazil, Turkey, India, and Taiwan.

However, founder Martynas Gudonavičius, in an interview with Amy Lewin in August 2020 for Sifted.eu admitted that at its start:

> . . . we never figured out a business model. About four years ago, we said we can't continue like this, we're not going to make money out of it.

Trafi decided to pivot and offer the tech it had built for its app to others. *'We talked to partners and ecosystem players and they liked it'*, says Gudonavičius. *'That was the click moment'*.

In Jelbi, the company boasts that it has the largest deployment of MaaS in the world. Berlin is a city of 3.6 million people and an estimated 10% of the population have downloaded the app.

It is part of a pivot common with MaaS technology (and other new mobility) providers – from B2C to B2G. Whilst Trafi continues to run its B2C app in more than 30 cities around the world, an increasing proportion of its business is with government and big tech. Trafi also licenses its software, including its multi-modal routing algorithm, to some of the world's biggest companies and is at work behind Google Maps, Apple Maps, Lyft, and the giant Indonesian delivery app Gojek, taking open data on public transport and enhancing it, for example offering more accurate predictions of when a vehicle will arrive than the authority's own algorithm can provide.

As the first of these deployments, Jelbi helped move Trafi to this business model – licensing off-the-shelf products to cities, tech giants, and transport companies, rather than offering a free product to consumers or building custom solutions for clients [30].

Increasingly, cities are licensing MaaS in the Jelbi/Berlin model rather than directly owning the technology. However, in an interesting demonstration that there is no single solution, the BVG announced in 2021 that it was commissioning a reboot of its trip planner app, Fahrinfo, combining 'all means of transport' including buses, trains, trams, ferries, e-scooters, shuttles and more, seemingly duplicates some of the Jelbi operations.

The partners for the development of Fahrinfo are Siemens Mobility subsidiaries Hacon and eos.uptrade, and user experience (UX) specialist Ray Sono. Significantly, however, the three partners are developing a new back-end system for ticket shops as well as a smart app design, which aims to provide seamless mobility for those travelling in Berlin.

Michael Frankenberg, CEO of Hacon and head of intermodal solutions, Siemens Mobility said, in an interview with Smart Cities World:

> Thanks to our long-standing partnership with eos.uptrade, we can seamlessly combine our trip planning with the booking process and contactless ticketing – a big step for BVG towards mobility-as-a-service [31].

Upgrading ticketing (and therefore the ticketing APIs) is a significant way of ensuring better-integrated mobility and, potentially, mobility accounts on a variety of platforms. In addition, the plans are to enable all BVG online sales channels to be connected to the new multi-client back end via an open interface. The BVG apps, Fahrinfo and BVG tickets, could potentially be integrated with the Berlin-Brandenburg transport association's (VBB's) Bus&Bahn app through the API, and additional applications may be connected in the future.

As the incumbent, the current BVG Fahrinfo app has far more downloads than Jelbi – 5.5 million downloads by 2021 – and is one of the most popular public transport apps in Germany. With such a huge user base, it is worth developing. Plans include a fully integrated ticketing and an intelligent, proactive travel companion. Fahrinfo aims to provide better predictive recommendations for selected routes and it will automatically consider individual mobility preferences, take into account possible forecasted disruptions along the route and transmit them via push to the user.

For numerous cities, the dilemma still revolves around the extent to which they should share information and ticketing data, how fully they should adopt third-party MaaS platforms, and how many mobility service providers they can attract under their umbrella. Like Vienna – but via a different route – the Berlin approach appears to be hedging its bets to some extent.

5.5.4 *Case study: The UK's Future Transport Zones*

Although some MaaS projects existed in the United Kingdom before the announcement of three Future Transport Zones in 2020, they were the first schemes expected to provide an insight into how MaaS could or would work in the United Kingdom. The three zones, Solent (south coast area incorporating the cities of Southampton and Portsmouth), West of England Combined Authority (covering Bristol and Bath in southwest England), and Nottingham and Derby were all required to create a MaaS platform and data hub as part of the funded projects.

Each of the Future Transport Zones in the United Kingdom took a different approach to the development of MaaS. Solent Transport, appointed developers Trafi relatively quickly in 2021, with the expectation that the platform would be built in parallel with plans and partnership development.

The ultimate ambition for the Solent app is to permit travel around the Solent region with nothing more than a phone. The aim is for it to unlock e-scooters for rent, enable purchase and scan tickets for trains, buses or ferries, and even rent cars from car clubs.

However, on the ground, those managing the process recognise that the technical work on the platform itself is not the biggest element of the workload. In 2022, Chris Hillcoat, Programme Lead for the Future Transport Zone said:

> Technology is by no means the only area of complexity. Nor are data standards so essential – [Bus Open Data Service] has definitely saved the developer time but essentially data integration is what technology developers do. The key learning is that the main focus has to be around partnership and stakeholders and that's more involved than the technical work.
>
> Technically we have many pieces of the jigsaw in place: the core platform, user authentication, ID verification, phone authorisation, a payments and refunds module, customer account set up, cloud hosting and a journey planner that is tailored to our region including specific data sources. Ticketing integration has begun with some operators. We have a new brand. But the biggest workload has been the work with partners [32].

As the app is selling tickets on behalf of operators, a bilateral agreement is needed for each ticketing integration. The contracts interact with technical issues ensuring that each one is different. Each operator has a bespoke agreement – so some activities will be on the MaaS platform and some on the operator's platform and the split will depend on factors including technology and commercial models. These agreements also cover revenue, data, and marketing terms.

This speaks to the complexity of transport ownership in the United Kingdom – where privately owned operators run most services. Privately owned operators do not necessarily see the opportunity or necessity of MaaS. They are not aligned with the authority in the same way as publicly owned transport systems are in cities like Vienna. With resources stretched on both sides, contracts are slow to be negotiated and signed.

Nevertheless, the Solent business case is expected to rest on a mix of revenue sources including sharing costs with other authorities (in a similar way to the Vienna model) and commission on ticket sales.

> We have put in the work and time to set this up. Once it's functioning, adding another authority would be a fraction of the time and cost. There would also be benefits to both travellers and transport operators if the platform includes wider regions – it could offer journeys across authority boundaries. Chris Hillcoat [32].

5.6 Resold to cities business model

With many open-access transport APIs available, it is possible for technology providers to create MaaS platforms that ingest available APIs and use them to plan journeys and resell tickets. The extent to which this can be achieved solely by a technology provider greatly relies on the accessibility of information and the comprehensiveness of open data. For instance, if transport data is radically open it may be possible for the platform to resell tickets as well as plan journeys. It is much more likely, however, that it will be necessary for the platform to negotiate locally for the rights to do this, and to negotiate either a discount with the transport operator on ticket prices or a fee for attracting customers (essentially for advertising the ride). This is particularly important for it to retain a positive cash flow as there are costs for card processing, servers, etc. Whim (created by MaaS Global) was early to market with this type of MaaS. It has been more successful when it is able to create subscription services so that people can access a certain amount of travel for a fixed monthly fee. This works where the public transport provider will provide bulk ticketing discounts or other incentives to sell tickets as well as operators providing some kind of revenue stream in return for bookings via the platform.

One of the earliest concepts of MaaS was that of a technology provider being the first mover (and intended financial beneficiary of MaaS). In the first decade of the 21st century, the possibilities of digitisation began to look limitless. The rise and rise of the mobile phone, renamed the smartphone was key. It was the cornerstone of interaction in social and, increasingly, economic spaces and was intertwined with start-ups promising new ways to share experiences, shop, and manage all manner of household functions and life stages – from travel booking to online meditation. These same start-ups claimed to be creating 'platforms' that promised investors 'scale' – access to global markets with one simple technology, and the accompanying vast potential revenues.

Unsurprisingly, transport was caught in the wave. The incredible growth of Uber (in terms of the number of people downloading the app and the global coverage it achieved) is probably the most high-profile and widely adopted of the 'platformisations' in transport. Its success perhaps speaks to the relatively simple concept of an electronic booking of a taxi and the compelling customer interface that kept travellers informed, giving them a semblance of control, whilst also managing payment digitally.

When Uber was founded in 2009, the riposte – in the form of a combined public transport and multi-modal service – was already formed in the minds of some of the first MaaS start-ups. Sampo Hietanen, founder of MaaS Global, conceptualised Maas in 2006 and published an influential paper for Eurotransport in 2014 setting out his vision of 'Mobility as a Service' whilst CEO of ITS Finland [33].

Meanwhile, in Sweden, Hans Arby, launched a pilot project MaaS in Gothenburg in 2013 for 6 months. The service, Ubigo was directly offered as a car replacement in which people trialled giving up their car for a month and using the app instead [34]. Some of the initial pilots were short-lived, however, they provided the foundations for many future developments that have furthered and further defined the concept of MaaS.

Whilst Uber gained enormous traction (a combination of a simple concept connecting people with a single mode with extraordinarily successful fundraising), MaaS technology platforms had a harder job. The basic mechanism of fundraising (as deployed by Uber) consists of pointing to an enormous consumer market and fundraising from investors based on its potential.

MaaS platforms have a more complex concept to explain, whilst managing a more complex service with multiple operators. This proved a harder (though not impossible) story with which to raise private capital.

The complexity of creating a MaaS was significantly reduced in 2017 when Finland decided to open up transport data, enabling third parties to use transport operators' data including timetables, real-time information, and ticketing data (see Chapters 3, 16, and 17).

This move provided a steer towards open data in several European countries. Importantly, it allowed MaaS technology platform providers to point to an environment in which they could demonstrate the viability of their offer.

5.6.1 Case study: Whim

Whim is the MaaS app operated by MaaS Global, the company launched by Sampo Hietanen after publishing his paper on MaaS in Eurotransport in 2014.

Whim was one of the first apps to provide digital access, payment, and mobility bundles. It was launched with full MaaS functionality – allowing users to view timetables, buy tickets, and plan journeys using public transport, taxis, car-share, and more, all in one place. In Helsinki, users can use the app to book public transport such as ferries and buses, over 2,500 taxis, and rent cars from more than 50 locations. It incorporated bike share and different types of access to car rental aiming to provide four different styles from flexible car clubs, bookable back-to-base car club rentals, daily to weekly rentals, and longer hires.

The Whim app was designed to simplify the travel experience and budgeting for transport. Users were offered a monthly subscription or pay-as-you-go plan. There was much made of the monthly subscription being equivalent to and replacing a car finance payment.

On the operations side, MaaS Global (as the operator of Whim) decides whether or not to include a mobility service in its Whim offers, according to its own criteria. MaaS Global takes into account the technical conditions of integration into MaaS (which it sees in terms of the existence of quality open data and the openness of the ticketing system), the relevance of the mobility packages offered, and the practicality of agreeing with the transport operator concerned. MaaS Global can use standard APIs (open through the Finnish Transport Act) by signing the standardised conditions of use of the APIs proposed by the transport authority. For other mobility services such as taxis or car rentals, MaaS Global signs agreements on a case-by-case basis that define the conditions of access to the service, fare systems, and payment terms.

Support from investors and service providers meant that MaaS Global could buy transport in advance and repackage it to sell to its customers via the different mobility subscriptions and payment plans it offers.

The first trip on Whim took place in 2016 in Helsinki, Finland. Insights from the initial tests show interesting travel patterns. MaaS Global was initially surprised to find that the use of public transport was higher than expected. It surveyed the modal split before and after people subscribed and found that use of public transport moved from 48% to 74% quite quickly. The team's expectation was a shift of just 10%.

The subsequent launch of Whim in the West Midlands in 2017 was less of a success story. The deregulated environment in Birmingham meant that integrating with bus ticketing was not straightforward for both technical and business reasons. On a technical level, the operator did not have a mobile ticketing solution at the time. On a business level, the agreements possible in Finland which included pre-buying tickets at a discount did not appeal to the operators. A workaround had to be created that combined the Transport for the West Midlands Swift card (which could be used to hold season tickets and other transport passes) with the app. The Swift card could hold a limited range of bus tickets with individual operators able to offer better 'single operator' tickets. This meant that overall the combination had a more limited audience than the very flexible offer possible in Helsinki. Despite the issues with public transport, the platform was able to build strong relationships with other providers including shared transport, ride-hailing, and car hire.

These other agreements covered much larger areas than the West Midlands transport zone, creating a powerful network. Car hire bookings could be made from any Enterprise in the United Kingdom. Wherever Gett taxis were available – both in the United Kingdom and internationally – the Whim app could book them. Likewise, Nextbike in Birmingham, Cardiff, Glasgow, and Milton Keynes (as well as 150 locations globally) was fully integrated with the app.

In many ways, adding a new 'supply' of transport was crucial to creating a network effect to make Whim more attractive than car ownership. The main barrier in the United Kingdom was around integrations with public transport ticketing on both a technological and regulatory level. Pricing of supply (in this instance public transport tickets) is also an issue – whilst some suppliers see an advantage to offering 'wholesale' or discounted rates, others do not.

Other launches, particularly where public transport was strong, were more straightforward. Whim was launched in Vienna in 2019 with the Wiener Linien (the public transport provider) plus a taxi company and the city airport train (CAT). It incorporated a micromobility provider later on. In Switzerland in 2021, the initial offer combined all forms of public transport across six cantons along with micromobility, later adding other shared transport partners [35].

Although not at the same kind of scale as Uber, MaaS Global has attracted investment from international companies wanting to understand how transport is being disrupted and what customers really want. Both Toyota, the world's biggest automotive manufacturer, and much smaller public transport innovator TransDev have invested in the company. According to Sampo Heitenan, MaaS Global CEO, as an independent player, MaaS Global has a better chance of creating partnerships to further the network, a feat that individual competitors might struggle with.

Investment is often in line with the territories Whim serves – with Swiss insurance group Vaudoise Assurances investing in the funding round before the launch in Switzerland. Companies including Ferrovial and Tesi (Finnish Industry Investment), NordicNinja, and Mitsui Fudosan were also involved in funding rounds. Mitsui Fudosan was one of the partners in a 2019 trial that incorporated MaaS into urban development [36]. In total, MaaS Global reported an investment of almost $65 million by 2022 [37].

Although MaaSGlobal's trajectory was generally positive, and it expanded to cover several global cities, its operations were affected by the COVID-19 pandemic. The pandemic and associated lockdowns reduced travel in many countries reducing revenue for all forms of transport. The energy crisis of 2022, following the Russian invasion of Ukraine, added to costs and uncertainty across the world. By mid-2022 some services were suspended and the company announced that it was 'reorganising'.

Whilst usage was in the process of growing again post-pandemic, and the company had hit €1 million recurring revenue per month for the first time, it was also experiencing what founder Sampo Hietanen described as a significant shift in 'the funding environment and investor expectations for start-ups like us'. These changes forced it to reorganise its team and find new partners to move forward.

This vulnerability to market forces is aligned with other evidence that a MaaS platform is not necessarily a license to print money. In parallel to MaaS Global and Ubigo, Lithuanian start-up Trafi was launched as a B2C offer. It offered a free journey planning app to consumers in Vilnius in 2013, expanding to other Baltic countries, Indonesia, Brazil, Turkey, India, and Taiwan the following year. After a couple of years, its free-to-use model was not generating revenue and it pivoted to licensing its software to cities to build MaaS platforms and also its multi-modal routing algorithm to other operations.

5.7 Resold to companies business model

This model allows new entrants to the shared mobility sector (bike share, scooters, shared cars) access to a white-label platform that helps them sell the use of their assets as part of a customer-facing multi-modal offer.

These providers often offer the building blocks of MaaS as services – whether APIs, SDKs, or SaaS. Ticketing companies have also started to offer software-as-a-service integrations into this market so that public transport can be offered via shared transport operators' apps.

Some of the early MaaS offers that fit this model could be described more as shared-fleet management or transport integration platforms. At their simplest, some platforms launched offering a 'marketplace' for connected vehicles (cars or micromobility) through which end-users could book and pay for their trips. However, in time they evolved their business models to broaden their appeal. Those platforms launched into this space may have started as a mobility marketplace or a full-service app aimed at specific markets but they have become a broad range of platforms and software – often enabling complex fleet management and integrations between different fleets and ticketing systems.

This model has become one that allows the broader on-demand or shared mobility sector (ride-hailing, bike share, scooters, shared cars) access to white-label platforms or integration layers that bring together multiple modes and connected vehicles. These providers often offer the building blocks of MaaS as services – whether APIs, SDKs, or SaaS – (almost to the point that they enable the rather iterative concept of MaaS-as-a-Service).

Some versions of this type of platform are 'thinner' layers which just enable a certain kind of integration, for instance, integration between providers on complementary territories. The benefit for the operator is that their customer can remain within their app but 'roam' beyond their normal area. Others are more full-service offers that enable operators (particularly new market entrants) to sell use of their assets as part of a customer-facing multi-modal offer.

In the fuller version, the platform helps manage the vehicles and customers, bookings and payments, very closely and in mapped areas. Where the same company offers its platform to multiple shared mobility providers, it can offer them the opportunity to integrate with each other to offer a bigger network or complementing modes. Operators using these platforms pay fees based on set-up costs and usage. Some of the platforms work with operators on hardware (in the shared mobility sector vehicle location, tracking, and IOT needs to align closely with the platform). Many operators are backed by these platforms, which include the likes of Wunder Mobility, Here, and Ridecell.

To understand the companies and technologies operating this business model and exactly what they facilitate, we need to break down MaaS into layers (which are further detailed in Section 3.3):

- Customer interface (sign up, data, id verification, account, journey planning, trip details)
- Mapping (base map)
- Transport mode (with geolocation, connectivity, locking, or some kind of token or ticket verification)
- Back office (event management (e.g., booking, locking unlocking, entry, egress), cost calculation/reconciliation, payment)

These companies offer some or a combination of these layers as services. For instance, Wunder Mobility offers the digital elements that allow a new operator to manage a fleet in a given area – for instance, a scooter operator could set up using the Wunder Mobility platform to manage its scooters and to provide an app for users. The platform can track the vehicles for operations (so they can be rebalanced or charged) as well as handle communications with customers from the basic booking and payment interface to special offers and incentives.

Platforms like Wunder Mobility are intended to provide robust technology support, able to handle the number of transactions per day (such as calls on the API) and scale up without the operator needing to invest in its own computing infrastructure or programming.

Ridecell provides fleet management services connecting vehicles into broader systems. By combining different business models for making vehicles available, it claims that it can offer mobility as a service in its broadest sense:

> With Ridecell Shared, you have a mobility solution that fits the widest range of customer needs—from free floating to station-based services and everything in-between. Let your customers rent by the minute, the hour, the day—or even longer, with options for multi-day and multi-week rentals and recurring monthly billing. And with real-time vehicle monitoring, and automated dispatch for sanitizing and servicing, you'll get more efficiency while your customers get peace of mind [38].

The more targeted versions are thin software layers that allow integration between operators or modes. This may mean the integration of providers on complementary territories where the benefit for the operator is that their customer can remain within their app but 'roam' beyond their normal area. It may also mean integration between types of transport – for instance, ticketing companies that have started to offer software-as-a-service integrations into this market so that public transport can be offered via shared transport operators' apps. They enable cooperation between providers either actively or passively.

5.7.1 Case study: Splyt

Splyt enables super apps, wallets, and travel platforms to fully integrate mobility services. For instance, Splyt offers two products that link on-demand services together:

- API: Operators gain access to Splyt's network through their interface. By connecting to the Splyt API the operator can give their end-user access to on-demand services from over 65 partners.
- SDK: A full software development kit enables the operator to integrate their app with up to six on-demand services with a full user interface that can be integrated 'out-of-the-box'.

Operators are motivated to invest in these integrations because they aim to establish a lasting customer experience, ensuring that users continue using their app instead of exploring rival alternatives.

The Splyt app has successfully enticed users to return to the app repeatedly instead of exploring competitive alternatives. This is achieved by being able to offer more services over a wider area (beyond the area the operator directly serves with its own vehicles). This is important for customer experience – once people are familiar with one app and have lodged their details and payment options in it there is a decided reluctance to keep adding more and more apps and repeating the process. By making their app more 'useful' – working over a wider area or for more modes of transport – the operator can keep the customer returning to their app rather than trying (and possibly migrating) to another service.

In terms of MaaS, these apps provide a more comprehensive travel planning, booking, and payments service. How comprehensive, however, depends on the transport options integrated.

5.7.2 Case study: Masabi

Masabi is a mobile ticketing platform that provides 'Software-as-a-Service (SaaS)' based fare collection and ticketing for public transport.

Early on in its development, Masabi created some stand-alone rail apps in the United Kingdom (where it originated). However, it developed its business strategy and created a single ticketing platform – its JustRide platform – which it then licenses to customers. The JustRide platform is used by Boston and New York commuter rail and several other public metro, bus, rail, and ferry services – including Preston Bus, Thames Clipper, and National Express in the United Kingdom.

Whilst the JustRide core is a full fare collection platform, Masabi can provide a software development kit to allow it to be embedded in other apps – for instance – trip planning apps. Operators using the JustRide platform can opt to allow the software development kit to be used to offer further opportunities to sell tickets more widely. The SDK allows third-party applications to request fare types, make payments, and deliver visual and barcode mobile tickets to a passenger through a secure ticket wallet. So a journey planner could add seamless payment for any participating operator on the JustRide platform – but this need not be limited to transport applications – in theory, a customer booking a cinema ticket could buy their travel at the same time if tickets were integrated.

Whilst Masabi is a mobile fare platform it works on an account-based payment model – and can provide integrations into gate systems and multi-format readers. This functionality means that although it has no aspirations to provide a MaaS, it can enable the integration of mobile payment options and ticketing into third-party apps.

In the United States, the integration of Masabi's ticketing platform with the Uber app enables Uber to provide public transit options in those locations where Masabi provides the JustRide fare collection and ticketing platform.

The broadest version of this B2C type of MaaS is the business travel MaaS provider. This business model taps into the enormous market of company travel requirements. In some cases, it addresses the issue of grey fleet – where employees use their own vehicles for work purposes. Grey fleet can be an issue for employers as they need to ensure that drivers using their own vehicles are properly insured and that the vehicles meet safety and environmental standards (which is not straightforward as, by definition, the employer does not own the vehicle). For some employers, platforms streamline the management of expense claims or facilitate keeping travel within a company ethos or sustainability standard.

5.7.3 Case study: Mobilleo/Fleet on Demand

In 2017, Mobilleo was the multi-modal platform of the UK business travel service Fleet on Demand. It combined both public APIs and private agreements with several providers to include their services and was marketed to businesses. It allowed business users to compare all forms of available transport and the cost of making the journey in a private vehicle, allowing them to find, book, and pay for, the most suitable mode(s) of transport for the journey, all in one place. As a business-facing app, it provided the facility to pay by invoice and simplify business travel accounting and employee expenses.

The platform claims its utility is in both cost savings and increased efficiency for business, for instance, to reduce the amount of time taken by travel bookers to plan journeys around Europe, and in another case to reduce a company fleet size [39]. Since 2017, the platform has diversified and developed to provide a public-facing MaaS solution. This diversification means that it has moved from B2B provision and has added government as a customer – shifting (as has often happened in MaaS development) to a B2G business model.

It is the engine behind MaaS platforms including a trial project aimed at university students in Derby, DerbyGo, and the large-scale public platforms commissioned by Scottish Transport Authority SESTRANS. In 2023 it was announced as the provider for the Transport for the West Midlands MaaS solution.

5.8 Resold to destinations model

This model is often sold to destinations to enable the inclusion of travel with events or other destination booking. It can also be sold to employers to manage business travel accounts, and commuter costs or for reward programmes to reward positive travel behaviours. This is also a model in which a local authority may be the customer of the technology provider (in some respects this model blends with city authority business models, depending on the procurement processes, transport availability, and degree of open data). Several companies launched as journey planners or even full MaaS providers but have pivoted to reselling their platforms, for instance, Skedgo, Meep, and Transit. See Section 5.4 and Chapter 13 for further details.

The concept of destinations taking on a more significant role in organising travel to their location grew in tandem with the surge in ride-hailing services. Many forecasts were made, suggesting that the cost of transportation would decrease due to automation and electrification, for instance, this RethinkX report, published in 2017 predicted:

> By 2030, within 10 years of regulatory approval of fully autonomous vehicles, 95% of all U.S. passenger miles will be served by transport-as- a-service (TaaS) providers who will own and operate fleets of autonomous

> electric vehicles providing passengers with higher levels of service, faster rides and vastly increased safety at a cost up to 10 times cheaper than today's individually owned (IO) vehicles. These fleets will include a wide variety of vehicle types, sizes and configurations that meet every kind of consumer need, from driving children to hauling equipment.
>
> The TaaS disruption will be driven by economics. The average American family will save more than $5,600 per year in transportation costs, equivalent to a wage raise of 10%. As a result, Americans will keep an extra $1 trillion in their pockets, potentially generating the largest infusion of consumer spending in history.
>
> The TaaS disruption will be both quick and inevitable on a global basis. Below, we lay out a baseline analysis of this disruption, followed by a study of its implications for the car and oil industries and a discussion of the choices that society will face [40].

At the time, the ease of summoning ride-hailing vehicles had captured the imagination of futurists and the idea that travel could be incorporated into the offerings of destinations from restaurants (book your table and your ride there at one inclusive price) to cinemas and shopping centres. Bundling on-demand transport into apartment rent was also seen in some housing These ideas figured in more than a few MaaS conferences of the time.

Whilst there are few examples where MaaS (or transport in general) is embedded in local destinations, the use cases where examples arise are around business trips and long-distance/high-value leisure where hotels, car hire, and ride-hailing are often bundled together. The other key area is high-profile event management. In this scenario, the large numbers of people gathering in one location necessitate the implementation of measures to alleviate traffic congestion and guide individuals effectively. Sporting events and music festivals in particular draw enormous crowds and have created congestion both on roads and on public transport.

There are several examples of low-car and car-free housing where providing shared mobility – from car club vehicles to e-cargobike share – to augment public transport enables the development to require a reduced land footprint. However, developing specific platforms for these developments has not been a major focus for the market. University usage is perhaps more advanced, where the DerbyGo app is being used as a behaviour change tool to try and reduce the pressure of student vehicles on parking spaces in the city.

5.8.1 Case study: 2017 UEFA Champions League Final, Cardiff

The 2017 UEFA Champions League Final in Cardiff drew huge numbers of fans to the city, which has limited transport and accommodation capacity. An app was developed by consultancy Steer Davies Gleave to help engage with fans converging on the city. With the dual purpose of making the fan experience more pleasurable and helping to run the event more efficiently, it contained both important and useful content for fans as well as giving the event team an opportunity to raise awareness and influence behaviours before and during the event.

The application developed included match information and a guide to Cardiff in five languages, as well as a tailored journey planner module, ticket/travel purchasing links (including Great Western Railway and Uber), and links to accommodation.

For fans travelling to the final, the app was able to provide targeted messages to fans depending on their locations and behaviour. The primary target was the travelling fans from the two finalist clubs, and with over 10,000 downloads captured a representative sample of these fans to engage with them directly in the days leading up to the event.

The app enabled real-time visualisation of fan movement and user audience segmentation based on geo-behaviour. These segments were used for targeted push notification messaging – enabling 30,000 messages to be sent. In total, there were 500,000 screen views and an average of seven sessions per user.

The situational awareness evidenced aggregated travel insights and supported the team in the Transport Coordination Centre by quantifying travel times on key corridors, arrival and departure profiles at key ports of entry, and unexpected travel behaviour [41].

5.9 Autonomous vehicle model

One of the early visions of MaaS was as a platform for summoning, managing, and optimising on-demand electric vehicles to replace car ownership. This vision has become somewhat blurred by the practical difficulties of developing reliable and safe autonomous vehicles and the conflicting views of what autonomous vehicles should be licensed for; as shared transport only or for single occupancy (or indeed single owner) use.

Ride-hailing apps like Uber and Lyft have (at points in their business growth) predicated profitability on future adoption of driverless vehicles (meaning that the app and the operator are one and the same). However, this has yet to result in anything other than limited pilot applications in the US cities with grid-type street networks.

Meanwhile, other small shared vehicles (mini-buses and large cars) are being trialled as on-demand shared transport in European and American cities. Both build detailed maps of their area and constantly collect data to augment the maps and enable their driving. This is very much at the trial stage, and their use and integration with MaaS will depend very much on the reality of operations within cities and on the types of businesses that emerge with viable autonomous vehicles.

Whilst autonomous vehicle-based MaaS has proved limited, platforms have also become important for the management of electrification (with journey planning for electric vehicles essential to enable timely charging) and also for the management of vehicles in cities – Clean Air Zones, Congestion Zones, and also for parking.

There is less integration in these services so far, but with an intense interest in better kerb management and using parking policy and charging to manage traffic, there is great potential for future platforms that plan vehicle journeys, access charges, and parking charges, help drivers (or connected and autonomous vehicles) comply with city requirements. These too could become instruments that enable drivers to leave their cars at optimal points and plan their onward journeys using public and shared transport.

Whilst 'autonomous vehicle MaaS' is currently in the slow lane, car-management platforms that enable people to drive less are likely to come into sharper focus in time.

In some of the earlier hype over mobility as a service, there was a distinct version of MaaS that anticipated fleets of driverless cars being summoned by apps, a constantly available swarm of robotaxis, dispensing with the need to own vehicles (and, potentially the need for mass transit as well). This version dates from early excitement that autonomous driving was achievable in the near term – around 2017 – when Waymo was able to trial autonomous taxis on the grid streets of Phoenix Arizona (with a safety driver) and Uber was developing its own autonomous vehicle, as were more traditional automotive brands. The progress in autonomy is still slow, with no fully autonomous vehicle on the roads (level 5 autonomy as defined in Section 4.5) as of mid-2023.

Whilst the vision of the robotaxi swarm has been undermined by modelling showing the increased traffic, congestion, and associated environmental impacts that would likely be created in this scenario, other elements of a personal vehicle-based MaaS have developed.

The automotive industry, broadly defined as organisations involved in the manufacture and selling of cars, either as units or as a service, has been taking an increasing interest in the development of Mobility-as-a-Service. Since the latter part of the twentieth century, the automotive business model evolved to involve financial services which has changed purchasing patterns. Rather than being paid in full for a vehicle once it leaves the showroom, customer behaviour has shifted increasingly towards leasing or buying cars on finance. Each car that rolls off the production line is now more likely to represent an income stream rather than an immediate full payment.

There is long-term interest in technology innovations along with the changing demographics and behaviour the industry faces. However, it is probably fair to say that the automotive industry is hedging its bets with respect to both MaaS and the roles it could play in it. Some have entered the MaaS market not by establishing their own MaaS, but by making strategic investments in MaaS start-ups and companies. For instance, Toyota was an early investor in Whim. Others developed their own models for on-demand services (for instance VW's MOAI) or carpooling (Porsche). Further, leasing, hiring, and 'cars as a service models' have been invested in by the automotive industry. This ranges from leasing models where the car is on long-term hire, to subscription models to car club (car sharing).

Examples include Land Rover, which offers a subscription service that involves paying a one-off joining fee and being able to choose a vehicle that is delivered and ready to drive with insurance and tax included. The subscription offers the option to pause 'if priorities change' after 3 months and subscribers qualify for a fresh vehicle after 6 months [42].

Whilst this is more like a luxury service-oriented lease, car club models provide more practical integration with MaaS. Both BMW and Peugeot PSA Group have invested in car clubs. One of the key innovations was BMW's launch of the DriveNow brand in London in 2014 with electric BMW i3s being run as a floating car share model in which users could pick up or drop off their vehicles anywhere within permitted zones. Although popular, BMW changed its brand and business strategy and partnered with competitor Mercedes on new mobility developments in 2018 [43]. Big ambitions for the cooperation, rebranded as Share Now, were short-lived, reversing from growth to pulling out of North America and reducing its footprint in Europe in 2019 [44]. The new mobility cooperation business was sold to Stellantis in 2022 [21], to augment the Stellantis Free2Move mobility brand (which was originally launched by Peugeot in 2016).

Whilst various headwinds have impeded the automotive industry developments in the area – notably the 2020 COVID-19 pandemic, followed by the energy crisis precipitated by the invasion of Ukraine, there is no less interest in autonomous vehicles and driving. The challenge is funding and financing within an industry that is under pressure.

Perhaps ironically, as the automotive industry interest is throttled, there is increasing interest from local authorities in how MaaS could be used to directly interact with motorists – through clean air zone charging, park and ride schemes, and electric vehicle charging integration. This is particularly true in the United Kingdom where the Bristol and Nottingham and Derby Future Transport Zones have attempted to specify the interaction between driving or parking within their MaaS platforms to enable behaviour change.

This fragmentation of the automotive model of MaaS is probably most beneficial for the integration of autonomous vehicles in public transport. The use cases for autonomous vehicles being tested across Europe focus on their integration into the public transport network. The last mile is the most expensive part of the public transport network, and yet it is essential to ensure that people are able to access transport and use rapid transit. In low-density suburbs, rural and peri-urban areas where high passenger numbers and economies of scale are unlikely, the opportunities to bring down the costs per passenger are limited. Reducing the driver cost is one possible option – which is where autonomy comes in.

An EU programme, Horizon 2020, trialled several different integrations of autonomous vehicles in public transport.

In Lyon, autonomous shuttles were set up to transport people from tram stops to a stadium development under construction. The development was expected to attract only a limited flow of passengers throughout the day whilst it was unfinished, hence the use of small vehicles to take people from the station to 'virtual stops' within the stadium. The shuttles were run as on-demand vehicles, booked and managed through a Demand Responsive Transport (DRT) platform created by Padam Mobility, integrated with the autonomous driving software.

A different type of population was served in the Île-de-France area. This area, surrounding Paris is different from the highly connected central zone. The population is less dense, with people often living and/or working some distance from the rapid transit network. An autonomous shuttle was tested to connect an end-of-line station – Saint-Rémy-lès-Chevreuse – with a campus 2 km away. Autonomous electric vehicles carry up to four people along an open road. The vehicle software had to obey road rules and take account of conditions and other road users.

The trials focused on incorporating autonomous vehicles into passenger transport, and as such, the vehicles had to be able to interact with passengers. The DRT platform managed the passengers' interactions with the vehicle from booking a ride through to safely leaving the passenger at their destination. It helped passengers understand their trip – from showing passengers their vehicle travelling towards them, tracking its progress in real-time to providing and helping them validate their ticket.

Booking information from the platform ensured that the vehicle arrived on time for each booking and opened the door for the passenger(s). The platform also confirmed when each person has boarded so that the vehicle can proceed. At the end of the journey, it determines where the person leaves the vehicle at the end of the booked trip.

In addition, the software optimises the vehicle route to serve all the trips requested, managing the pattern of picking up and setting down other travellers en route and optimising the use of the fleet.

These proof of concept trials were successful and the second iteration was launched in three new locations through the EU's ULTIMO programme in 2022.

Although small shuttle trials have been relatively rare in the United Kingdom, there is one notable project: a full-sized autonomous bus was tested in 2022. The service, operating over the Forth Road Bridge, was launched carrying passengers in May 2023 [45]. Fitted with a ground-breaking sensor and control technology, these buses can run on pre-selected roads without the safety driver having to intervene or take control. The buses can carry up to 36 passengers over the 14 miles across the bridge, with capacity for over 10,000 passengers a week.

The investment required to get autonomous vehicles in use is huge – the ULTIMO project alone has budgeted for €40 million (combining EU and Swiss resources) across three test sites for 4 years.

As development continues, involving enormous budgets, there is the possibility that the investment can only be justified by the development of highly utilised vehicles rather than private cars. As resources are constricted by the limits of planetary resources and carbon budgets, the constraints point to future returns on this investment only being achieved through the kind of utilisation that public transport achieves.

5.10 The consolidation of business models

There have been numerous attempts by a variety of authors to understand the degree of consolidation that is taking place across MaaS now that many of the original trials are reporting. This depends to a significant degree on definitions of MaaS and is based upon the technical characteristics of the MaaS offering [46]. Our observations presented here are intended to be general observations in relation to the commercial offering by MaaS providers.

5.10.1 Evolution vs. revolution

The business models of MaaS have not remained static. Whilst some of the launches of the apps and platforms have been described as revolutionary, harnessing the language of paradigm shift, and the business models of the tech giant which scaled through millions if not billions of users, there has been a tempering of MaaS business models on contact with reality.

Transport systems are more constrained than the world of social media, bounded as they are by vehicles, road, and rail space, which limits somewhat what can be offered. More so because transport also depends on a degree of investment in physical infrastructure and localisation at odds with the (almost) limitless potential of the virtual world. It is not possible to offer infinite transport options: only those that are bounded by space, time, and the infrastructure are available. At the same time, within these constraints, people have demonstrated a keenness to adopt better systems for accessing transport with better information and automatic payment processes.

The business models of MaaS have been iterated between the available technologies, the available infrastructure, public policy and investment, private investment and business models, and the willingness of people to adopt new ways of travelling and means of planning, booking, and paying for them.

Different national frameworks have provided different baselines and shaped the development of MaaS in their contexts. In the course of the book, we will see how these different business models have played out in contact with reality in greater detail. There has been some convergence of MaaS business models over time, and this is discussed at the end of this section.

5.10.2 The business to government shift

The search for a business model in MaaS has followed many paths – and not all are linear. One strand of MaaS development has been that where a provider has started out as a private enterprise and then developed into a MaaS-type offering as a means of diversification or business model development. Another is where the enterprise has pivoted from being largely consumer-facing to seeking markets with business or government. This pivot has been driven partly by commercial realisations and partly by the changing global financial situation. Whilst most MaaS development has happened post the 2008 financial crash in a period of growth, the COVID pandemic created economic headwinds during 2020 and into 2021. The anticipated recovery has been undermined or reversed by further turmoil due to war at Europe's borders, the related energy crisis and friction in global trade. Despite the most severe impacts being manifest in Europe, the entire global economy showed signs of slowing throughout 2022. In uncertain times consumer-facing businesses often must rethink their offer, their markets, and their business models to survive. It has been no different in the world of transport technologies and MaaS.

Uber is an interesting case in point. Founded in 2009 as UberCab, Uber started life offering 'ride-share' via a slick app. With glossy marketing and some dubious business practices it quickly snowballed, despite contention with traditional taxis and regulations – and indeed, its own mounting losses.

Early on, in line with a start-up culture aiming to gain market share rapidly, Uber spent a great deal of money acquiring both customers and growing its capacity – the drivers that enabled it to offer the new service. Whilst taxis were in no way a new service, the customer experience of ride-hailing was. Uber also paid excellent rates to drivers to establish itself in territories and ensure there was capacity to meet the demand it had created.

Uber also claimed that its algorithms could help predict rides and best channel resources towards anticipated pick up locations. The predictive capacity and smart routing capabilities were touted as ways to reduce the environmental impact on cities by maximising vehicle utilisation. Whilst its technology was undoubtedly advanced and data-rich, the claims did not bear academic analysis. Both Uber and its competitor Lyft have been found to be contributing to additional congestion rather than reducing it [47].

Since its inception in 2009, Uber has grown astronomically and struggled to make a profit. This tension probably reflects a theme common to much of the transport sector – the ubiquity of travel and the public reluctance to pay for it.

Key milestones have accelerated over the years, with the service reaching 1 billion trips by the end of 2015, 5 billion trips by mid-2017, and doubling to a total of 10 billion trips by June 2018. By 2021, annual trips had reached 6.3 billion per year.

However, for Uber, its technology and its reach have enabled it to experiment with multiple business models. Along the way, forays into freight, delivery, autonomous driving, bike, and scooter share rank among the attempts to create a profitable business model. Whilst some areas are more profitable than others, overall, by the end of 2022, the company had only turned a profit in 1 year of its existence – 2018.

The growth of Uber as an 'anything app' is where it intersects with Mobility-as-a Service. As a widely used, consumer-facing app, with 118 million people using Uber (or Uber Eats) at least once a month in 2022 there is a huge opportunity for Uber to effectively sell other forms of transport. Capitalising on this to become 'a one-stop shop for transportation' has become a stated part of the company's vision.

In 2018, Uber started offering trips by bike directly in its app, via its acquisition of the electric bike-share start-up JUMP. However, the acquisition was unwound and JUMP sold to Lime in 2020 (possibly due to the difficulty of integrating bike-share – which relies on exceptional hardware management – into Uber's very tech-focussed operation). The 2 years of operation did give the world an insight into what happens when bikes are offered alongside cars. In San Francisco, JUMP bikes were integrated into the Uber ride-hailing app in 2018. Once the full integration had taken place, Uber's analysis indicated that users'

Uber trips went down 10%, whilst their overall trips (comprising both JUMP and Uber) rose by 15%. During the weekdays when the congestion was at its worst, the decline in Uber trips was higher at 15%. In a 2018 Medium post about the integration of 'Uber under the Hood', Santosh Rao wrote:

> To sum up, eBikes were popular with these early adopters and some Uber trips, especially during congested periods, were replaced by JUMP trips [48].

Similarly, Uber has tested offering peer-to-peer car-sharing via a partnership with Getaround. This limited trial allowed Bay Area Uber users to book and drive 'Uber Rent' vehicles by the hour or day at prices set by the individuals who own and list them on the Getaround platform.

By 2019, Uber was also integrating public transport information in its app in some territories, and in some places, public transport ticketing. The ticketing feature is available via ticketing software as a service supplier Masabi, which facilitates mobile transit ticketing in over 150 agencies or authorities worldwide by mid-2023 [49].

The extent of Uber's ambitions was set out in its IPO filing in 2019, which clearly identifies miles travelled by public transport as part of its potential market – the 'Total Addressable Market' (TAM).

> Our Personal Mobility TAM consists of 11.9 trillion miles per year, representing an estimated $5.7 trillion market opportunity in 175 countries. We include all passenger vehicle miles and all public transportation miles in all countries globally in our TAM, including those we have yet to enter, except for the 20 countries that we address through our ownership positions in our minority-owned affiliates, over which we have no operational control other than approval rights with respect to certain material corporate actions. We estimate that these 20 countries represent an additional estimated market opportunity of approximately $0.5 trillion [50].

What is less clear is how much of this is direct with ride-share and how much of it is by facilitating the integration of modes (and monetising that via commission on ticket sales or other fees).

However, from the interviews around the IPO in 2019, it seems that Uber was increasingly pivoting from its B2C model – providing individuals with rides – to targeting more B2G or business-to-government business. By this point, far from presenting itself as an innovative start-up chafing against regulation and championing individualism, Chief Executive Officer Dara Khosrowshahi was keen to speak the language of MaaS and repeat aspirations to end car ownership.

This redirection towards winning business from government authorities became clearer still in 2020 when Uber published a paper on public transport: Transit Horizons, Towards a New Model of Public Transportation [51]. Rather than fighting public transport agencies for a share of rides, in this document, Uber sets out its stall to make 'our technology, our app, our network, and our expertise available to public transportation agencies to use as their own'.

In this, Uber has capitalised on early experiments in providing its tools and services to local authorities. The B2G pivot was heralded by the 2017 implementation of a subsidised shared Uber service in the Canadian town of Innisfil. Rather than procuring a fixed-line bus service, the town authorities subsidise Uber to provide on-demand shared rides with fixed fares for certain trips and discounts on others. The case study (see below) has provided a wealth of data and – as it evolved – highlights key philosophical questions about the purpose of transit.

In 2020 Uber announced its first Software as a Service (SaaS) partnership. This puts Uber in charge of an on-demand service along a corridor route in Marin County in the San Francisco Bay area. The deal involves the Transportation Authority of Marin (TAM) paying Uber a subscription fee to use its management software to facilitate requesting, matching, and tracking of its high-occupancy vehicle fleet in the area. The vehicles – six-seater and wheelchair-accessible vans – are owned and operated by Marin Transit but are visible in the Uber app. The fares are charged per mile at $4 per mile or $3 per mile for Marin Access. Revenue from fares is directly paid to Marin Transit. Uber's income is derived from software subscription fees.

As the company has emerged from changes and challenges thrown up by the COVID-19 pandemic – in which it relied heavily on its food delivery business for survival – it has renewed its focus on public transit. This model shift means that Uber is in competition with other services that either started with the aim of providing public transport-based MaaS or pivoted towards B2G business models earlier.

5.10.3 Case study: Innisfil

Innisfil is a town in Ontario, Canada 8 km north of Toronto. Its population of around 40,000 people is spread across 200 km^2, giving rise to a low-density urban area of around 165 people per km^2. The suburban area (its road network largely based on a typical North American grid layout) is surrounded by rural farmlands.

Innisfil Transit is built on the uberPOOL service which offers users a discount if they are willing to share their ride. It matches riders going in the same direction so they can share a vehicle and the cost of the ride. To use the Innisfil Transit system, people download the Uber app and define the trip they would like to make. If they are within the town boundaries they will be offered 'Innisfil Transit' as an option, which is lower cost than the standard Uber offer. Fares to popular destinations, such as the train station, the recreation complex, or the Innisfil Employment Area are flat fares, and a discount ($5 at the launch) is applied to any other rides beginning or ending anywhere else in Innisfil. Rides booked via 'Innisfil Transit' may be 'pooled' with others going in the same direction and may include additional pick ups or drop-offs along the way.

In 2016, the town had no public transport system, and the local authority commissioned a feasibility study into setting up a public bus system. The initial report from the consultants suggested a fixed-line bus service running between 7 am and 5 pm. The price tag for a single bus service was reportedly $231,000 capital costs and $330,000 annual operating costs, rising to $439,000 capital and $541,000 operating costs for two buses.

In an interview with *The Guardian* [52], Paul Pentikainen, Innisfil's senior planner suggested a three-bus route system would cost the council nearly $1 million. The low-density topography and costs led the council to look at alternatives.

> '[We wanted to] connect all the drivers in the town with those that need these rides: that could be the basis for our transit system', said Pentikainen. 'And because Uber was already doing that, it just seemed to make sense for us'.

The town authorities approached Uber at the end of 2016 to explore alternative solutions. The process that followed designed what Uber terms 'a custom transit solution' and which was named Innisfil Transit.

Innisfil Transit is available 24/7 and within the first 8 months, Uber reported that 3,400 users made 26,700 trips. The average wait for transport was less than 10 min. This early success continued and during 2018 there were 85,943 trips, with many residents embracing the service. However, the linear relationship between cost and ridership has proved difficult for the town. As a subsidy is paid per ride, costs rise in direct proportion to the number of people using the service.

Early on, Uber boasted that they were saving the town money:

> By partnering with us to create Innisfil Transit, the Town estimates it is saving more than $8 million per year, based on the cost of implementing a public transit accessible by every household. Rather than absorbing these significant capital and operating costs to provide bus coverage across town, the Town has only spent $165,535 on Innisfil Transit rides after eight month.

However, this calculation has become more fraught. In 2019, a Guardian report on the project noted that 'Because Innisfil subsidises each ride, the more successful it is, the more the town pays to Uber. That figure is now projected to reach $1.2 million for 2019 – more than the bus programme would have cost, and well above the $900,000 the city allocated. With ridership increasing each year, costs will only rise'.

In 2019, to limit escalating costs, Innisfil Transit changed its structure. From April 1, flat fares rose by $1. Trip discounts dropped to $4, and a 30-ride monthly cap was implemented. Town leaders predicted that this would allow Innisfil to continue to cover the scheme's costs. Residents are allowed to apply for an exemption on a case-by-case basis and the Innisfil Transit site notes (in 2022) that:

Residents can apply to increase their monthly trip limit. We will give special consideration to seniors, students, and those who depend on Innisfil Transit for work. If we receive your application after the 10th day of the month, you will not be eligible for a trip increase until the next month.

As far as limiting the cost of the services, it is not clear how successful this has been. In a report to the mayor in May 2019, Innisfil authorities said few cost savings had in fact been made. Additional revenues from the fare increase were offset by increased ridership costs, as more residents successfully petitioned for exemptions [53].

Whilst costs mount, the social benefits of the service have been well-documented. An extensive survey published in 2020 showed that Innisfil Transit was being used for a broad variety of trip purposes from medical appointments to commuting – and that the service is likely to be used for a broader variety of trip purposes than Uber by itself.

In addition, one-quarter of Innisfil Transit users indicated that they would not be able to get to or from their jobs without Innisfil Transit. This trend was very marked in lower-income households, who are significantly more likely to depend on Innisfil Transit to access work. In addition, Innisfil Transit appears to be significantly more likely to serve low-vehicle households in meeting work and shopping trips.

Despite expectations of on-demand services providing a low-cost service being confounded, there were other sound topographical reasons for creating an on-demand system rather than fixed routes.

The low-density housing in the town and grid-based road layout make it hard to create fixed-line bus routes and services that are within a sensible walking distance of sufficient numbers of people without exorbitant costs. The urban area is about 20 km long and up to 10 km wide, with building type and density spreading the population out in such a way that makes a demand for high frequency/high capacity routes less likely than in higher density areas.

This observation was supported in the 2020 report on Innisfil Transit and Social Outcomes. When Innisfil Transit users were surveyed, only a third of users indicated that the fixed line service hours originally proposed in 2015 (0700–1700) would have met their needs, and a similar proportion would have been willing to walk to one of the 'transit stations' in the fixed line service design [54].

However, at the same time, the launch of the service revealed a level of suppressed demand that clearly exceeded expectations. The budgets set aside for the project – based on anticipated ridership – were rapidly exceeded.

In this model of DRT, costs grow in direct proportion to ridership. This is because Uber's drivers are paid per ride, the pool of drivers is potentially unlimited and the subsidy per ride is fixed, there are no economies of scale available to the city as ridership grows. Any savings gained by pooling rides (for instance if a car is fully occupied for much of the day) will accrue to the service provider. And for small vehicles, the vehicle capacity limits the potential for this in any case.

This is a different business model to traditional public transport bus services where start-up costs are high, but the cost per person falls as vehicles – buses – are filled. In addition, with set numbers of vehicles and service specifications, it is easier to forecast the total costs of the service (and therefore subsidy required). Accurate prediction of use (or at least accurately predicting 'break even' use) is important in these business models but for higher numbers of passengers, the additional costs are marginal.

5.10.4 B2G – the trend for technology

Uber is not alone in its B2C to B2G pivot. Via Transportation, originally set up as a ride-hailing service in 2012, rapidly morphed into a ride-sharing platform and thence to an on-demand shared transport service that had, by 2022, completed its shift away from a consumer focus. In December 2021 it announced it was shutting down its ride-hailing services to focus on public transit.

By the end of 2022, it claimed 500+ partners (including local authorities, transport operators, corporations, and educational institutions) operating in 35+ countries with over 115 million trips completed. Unlike Uber, Via has remained focused on providing on-demand transport rather than broadening the types of transport or providing access to multiple forms of shared and public transport. However, like Uber, its target market has shifted from consumers to the government. In an interesting move, Via acquired CityMapper in 2023.

Similar stories recur across MaaS technologies and the broader transport technology field.

Moovit was founded in 2012 as a free journey-planning app for iOS, Android, and the web. Initially based on publicly available transport, within 2 years it moved to incorporate crowd-sourced data, allowing volunteer editors to generate schedule and map data in areas where there is no official data available to app developers. It also derives information by crowdsourcing from passively shared data gathered from app users as well as active reporting.

As a journey planner, Moovit was named the official mobility partner for several sporting events between 2016 and 2018. Managing journeys around large sporting events has proved a headache for venues and local authorities. Journey planners were quickly adopted with enthusiasm to reduce traffic congestion and help people travel safely within unfamiliar environments. In this way, event journey planners and travel platforms have prefigured MaaS applications or offered bounded MaaS solutions to specific events and tournaments.

By 2019 Moovit's transport APIs were being integrated by Uber to enable public transport information provision in the Uber app in Denver Colorado. At the same time, these APIs were also integrated with driving and parking information using TomTom's APIs, to show the options for driving a car from a suburb to a parking lot near a transit station, riding transit into the city and completing the journey on transit, on foot, via car-sharing or on a scooter or bike. In addition, the Moovit API was integrated with Azure Maps, Microsoft's location intelligence platform for the Azure cloud to enable developers to integrate multi-modal trip planning into their IoT, mobility, smart city, and logistics solutions.

The provision of transit APIs into other ecosystems is a powerful way to embed the business into a variety of applications and build a business model around that. Moovit APIs are now fundamental to many applications in use, securing revenues for the business.

After this point, Moovit was able to develop from a journey planner serving individuals to providing MaaS to authorities relatively quickly. In 2020 Moovit and Cubic Transportation Systems entered into a partnership to incorporate Moovit's MaaS functionality with Cubic's Traveler App. This means its customers can pay seamlessly for the journeys they plan via the Moovit data. In the same year, Moovit was acquired by Intel [55].

In 2021 Moovit announced a partnership with Arriva to power the Netherlands' national MaaS app 'Glimble'. Tapping into Moovit's technology,

Glimble combines official information from all Dutch transport agencies, as well as crowdsourced information to calculate the best route for each journey with urban mobility options like bus, rail, tram, underground, ferry, taxi, Uber, carsharing, scooters, and bikes.

Glimble enables 'personalised MaaS' that ensures the routes meet users' needs and priorities. It incorporates accessibility features, empowering people across the disability spectrum to use public transportation with more assurance. The app is optimised with screen-reading features for low-vision users, including talkback/voiceover capabilities. It also identifies wheelchair-accessible routes and stations and calculates step-free routes. Glimble is also designed with optimised menus and buttons for those with hand-motor disabilities. In addition, users can also sort Glimble's suggested routes via the cost of the journey or by the total level of CO_2.

Moovit has developed from B2C to providing government and transport agencies with technology, but their model also encompasses B2B activity and as such has the potential to derive a broad set of revenue streams. Through Glimble, Moovit is also providing Arriva with a web-based employer portal. The portal enables companies and organisations to provide their employees with budgets for commute, business, and private trips. Employers can view the historical business trips of each employee and can view and allocate additional funds to employees if they wish.

A further diversification means that Moovit offers on-demand transport as well as MaaS. The on-demand services are powered by Moovit and can be integrated with MaaS as part of the mobility ecosystem. This means that Moovit can offer on-demand transport to a variety of government and transport authority partners – and indeed a suite of options and solutions.

Companies focused entirely on MaaS have not proved immune from the contractions experienced in 2022/3. One of the earliest commercialisations of the concept of MaaS, the Whim app, launched by MaaS Global in 2015 has not been unaffected by this process. Capitalising on the Finnish Data Law, which made transport data open, MaaS Global was able to design a service based on that data (covered in Section 5.6). Whim provided mobility subscriptions, transport bundling, and pay-as-you go targeted at consumers as its original concept. It relied largely on open data to deliver this. Over the subsequent years, it has provided its services in cities in Finland, Belgium, Vienna, Switzerland, the United Kingdom, and Japan. Whilst it has partnered with transit and local authorities, they have not been its key markets. The demand for personal transport dropped dramatically during COVID lockdowns in 2020-21, affecting revenue across the transport industry in general. Virtually no part of the industry, whether operator or digital was unaffected.

Despite this, by May 2022, the company announced that it had hit €1 million in monthly recurring revenue for the first time. However, despite regular travel behaviour returning and 'strong growth' in Whim app usage, the company announced that reorganisation was necessary as *'the funding environment and investor expectations for start-ups like us shifted significantly'*.

In an interview Hietanen gave around mid-September 2022 in Brazil, he said MaaS Global would 'pivot its business model', using its technology, products, and market experience. '*MaaS Global will partner with companies and partners who want to deliver MaaS locally*'.

With no likely remission in the global financial situation venture capital firms that have become more risk averse are likely to remain so for some time – a situation likely to affect the sector beyond MaaS Global and further accelerate the pivot within the sector from consumer facing to government and business facing.

For many years, the question of 'who should provide MaaS?' has been debated by transporter planners, national, city and regional authorities, technology providers, transport operators, and other companies.

Whilst the debate is interesting and there are valid points in favour of provision by commercial, government, and third-sector organisations (often flavoured by local circumstances), there is a quiet tide within the market to lay MaaS at the door of government and transport authorities.

5.11 Summary of MaaS business models

The development of MaaS has resulted in several business models being developed and delivered, with varying degrees of success. There have been numerous analyses of the different types of business models developed. Based upon our analysis of the MaaS propositions currently available, these can broadly be defined as seven types. The key features of these types are summarised in Table 5.1.

Table 5.1 Summary of business models of MaaS

Business model	Key characteristics
Single Mode Operator Business Model	Ownership of app: Private Ownership of modes: Operator Payment is to the platform which retains money for its own operations. Revenue is from fares paid by traveller mediated by app (fares disbursed to driver/hire fees to other operators less costs) MaaS platform and App is a cost of doing business for the operator.
Single Function Business Model	Ownership of app: private Ownership of modes: public and diverse private operators Revenue from subscriptions paid to platform which pays external operators for travel consumed. Revenue is derived from people not fully utilising fare cap. Revenue could be derived from location-based advertising. MaaS platform and App is essential for access to customer and therefore revenue.
City Authority Business Model	Ownership of app: public authority Ownership of modes: public and private operators Revenue is derived from payments to the platform (which is usually also the public transport ticketing authority) and which disburses payments to other modes unless payment is direct to

(Continues)

Table 5.1 (*Continued*)

Business model	Key characteristics
	shared/PAYGO operators through deep links. The app is part of the cost of doing business for public transport operators and any partners sharing costs.
Resold to Cities Business Model	Ownership of app: Technology company Ownership of modes: public and private operators Payment is to the platform which then disburses payments to other modes (unless payment is direct to shared/PAYGO operators through deep links – however this is commercially difficult without some form of introduction/affiliate fee). The MaaS platform and app is a commercial venture of the technology company.
Resold to Companies Business Model	Ownership of app: Shared mobility operators (on licence from technology company) Ownership of modes: One or more shared transport providers Payment is to operators but mediated by the platform. The app is a business cost to shared transport operators.
Resold to Destinations Business Model	Ownership of app: Destination/authority/company (on licence from technology company) Ownership of modes: Public transport and shared mobility operators Payment is to the licence holder for combined travel and ticket offers. In this case, the licence holder settles the fare or ticket costs in line with the agreements. Where the platform is used to manage business travel the license holder may settle ticket costs directly (without requiring the traveller to pay). The app is a business cost to destinations or businesses. It may also be a cost to a local authority seeking to provide better transport options.
Autonomous Vehicle Model	Ownership of app: Various versions could include the autonomous transport operator or the public transport provider or the infrastructure owner/manager. Ownership of modes: Potentially ride-hailing operators, public transport operator or/and shared mobility operators Payment is to the operator, if it is combined with other modes the platform disburses payments to other operators. The MaaS platform and app may be a commercial venture of the technology company (e.g., in the ride-hailing scenario), and it could also be part of the business costs of the operators. It may also be a cost to a local authority seeking to provide better transport options.

As trials have been undertaken and services have become more mature, there has been a gradual evolution of business models over time. The most notable shift has been from selling directly to businesses and consumers to selling directly to the government. This shift has primarily been for commercial reasons but has been enabled by governments seeking to deploy MaaS-style solutions in line with policies.

References

[1] Can Mobility Credits Help Resolve the Urban Mobility Challenge? [Internet]. CURB. 2021 [cited 15 Aug 2023]. Available from: https://blogs.coventry.ac.uk/researchblog/can-mobility-credits-help-resolve-the-urban-mobility-challenge/

[2] Start Selling Public Transit Tickets via MaaS Apps in Just a Few Weeks [Internet]. Masabi. 2022 [cited 16 Aug 2023]. Available from: https://www.masabi.com/2022/04/20/start-selling-public-transit-tickets-via-maas-apps-in-just-a-few-weeks/

[3] Kubitz, B *et al.* (2018) Mobility as a Service: The Second Annual Survey of MaaS in the UK, Landor LINKS, ISBN: 978-1-899650-85-9

[4] Bosa LF. Uber Leads $170 Million Investment in Scooter Company Lime [Internet]. CNBC. 2020. Available from: https://www.cnbc.com/2020/05/07/uber-leads-170-million-investment-in-scooter-company-lime.html

[5] Lime. Lime Locations | Bring Lime Scooters and Bikes to Your City or University [Internet]. www.li.me. 2019. Available from: https://www.li.me/locations

[6] Car Hire by the Hour, Day or Week in the South West, Exeter | Co Cars [Internet]. www.co-cars.co.uk. [cited 16 Aug 2023]. Available from: https://www.co-cars.co.uk/

[7] Discover Renfe [Internet]. www.renfe.com. Available from: https://www.renfe.com/es/en/renfe-group/renfe-group/get-to-know-us/the-company

[8] May T. Renfe-as-a-Service to Contribute to a Sustainable Railway Network [Internet]. Railway-News. 2022. Available from: https://railway-news.com/renfe-as-a-service-to-contribute-to-a-sustainable-railway-network/

[9] Citymapper – The Ultimate Transport App [Internet]. Citymapper. 2019. Available from: https://citymapper.com/london

[10] Introducing the Citymapper Smartbus [Internet]. Citymapper. Available from: https://citymapper.com/news/1800/introducing-the-citymapper-smartbus

[11] 2.1.2 Case Study – TfL Journey Planner [Internet]. MENTZ. Available from: https://www.mentz.net/en/articles/tfl-journey-planner/

[12] Pricing and Plans [Internet]. SkedGo. Available from: https://skedgo.com/pricing-and-plans/

[13] Francisco SN in S. Google hits Uber, Lyft [Internet]. www.theregister.com. [cited 16 Aug 2023]. Available from: https://www.theregister.com/2016/05/16/waze_opens_carpool/

[14] Quach K. Google Says There's No Waze Forward, Carpool App Axed [Internet]. Theregister.com. The Register; 2022 [cited 16 Aug 2023]. Available from: https://www.theregister.com/2022/08/25/google_waze_carpooling_closes/

[15] Waze. A Living History of the Waze Map [Internet]. Medium. 2021. Available from: https://medium.com/waze/a-living-history-of-the-waze-map-bec8da549659

[16] Hanover's "One Stop Mobility Shop" [Internet]. Intelligent Transport. [cited 16 Aug 2023]. Available from: https://www.intelligenttransport.com/transport-articles/69062/hanovers-one-stop-mobility-shop

[17] ktv_tkiebl. Vienna in Figures – Urban Area, Population, Education, Economy, Transport, Public Administration [Internet]. www.wien.gv.at. Available from: https://www.wien.gv.at/english/administration/statistics/

[18] Development of Passenger Numbers in Vienna [Internet]. 2019. Available from: https://www.wienerlinien.at/media/files/2020/wl_betriebsangaben_2019_englisch_358275.pdf

[19] Vienna Mobility Report [Internet]. [cited 16 Aug 2023]. Available from: https://www.wienzufuss.at/wp-content/uploads/sites/3/2020/04/Mob_Report_EN_2019_RZscreen.pdf

[20] wegfinder [Internet]. Wegfinder.at. 2023 [cited 16 Aug 2023]. Available from: https://wegfinder.at/faq/de/

[21] Mercedes and BMW Offload Car-Sharing Venture to Stellantis. Financial Times [Internet]. 3 May 2022. Available from: https://www.ft.com/content/5a5351eb-3007-4268-8636-6a4f42ee9a57

[22] Graph Integration Platform GIP: The Reference System of Austrian Public Authorities for Transport Infrastructure Data [Internet]. gip.gv.at. [cited 16 Aug 2023]. Available from: https://gip.gv.at/en/index.html

[23] Directive 2010/40/EU Progress Report 2020 Austria [Internet]. European Commission. 2020 [cited 19 Aug 2023]. Available from: https://transport.ec.europa.eu/system/files/2021-06/2021_at_its_progress_report_2020.pdf

[24] MaaS in Europe: Lessons from the Helsinki, Vienna and Hanover experiments [Internet]. Available from: https://www.cerema.fr/system/files/documents/2020/04/cerema_parangonnage_maas_rapport_complet_eng.pdf

[25] Scotland M. MaaS Scotland Annual Conference 2019 [Internet]. maas-scotland.com. [cited 22 Dec 2022]. Available from: https://maas-scotland.com/maas-scotland-annual-conference-and-dinner-2019

[26] Busvine D. From U-Bahn to E-Scooters: Berlin Mobility App has It All. Reuters [Internet]. 24 Sep 2019. Available from: https://www.reuters.com/article/us-tech-berlin-idUSKBN1W90MG

[27] Berlin Launches Single App for Booking Private and Public Transport [Internet]. Yahoo Finance. 2019 [cited 16 Aug 2023]. Available from: https://uk.finance.yahoo.com/news/berlin-launches-its-onestop-mobility-app-070000776.html

[28] Jelbi – Mobility Partners: More Mobility than Ever Before [Internet]. Jelbi. [cited 16 Aug 2023]. Available from: https://www.jelbi.de/en/mobility-partners/

[29] GetNews. Trafi Selected by BVG and Wins Tender: Jelbi Platform in Berlin to be Operated by Trafi Starting in 2022 [Internet]. Digital Journal. 2021 [cited 16 Aug 2023]. Available from: https://www.digitaljournal.com/pr/trafi-selected-by-bvg-and-wins-tender-jelbi-platform-in-berlin-to-be-operated-by-trafi-starting-in-2022#ixzz7Xmu1U0Nd

[30] Has Trafi Cracked Mobility-as-a-Service Where Others Have Failed? [Internet]. Sifted. 2020 [cited 19 Aug 2023]. Available from: https://sifted.eu/articles/trafi-series-b

[31] Companies Team-Up for Next-Gen Trip Planner in Berlin [Internet]. Smart Cities World. [cited 16 Aug 2023]. Available from: https://www.smartcitiesworld.net/news/news/companies-team-up-for-next-gen-trip-planner-in-berlin-7060

[32] Kubitz B. Smart Transport Journal [Internet]. Aug 2022 [cited 19 Aug 2023];15. Available from: https://www.smarttransport.org.uk/members/resource-hub

[33] ITS Finland [Internet]. ITS Finland. Available from: https://en.its-finland.fi/ Sampo Hietanen. CEO, ITS-Finland, 'Mobility as a Service' the New Transport Model?. *Eurotransport*, Vol 2, Issue 12, 2014. Available from: https://silo.tips/download/sampo-hietanen-ceo-its-finland

[34] UbiGo: Swedish Mobility-as-a-Service App Brings Smart Mobility [Internet]. Fluidtime. Available from: https://www.fluidtime.com/en/ubigo/

[35] MaaS App Gives Whim New Swiss Role [Internet]. ITS International. [cited 16 Aug 2023]. Available from: http://www.itsinternational.com/its17/news/maas-app-gives-whim-new-swiss-role

[36] MaaS Global Brings Mobility on a Whim to Japan [Internet]. www.goodnewsfinland.com. [cited 16 Aug 2023]. Available from: https://www.goodnewsfinland.com/en/articles/breaking-news/2019/maas-global-brings-mobility-on-a-whim-to-japan/

[37] MaaS Global – Crunchbase Company Profile & Funding [Internet]. Crunchbase. [cited 16 Aug 2023]. Available from: https://www.crunchbase.com/organization/maas-finland

[38] Digitally Connect and Control Your Fleets I Ridecell [Internet]. Ridecell | Powering Digital Transformation for Fleets. 2023 [cited 19 Aug 2023]. Available from: https://ridecell.com/

[39] Kubitz, B. *Mobility as a Service, First Annual Survey of MaaS in the UK*. July 2017. Stonham, P, editor. London: Landor Links; 2017.

[40] Arbib J, and Seba T. A RethinkX Sector Disruption Report [Internet]. 2017. Available from: https://static1.squarespace.com/static/585c3439be65942f022bbf9b/t/59f279b3652deaab9520fba6/1568193053392/RethinkX+Transportation+Report

[41] Beate K. *Mobility as a Service, First Annual Survey of MaaS in the UK*. July 2017. Peter S, editor. London: Landor Links; 2017.

[42] Pivotal: Vehicle Subscription by Jaguar and Land Rover [Internet]. Pivotal. [cited 16 Aug 2023]. Available from: https://www.drivepivotal.com

[43] Boeriu H. Car Sharing by BMW and Mercedes Plan to Headquarter Joint Company in Berlin [Internet]. BMW BLOG. 2018 [cited 16 Aug 2023]. Available from: https://www.bmwblog.com/2018/09/20/car-sharing-by-bmw-and-mercedes-plan-to-headquarter-joint-company-in-berlin/

[44] Miller J, and McGee P. BMW, Daimler Reverse Out of Car-Sharing Venture. *Financial Times* [Internet]. 18 Dec 2019 [cited 19 Aug 2023]. Available from: https://on.ft.com/3sl0KXu

[45] Urquia BV de. UK Unveils First Fully Autonomous Bus Service [Internet]. eandt.theiet.org. 2023. Available from: https://eandt.theiet.org/content/articles/2023/01/uk-unveils-first-fully-autonomous-bus-service/

[46] Esztergár-Kiss D, Kerényi T, Mátrai T, and Aba A. Exploring the MaaS market with systematic analysis. *European Transport Research Review*. 2020;12(1).

[47] UNSUSTAINABLE? [Internet]. 2017 [cited 16 Aug 2023]. Available from: http://www.schallerconsult.com/rideservices/unsustainable.pdf

[48] Rao S. Understanding Multimodality: An Analysis of Early JUMP Users [Internet]. Medium. 2018. Available from: https://medium.com/uber-under-the-hood/understanding-multimodality-an-analysis-of-early-jump-users-4a35d647b7e6

[49] Why Justride is the Leading Fare Payments Platform for Public Transport [Internet]. Masabi. [cited 16 Aug 2023]. Available from: https://www.masabi.com/why-choose-justride/

[50] S-1 [Internet]. www.sec.gov. Available from: https://www.sec.gov/Archives/edgar/data/1543151/000119312519103850/d647752ds1.htm#toc647752_1

[51] Mihov, I., Pangilinan, C, and Reich, D, Transit Horizons: Toward a New Model of Public Transportation [Internet]. [cited 19 Aug 2023]. Available from: https://www.uber.com/us/en/transit/horizons-paper/

[52] Cecco L. The Innisfil Experiment: The Town that Replaced Public Transit with Uber. *The Guardian* [Internet]. 16 Jul 2019. Available from: https://www.theguardian.com/cities/2019/jul/16/the-innisfil-experiment-the-town-that-replaced-public-transit-with-uber

[53] Staff Report Executive Summary [Internet]. Innisfil Civic Proceedings. 2019 [cited 19 Aug 2023]. Available from: https://innisfil.civicweb.net/FileStorage/36706CEF3729468E9ECCF1C073DD3B98-Innisfil%20Transit%20%E2%80%93%20Pilot%20Program%20to%20Support%20Acc.pdf

[54] Matthias Sweet, Raktim Mitra, and Sarin Chemilian, Innisfil Transit and Social Outcomes, Final Report, 2020, *Ryerson University School of Urban and Regional Planning*, Faculty of Community Services

[55] Confirmed: Intel Is Buying Urban Mobility Platform Moovit in a $900M Deal [Internet]. TechCrunch. Available from: https://techcrunch.com/2020/05/04/confirmed-intel-is-buying-urban-mobility-platform-moovit-in-a-900m-deal/

Chapter 6
MaaS in cities

6.1 Introduction

The urban core is usually an area where the highest concentrations of trips at the highest densities take place, so it is no surprise that many of the early trials of MaaS have taken place in cities of varying scales. From a commercial perspective, the benefits of cities are clear: a high concentration of trips combined with a variety of potential mode choices, as well as a greater propensity of the population to use alternatives to the car. When combined with a supportive city authority, there is significant potential for MaaS to become established and successful in cities.

In the last 20 years, there has been an increasing focus on cities as testbeds and living labs for technological innovation and greater experimentation in policy making. It is within this fast-changing context, and the inevitable friction with legacy systems, processes, and the value of established public transport networks, that MaaS has had to develop and adapt.

6.2 MaaS in city transport policy

Across the last 15 years, and even during the time within which MaaS has developed as a service offering, there has been a significant shift in city policy making when it comes to transport. Previous approaches to such policy making have been predicated on a model commonly referred to as 'predict and provide'. That is the continual expansion of transport services and infrastructure to meet demand [1]. This was further complimented by a focus on enabling the market in transport services and towards improvements in speed, technical efficiency, and safety.

The paradigm has shifted within city policy making over the last 10 years to a model that emphasises visioning, setting a desired future, and developing strategies and action plans that seek to progress towards such a desired future. Such a policy approach is referred to by several terms, including vision and validate [2] and decide and provide [3]. This is also underpinned by a greater consideration of uncertainty as part of the policy-making process [4], and even processes by which schemes are assessed [5].

These changes in the process of policy making have also shifted the emphasis of transport policy within cities towards achieving more significant policy objectives traditionally considered to be outside the traditional remit of transport services. And it is such policy objectives that MaaS are increasingly being judged by. Common thematic areas for such objectives in city transport policies include:

- Tackling climate change and reducing carbon emissions, notably through reducing the number of trips and modal share of the highest polluting modes of transport;
- Achieving wider economic development goals, but primarily focusing on improving access to employment, providing access to opportunities to achieve new skills, and encouraging new innovations in the transport sector;
- Tackling social exclusion and its contributory factors, which is often considered in terms of journey time to key public services but is increasingly being considered in terms of social justice and tackling embedded injustices in society;
- Supporting wider policy initiatives to boost the livability of the city, mainly considered in terms of improving the quality of streets and mitigating the environmental impacts of the car, as well as boosting health through improving active travel.

In early cases of MaaS, especially when the concept was relatively new, the deployment of MaaS was seen as a policy objective in of itself. This was often closely related to a policy objective of enabling innovative companies and deploying new services to realise an economic opportunity associated with it [6].

An example of this is in the West Midlands in the United Kingdom, a major urban region covering the cities and towns of Birmingham, Wolverhampton, and Coventry (among others). The Local Transport Plan sets the overall policy context for the region, with a significant focus on six 'Big Moves' of [7]:

1. Behaviour change;
2. Walk, wheel, and scoot;
3. Accessible and inclusive places;
4. Public transport and shared mobility;
5. Safe, efficient, and reliable network; and
6. Green transport revolution.

The deployment of MaaS, along with other transport innovations, is considered under the 'Green transport revolution' Big Move. The strategy identifies opportunities and threats associated with the deployment of MaaS in the West Midlands, taking the learnings from the experience of systems previously deployed in the region such as Whim. A noted opportunity is the economic growth potential of MaaS (Table 6.1).

Meanwhile, the strategy summarises its approach to the deployment of MaaS [8]:

> We have the experience of a previous MaaS system, Whim, to draw from. For example it would suggest that a subscription system is likely to be less popular than pay-as-you-go. Currently existing or incoming systems can also support MaaS. A key system which will bring all of this together is

Table 6.1 Opportunities and threats to the deployment of MaaS in the West Midlands [8]

Key opportunities	Potential threats
Increased mode share of sustainable transport, improved air quality, reduced congestion, improved regional health and well-being	Different actors creating and focusing on developing their own solutions, meaning that no one solution has the user base required to deliver the outcomes required
Economic growth and improved data for those travelling within the West Midlands	Whilst research highlights the demand for a platform that gives access to all services, a full release has not yet been delivered, so the expected outcomes are not yet confirmed

> the Swift ticketing system. Not only does the simple technology support a MaaS system, but it gives the precedent and pre-formed relationships of bringing different companies together with TfWM under a single banner. There is also the incoming project of Mobility Hubs which will provide a physical coming-together of the different forms of transport.
>
> Any new transport systems and projects can be brought under the same banner to encourage using the MaaS system. As the system develops through the trial we will add more features and cater to the particular needs and desires which develop.

Also within this strategy, a key delivery organisation in relation to future transport is Midland's Future Mobility. This organisation is intended to build on the area's automotive heritage and establish the West Midlands as a testbed for connected and automated mobility solutions. A key driving force behind the establishment of this testbed facility was to enable the West Midlands region to realise the economic value of next-generation mobility services, estimated by the organisation to be worth up to £62 billion to the UK economy by 2030 [9]. Whilst much of MaaS strategy focuses on the move away from personal vehicles, many areas – like the West Midlands – with long-standing connections to the car industry have maintained investment in potential future iterations of automotive technology.

MaaS has also been deployed in other cities in the United Kingdom as part of Future Transport Zones, a series of 3-year projects ending in 2023 seeking to trial new technologies and service offerings at scale in a number of urban areas. In this case the West Midlands, the West of England (covering Bristol and Bath), Nottingham and Derby, and the Solent (primarily the cities of Portsmouth and Southampton). As part of the business case for funding for these Future Transport Zones, the deployment of MaaS was considered to align with several strategic transport policy objectives (Table 6.2).

The transition of MaaS from a service concept to a trial to a potential service offering, although one that is still early in its maturity compared to similar policy initiatives such as integrated ticketing, poses a challenge to city authorities in terms of how to judge the success of MaaS policies. This also reflects a wider challenge

Table 6.2 Alignment of MaaS and other Future Transport Zone initiatives to regional objectives in the West of England region [10]

Regional strategic objectives	**Data hub**	**MaaS**	**Mobility stations**	**Micro-mobility**	**Dedicated demand responsive transport**
Sustainable and inclusive economic growth, increasing productivity	√√√	√√√	√√√	√	√√√
Equality and accessibility	√	√√√	√√√	√	√
Better health, well-being, safety, and security	√	√√√	√√√	√	√
Enable independence, reduce health inequalities	√	√√√	√√√	√	√√√
Access to public transport which reduces reliance on cars	√	√√√	√√√	√√√	√√√
Access to jobs	√√√	√√√	√√√	√	√√√
Reducing congestion and more resilient road network	√√√	√√√	√	√√√	√√√
Improving air quality	√	√√√	√√√	√√√	√√√
Well-connected with digital services	√√√	√√√	√	√	√

for trialling new transport services and technologies. Is success an offering that is commercially viable? One that achieves policy objectives, and if so which of those policy objectives is the highest priority? Or is it a solution that is universally adopted?

6.3 Assessing the impact of MaaS

Within transport policy, effort has been made historically to understand the impacts of a wide variety of transport interventions in terms of the impact on the transport network itself. This is often measured in terms of changes in the use and utilisation of different transport modes and in how people and goods get around a city. For example, a city may wish to understand how a MaaS solution may change the number of people using public transport, or change how frequently people drive in the city.

Policymakers now are increasingly seeking to understand how their interventions are having an impact in terms of outcomes. This implies that MaaS trials are exerting a broader influence on society, the economy, or the environment beyond their immediate impact on transportation. One example of this is the use of logic mapping, a systematic and visual way of presenting the key steps required in order to turn a set of resources or inputs into activities that are designed to lead to a specific set of changes or outcomes (Figure 6.1).

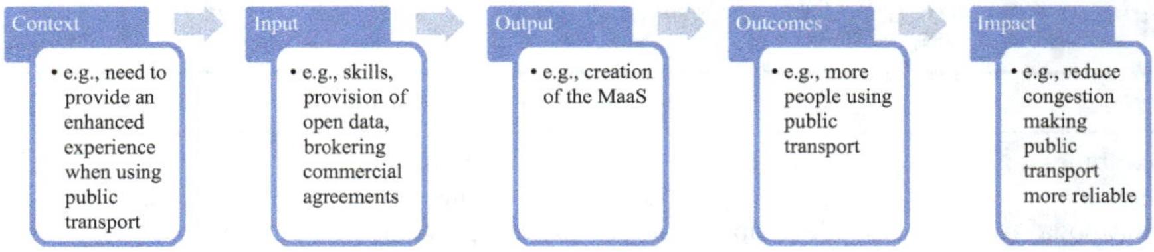

Figure 6.1 Simple logic map

Efforts are being made in academic literature to develop city-level policy frameworks against which the success of MaaS pilots can be judged. Some such frameworks also take account of the need to judge the success of MaaS initiatives in furthering the understanding of MaaS. For instance, an approach that has been applied to MaaS trials in New South Wales, Australia, has been to consider the success of MaaS against several areas (Table 6.3) and their impact at a number of levels:

Table 6.3 A framework for analysing MaaS programmes, utilised for New South Wales, Australia [11]

	Strategic	**Tactical**	**Operational**	**Reflexive**
Terminology (Establishing a shared language on MaaS)				
Objectives (Crafting inspiring visions and targets for MaaS)				
Rules (Lowering institutional barriers and steering the trajectory towards contributing to policy objectives)				
Technologies (Developing front end, back end, and support systems for MaaS)				
Business models (Assisting the development of viable and sustainable business models for MaaS)				
Modes (Pushing the diffusion and availability of transport services to include in MaaS)				
Alternatives (Supporting the relative attractiveness of MaaS, compared to the alternative options)				
Partnerships (Assisting inter-organisational collaboration, experimentation, and mutual learning)				
Diffusion (Accelerating the adoption and use of MaaS)				

Table 6.4 Measures of success for alternative mobility systems in Tunis [12]

Criteria	Measure
Awareness	Peoples' identification with public mobility systems
Security	Increasing security by decreasing overcrowding in buses and metros.
Intermodality	Developing an intermodal approach by improving connections between individual and collective transportation systems
Common vision and cooperation	Common vision and structure of Greater Tunis urbanisation, allowing assessment and foresight of the population's mobility needs. Cooperation of the governorates on mobility issues
Electronic information systems	Electronic information system to analyse mobility behaviour.

- Strategic – activities focus on culture and address the long-term development (15+ years) of the personal mobility regime (sectorial level) or broader regimes;
- Tactical – focus on structures and address the mid-term development (5 – 15 years) of personal transport service regimes;
- Operational – focus on practices and address the short-term development (0 – 5 years) of concrete personal transport services; and
- Reflexive – focus on monitoring and evaluation actors, agendas, and processes involved in the transition towards increased alignment between transport service regimes.

The academic literature has also extensively used the concept of expert scores. This method provides us with information about transportation activities and urban mobility implications. In terms of quantitative assessments, an example was deployed in the assessment of the social acceptance of alternative mobility systems in Tunis, Tunisia, where a series of local experts in transport were interviewed to identify key issues and opportunities associated with the delivery of new mobility services. This identified a series of potential interventions that could be used to increase social acceptance of new mobility services (Table 6.4) [13].

Whilst several trials of MaaS-style services have taken place within cities, with some research published on the impacts of these trials on specific transport modes, there currently does not exist a comprehensive framework for measuring and assessing the impacts of these trials in cities. Cities will likely be required to develop their own impact assessment frameworks, either as part of their own transport strategy work or for specific projects, to understand the scale of the impact of new MaaS on their transport networks and services.

6.4 MaaS in the context of smart cities

Many city authorities have been developing their capabilities in smart cities. Smart cities are as loosely defined as MaaS is. A good broad definition is that provided by the British Standards Institute which is the effective integration of physical, digital, and human systems in the built environment to deliver a sustainable, prosperous, and inclusive future for its citizens [13]. It should be noted that there are other definitions of smart cities that take a technological or citizen-centred focus [14], however, as in Chapter 4, the emphasis is on technologies working together for the benefit of the city inhabitants and visitors.

Smart Mobility is one of the defining pillars of Smart Cities and is a policy area that is commonly present in the application of Smart Cities, reflecting the fact that challenges associated with transport are common across many urban areas. A key part of this is the implementation of Intelligent Transport Systems. This is a suite of technologies and applications aiming at improving safety as well as increasing labour productivity and reducing the negative impacts of traffic [15]. Although such a definition has expanded to incorporate not only environmental and social objectives but has come to define a planning process that makes use of technological supports in the simulation, use, and monitoring phases of individual and shared transport systems to guarantee safety standards, functionality, and sustainability [16].

Different cities are at different levels of technological and institutional maturity in the deployment of Intelligent Transport Systems (ITS). Any definition of this maturity should allow for enhancing the awareness and understanding of the capability maturity levels, not only for the government institution but also for all the stakeholders throughout the mobility ecosystem, with the final scope of taking concrete actions on granting strategies and resources for moving forward in addressing the issues of ITS. Depending on the common understanding, agreed priorities, and available resources, the movement between lower to higher maturity levels is faster in a collaborative economy, when the government regulates and plans the necessary investment, and provides a stable implementation of ITS processes (Figure 6.2) [17].

Within this context, the development of Smart Mobility within Smart Cities is complex. It requires new measures of service quality, skills, management, and planning methods [18] on behalf of all actors within the smart city. It is within this context that MaaS is being deployed within smart cities.

The development of MaaS has aligned well with the development of Smart Cities due to the latter's requirements for the closer integration of various forms of intelligence, including mobility data. This ultimately resides in the increasingly effective combination of digital telecommunication networks (the nerves), ubiquitously embedded intelligence (the brains), sensors and tags (the sensory organs), and software (the knowledge and cognitive competence) [19].

Transport and mobility have often been one of the key areas of focus in work on smart cities. Relating to this is the variety of focuses that smart city technology deployment in the realm of transport that have been delivered to date. These can be summarised as so:

- A focus on **improving operational efficiency**. This is where sensors and data analytics are deployed to achieve improvements in the operational efficiency of public and private sector services. Examples may include remote monitoring of transport infrastructure assets for maintenance purposes, or real-time monitoring of buses and trains to identify patterns of delay.

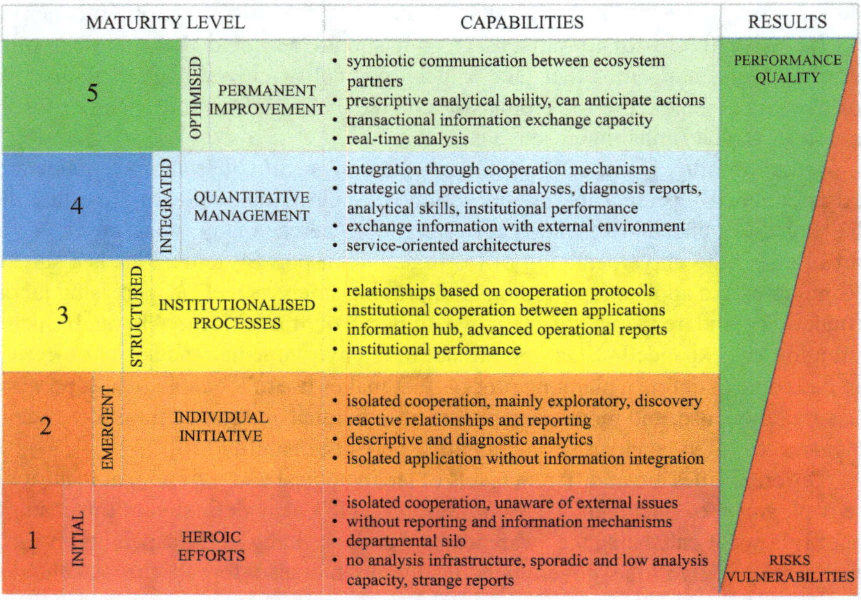

Figure 6.2 Mobility ecosystem's maturity levels, specific capabilities, and results [17]

- Increasing the **degree of servitisation**. This is primarily achieved through providing services through digital interfaces and platforms, complementing and sometimes replacing face-to-face and telephone-based service offerings. Examples can include the provision of parking permits, applying for passes to use local public transport and a significant number of public services.
- Seeking to **improve citizen participation in public services**. Digital platforms provide the means and the mechanisms by which citizens can participate in citizen engagement processes, such as public consultation and engagement. More radically citizen insight can be crowdsourced, and citizen science can be used to collect data on a specific policy matter.

As much as Smart Cities focus on a desire to deploy new technologies, an equally important desire is the focus on policy experimentation (Figure 6.3). Within this context of government bodies seeking to trial new ways of delivering on transport goals, MaaS has been both the result of a significant change and an instigator of it. For the former, this has often been in unexpected ways, with the most notable example being that of one of the pioneers of MaaS, Finland. Whilst the link between policy experimentation in cities and institutional change is a vague one, there is an expectation that favourable changes will become institutionalised over time [20].

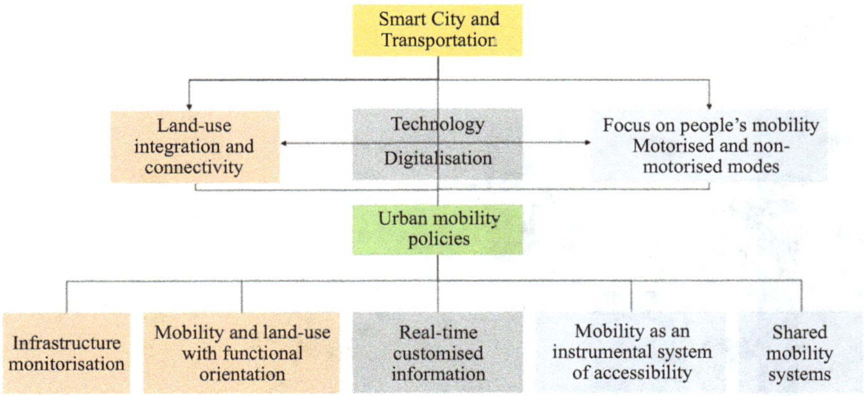

Figure 6.3 Key relationships in a smart city

In Finland, in addition to directly funding MaaS trials and supporting the development of new mobility concepts, a series of institutional changes were delivered to provide the basis by which MaaS could be delivered as a new and innovative transport policy. The most significant institutional changes were identified as [21]:

- Agencies for aviation, road, rail, and marine transport were merged into new multi-modal transport agencies, the Transport Agency and the Transport Safety Agency (Trafi), enabling more coordinated planning of different transport modes with potential benefits to low carbon transition. This led to a shift from 'solutions first' to 'problems first'.
- In 2016, the Ministry for Transport and Communications had an organisational change, where the separate transport and communications departments were merged and divided into four new departments: ministerial governance, networks, data, and services; and these comprised all transport modes and communications. The main driver for this reorganisation was a change in thinking in the ministry towards a more holistic policy.
- New framework legislation was created in 2017, the Transport Service Act. Its idea was to proactively support new transport services and innovation through significant regulatory change. The Act was influenced by developments in intelligent transport systems and issues explored in the policy experiments and was prompted by discussions on MaaS and the need to open access to information and ticket sales to third parties.
- A climate and environmental unit was established in the Ministry's network department and the administrative agencies were reorganised. The Transport Agency became the Finnish Transport Infrastructure Agency, responsible for road, rail, and marine transport and integrating transport and land use. More importantly, a new integrated Transport and Communications Agency (Traficom) was created.

Such significant changes have a number of impacts in terms of making significant policy changes such as MaaS 'stick' (Figure 6.4). Whilst the experience of

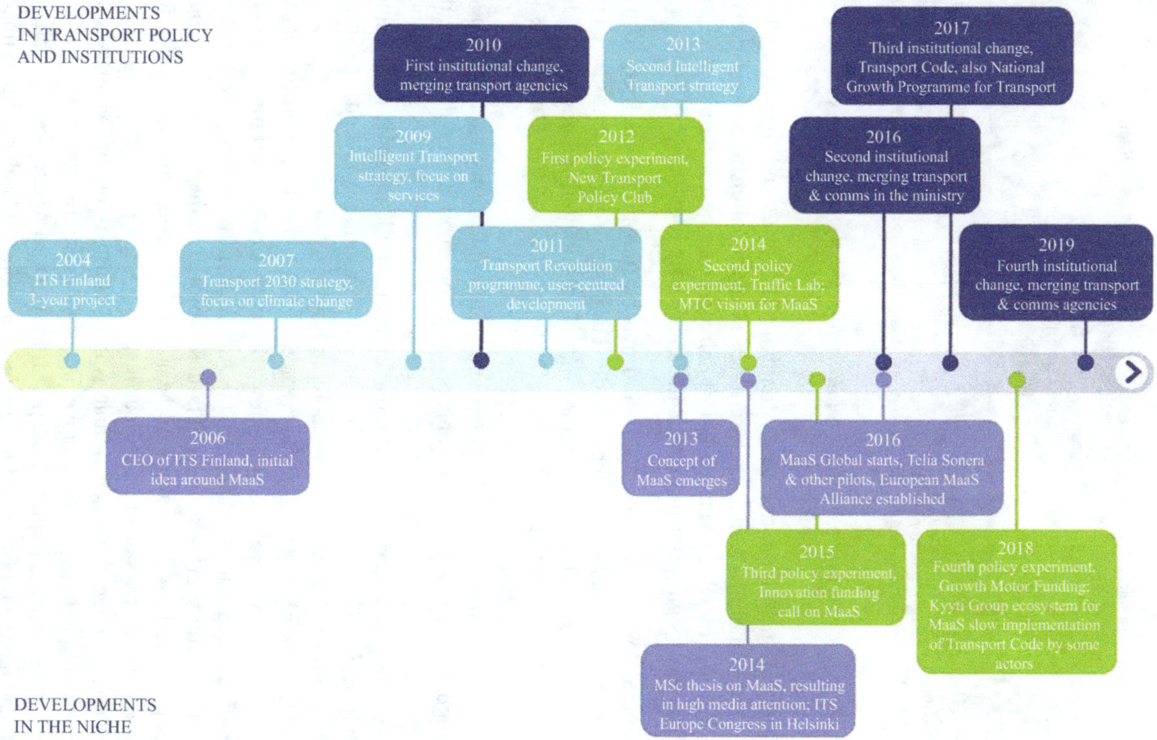

Figure 6.4 Developments in transport policy and institutions in Finland [21]

Finland shows that where institutional change is already occurring, the chances of policy experiments being normalised are greater, political backing and the ability to disseminate learning are critical ingredients.

6.5 MaaS in land use planning and development of cities

By 2050, an estimated 68% of the global population will reside in cities [22]. Significant urban development is planned across all regions of the globe to cater for this significant growth in the urban population, though the degree and nature of this growth varies from region to region. This means that a market for MaaS that already has significant potential may have an even greater potential for take-up in the future.

Land use planning and urban development also present a potential opportunity for the adoption of MaaS-style services. A commonly observed phenomenon is the impact of the establishment of new lifestyle choices, such as moving home on travel habits, a matter not always related to the density of development [23]. Consequently, the provision of a MaaS-style service as part of the process of moving home into a new development, when aligned with development priorities focussed on sustainable transport and prioritising walking, cycling, and public transport, may present an opportunity for securing new customers.

Data from the United Kingdom has shown that in any given year, around 18% of people changed their commuting mode on average. However, this percentage increased to 28% for those who moved house, 33% for those who changed their job, and 45% for those who did both [24]. Changes in commuter modal choices over a 3-month period were affected by job-related characteristics, access to mobility resources, satisfaction with current commuting, awareness of sustainable transport measures, and changed life circumstances [25].

The most significant challenge facing the coordination of land use planning and actions relating to MaaS is the alignment of timescales. The process of planning and delivering land use and transport infrastructure is often extended, with many cases taking more than 5 years from the initial planning stage. Build-out rates for developments that have secured permission can be many years, and though this may give a steady stream of new potential passengers each year, for services seeking quick financial returns this presents a challenge if a critical mass of users is needed to ensure financial viability.

The adoption of MaaS within the realms of land use planning is still at an early stage, with much of the technical literature focussing on the potential alignment as opposed to the actual alignment. Several areas where there was potential alignment between what MaaS can achieve and land use planning goals have been identified in the research, however much of this has been applied to what MaaS could theoretically achieve, as opposed to relating to the practice experience from MaaS trials to date (Table 6.5).

Exploratory research from Finland has also indicated the potential presence of feedback loops between the adoption of MaaS and land use planning. Interviews

Table 6.5 Potential alignment of MaaS and land use planning [26]

Increased cooperation at the ministerial level (including authorities) to make the transport system cope with new concepts
Greener transport including vehicles
Improved efficiency of transport system
Increased use of public transport
Boost to new technology (e.g., automated vehicles)
More efficient use of transport infrastructure (including maintenance and improved quality)
Changes in transport hubs (e.g., stations, terminals)
Facilitates introduction of the sharing economy in transport
Decrease in the number of private cars
Reduces needed parking spaces and transport infrastructure in general (frees space for other functions like complementary construction)

with MaaS and land use planning experts have identified the below examples of this, and the presence of these feedback loops is yet to be validated by research. However, this indicates that MaaS and land use have the potential to have a mutually beneficial relationship outside of a simple relationship between increasing the number of passengers and providing an easy means to meet modal shift targets (Table 6.6).

Changes in land use planning can provide a number of critical technology enablers for MaaS to become an attractive service offering. Across many areas, highway and public transport operators are heavily involved in urban developments and increasingly specify that where new infrastructure is provided or existing infrastructure is upgraded as part of such developments, new intelligent transport systems, sensors, and other Internet of Things devices should be installed as standard. Much of this is primarily focused on asset management, real-time detection of movements of vehicles, measuring running times, and detecting issues in real-time [27].

There has been significant effort on behalf of many cities to focus new developments on wider goals of sustainability, low-carbon lifestyles, and supporting local economies in line with the UN Sustainable Development Goals. Local Development Plan policies, by which such developments typically should accord, often have such requirements as policy tests which developments need to deliver against to secure appropriate authorisations. New innovations in urban mobility have been experimented with as a means of delivering against these policy goals.

In an assessment of four case study cities (Birmingham, Melbourne, Stavanger, and Milton Keynes), a number of factors were considered when practice on new developments was used to overcome inherent biases towards the car [28]. Namely the diversification of travel options, a shift from individual towards shared forms of mobility, and whether these impacts are implemented at scale. Within this context, the deployment of MaaS – in this case, the deployment of Whim in Birmingham – was considered to have promise in terms of diversifying travel options, potentially shifting towards shared use of assets and trips, but uncertain of its scalability.

The overall conclusion of this work was that as many deployments of new urban innovations are based on a commercial imperative, these innovations cater to

Table 6.6 Plausible impacts of MaaS on land use [26]

1st level impacts	Perspective of land use professionals	Perspective of MaaS professionals
Legislation	Increased cooperation between ministries (in legislation preparation), statutory transport services are organised more efficiently (and in increased service level), joint projects to find optimal solutions, traffic planning to be integrated into land use	Cooperation between legislation branches, enabling (not preventing) legislation must be developed, piloting of new services as a first step, cooperation within EU and internationally, goal to deregulate
Changing transport infrastructure	Current capacity adequate, re-use of vacant road infra, redundant parking spaces for other means (e.g., complementary construction), savings on infra spending, fringe areas better served, main changes in road infra (rail infra fixed), no new reservations on planning	New requirements for infra quality, current road infrastructure capacity adequate, maintenance must be ensured, infra steers land use, more efficient use of current (adequate) infrastructure, quality over quantity, re-use of vacant road infrastructure, redundant parking spaces for other means, savings on infrastructure spending, fringe areas better served
More efficient transport system	New mobility services, tools for transport planning	Transport flows become more fluent and commuter traffic increases, mobility for all, the land value between areas may alter, leisure mobility increases – compulsory decreases
Increased attractiveness of public transport	Improved utilisation, better service, on-demand public transport, and more options for organisers, pedestrians, and cyclers must be considered as part of public transport	Seamless integration of all transport modes, responsive services, connection parking facilities, improved service level, on-demand services, door-to-door transport, traditional line transport diminishes
Sharing of transport assets	Better utilisation via sharing, sharing economy more common	Sharing economy increases, less parking spaces needed, re-use of infrastructure, less congestions, pedestrians and cycling increases
Less cars	Less emissions, less congestion in urban areas, fewer parking spaces needed, increased utilisation of infrastructure, safer traffic	Services concentrate on hubs (decreased services in other areas), hubs need to be re-engineered for MaaS, multi-modality important
Changing hubs, nodal points, and centres	Terminals are needed, connection parking must be organised, services concentrate on hubs, traffic flows change hubs, smaller hubs than today, accessibility must be addressed	Leisure mobility increases, compulsory mobility decreases, remote work becomes more popular
Mobility volume	Leisure mobility increases,	Increased satisfaction and

(Continues)

Table 6.6 (*Continued*)

1st level impacts	Perspective of land use professionals	Perspective of MaaS professionals
increases or remains	compulsory mobility decreases, remote work becomes more popular	mobility, new business opportunities, more options in rural and urban areas, home delivery increases, mobility increases, digitalisation bundles services, new applications, internet of mobility, more options, virtual reality changes mobility (eCommerce/eWork), new service concepts
Regional aspects	Rural areas are more vigorous, fringe areas increase attractiveness, and urban areas are more functional	None
Automation of transport	None	Increased mobility, new requirements for infrastructure, resources in more efficient use, freedom to choose the place of living, terminals need space, autonomous driving, cleaner traffic
Services for new users	More options, on-demand services, optimised efficiency, and travel chains are served, services need to be re-engineered	

unquestioned levels of individual mobility. Accordingly a key policy failure of the past – narrowing down transport options to the private car – is being reproduced through such innovations. Whilst for this study, the evidence of impact in terms of sustainable travel behaviour in new developments was unproven, some older localised case studies of mobility hubs and car-free developments have more positive data (these are detailed in Chapter 15).

6.6 The role of Mobility-as-a-Service in city transport decision making

Much of the work on MaaS to date in cities has focussed on cities acting as enablers to the delivery of MaaS. This is achieved through investment in infrastructure, supporting data capability, and providing a supportive policy environment. But smart city literature indicates a vision whereby MaaS are delivered as a mutually beneficial ecosystem. Whereas MaaS providers may benefit from a supportive policy environment and investment from cities, MaaS providers should in turn provide insight to support decision making.

Some early work has been undertaken in the world of data standards to enable insight from new mobility services to inform policy and operational decisions at a city level. The Mobility Data Specification is the most advanced in undertaking this work (Figure 6.5). It gives cities insight into how their public realm is being navigated, and the tools to improve that experience. A suite of case studies have identified a number of city use cases where data from private mobility companies is shared in a manner that is meaningful to city authorities:

- Real-time and historical data allows better planning and programme management
- Digital management greatly reduces operating costs and staff time spent monitoring mobility programmes and service providers
- Allows real-time policy changes to be made to adapt to planned events and emergencies
- Supports policies that allow dynamic pricing, equitable access, and safety initiatives

A key distinction that should be considered is the nature of the decision-making framework within the city itself. In the deployment of smart cities, the use of multi-criteria analysis as a decision support model, to test a variety of types of interventions against a variety of criteria. This can also be applied when comparing the decision-making mechanisms for smart cities. For example, in a review of a number of smart cities and applying a fuzzy-based multi-criteria decision-making model, the results revealed London as the top-ranked smart city that promotes sustainability, resilience, and livability in its current urban development model. Dusseldorf, Zurich, Munich, Oslo, Dublin, Amsterdam, Hamburg, Rome, Moscow, and Stockholm were no exemption from addressing the tritactic goals of sustainability, urban resilience, and livability well into their urban development plan and were placed in the high-performance cluster [30].

In practice, the deployment of MaaS as part of transport decision making within cities is a relatively novel idea with little in the way of practical deployment

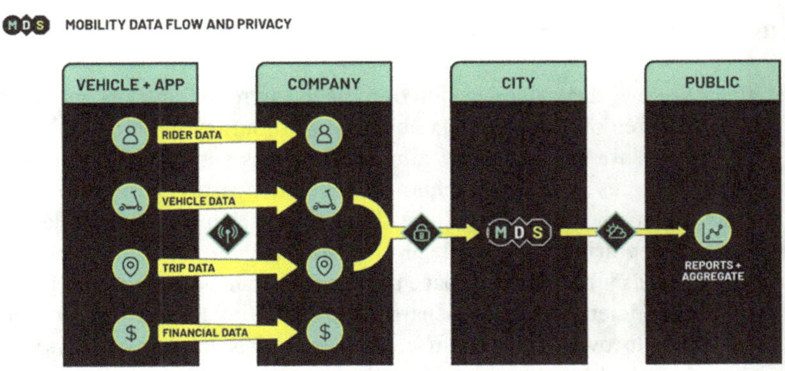

Figure 6.5 Mobility data specification, mobility data flow, and privacy [29]

so far for which there are verified results. However, there is some experience from the deployment of intelligent transport systems that may offer some insight into how this would work in practice.

In deployment across traffic management, intelligent transport systems have primarily aimed decision making in terms of making tackling congestion at a network scale an automated decision (e.g., changing traffic signals from red to green), providing decision makers with that same information so that they can make that decision (e.g., at traffic control centres), or aid in scenario planning significant events (e.g., a road closure) [31]. MaaS systems could complement this by providing data on travel behaviour changes of individuals at a transport system level.

In deployment in public transport, the decision-making aim of intelligent transport systems is often centred on improving experiences for the user, with outcomes judged upon key factors that influence this – notably reliability and the ability to help customers plan their journey. Integration with numerous technologies, including the smartphone, can make this both a highly automated decision to provide relevant information to people and also enable decision makers to highly tailor this information to end users to support wider transport network objectives, for example helping to manage the impacts of disruption [32]. MaaS offerings could be an additional method of communicating this information.

With many decisions being made through public-private collaboration, the correct institutional conditions are needed so that both parties understand their common goals, and the goals of each party, and work collaboratively through existing decision-making processes. When analysing the performance of four intelligent transport systems projects (TS Vienna Region, a dynamic intermodal traffic information service; compano, a novel ride-sharing scheme in Austria; SIS, a real-time traffic information services, displays and management system in Oslo; and Flexus, an electronic ticket system for public transportation in Oslo), this understanding was found to be critical to the success of each project in achieving its objectives [33].

6.7 Summary of MaaS in cities

MaaS in cities is seen both as an extension of tools to improve transport options for citizens and as a source of valuable data about citizen journey patterns. It is often seen as a policy tool for public policy aims as broad as sustainability, equality, health, access to services and employment, improved congestion, and air quality.

It also has a valuable role in land use distribution. Helping people plan and make trips without requiring a personal car is seen as having particular benefits in urban environments, as it reduces the space required for transport purposes, shifting people onto space-efficient public and shared transport. In addition, with a reduction in private vehicle ownership and use, the required space for car parking is reduced freeing up valuable urban space for other purposes. Although direct results from MaaS are not yet mature, early schemes such as mobility hubs which include many elements of MaaS have shown an impact on reducing urban car ownership and space requirements.

Importantly, however, MaaS enables cities to understand where people are travelling to and from and have greater insight into their modal choices. MaaS is integral to Smart City planning, as transport is key to effective urban spaces and economies. The data associated with MaaS presents a huge opportunity as a key data generator for Smart City insight and modelling.

References

[1] Goulden M, Ryley T, and Dingwall R. Beyond "predict and provide": UK transport, the growth paradigm and climate change. *Transport Policy* [Internet]. 2014;32:139–47. Available from: https://www.sciencedirect.com/science/article/pii/S0967070X14000213

[2] Jones P, Anciaes P, Buckingham C, *et al.* Preparing for the Future, Learning from the Past with Contributions from: Project Summary and Recommendations for Cities [Internet]. Available from: https://discovery.ucl.ac.uk/id/eprint/10058850/1/Jones%20et%20al%202018%20Preparing%20for%20the%20future%2C%20learning%20from%20the%20past.pdf

[3] Ngā kaupapa here whai rautaki mō ngā waka 2014–2016 | Strategic Policy Projects for Transport 2014–2016 [Internet]. Ministry of Transport. [cited 17 Aug 2023]. Available from: https://www.transport.govt.nz/area-of-interest/strategy-and-direction/strategic-policy-projects/#stageaccordionitem-412

[4] Benefits of a "Decide and Provide" Approach to Transport Planning | Insight [Internet]. Transport for the North. [cited 17 Aug 2023]. Available from: https://transportforthenorth.com/blogs/benefits-of-decide-provide-approach-transport-planning/

[5] New Guidance on Practical Implementation of Decide-and-Provide Approach Published | CIHT [Internet]. www.ciht.org.uk. [cited 17 Aug 2023]. Available from: https://www.ciht.org.uk/news/new-guidance-on-practical-implementation-of-decide-and-provide-approach-published

[6] Pangbourne K, Mladenović MN, Stead D, and Milakis D. Questioning mobility as a service: Unanticipated implications for society and governance. *Transportation Research Part A: Policy and Practice*. 2020;131:35–49.

[7] 6 Big Moves [Internet]. www.tfwm.org.uk. [cited 17 Aug 2023]. Available from: https://www.tfwm.org.uk/who-we-are/our-strategy/local-transport-plan/6-big-moves/

[8] Green Transport Revolution [Internet]. www.tfwm.org.uk. [cited 17 Aug 2023]. Available from: https://www.tfwm.org.uk/who-we-are/our-strategy/local-transport-plan/6-big-moves/green-transport-revolution/green-transport-revolution/#transport-innovation

[9] About Us – Midlands Future Mobility – Accelerating Future Mobility Solutions [Internet]. Midlands Future Mobility. Available from: https://midlandsfuturemobility.co.uk/what-is-mfm/

[10] Future Mobility Zones Fund Application Form – Final Proposal Applicant Information [Internet]. [cited 17 Aug 2023]. Available from: https://www.

westofengland-ca.gov.uk/wp-content/uploads/2019/10/WECA-FMZ-Bid-Submission-website-version.pdf

[11] Smith G, and Hensher DA. Towards a framework for Mobility-as-a-Service policies. *Transport Policy*. 2020;89:54–65.

[12] Storme T, Casier C, Azadi H, and Witlox F. Impact assessments of new mobility services: A critical review. *Sustainability*. 2021;13(6):3074.

[13] PAS 181 – Smart Cities Framework | BSI Group [Internet]. Bsigroup.com. 2015. Available from: https://www.bsigroup.com/en-GB/smart-cities/Smart-Cities-Standards-and-Publication/PAS-181-smart-cities-framework/

[14] What is a Smart City? [Internet]. Centre for Cities. 2016. Available from: https://www.centreforcities.org/reader/smart-cities/what-is-a-smart-city/

[15] Tran CNN, Tat TTH, Tam VWY, Tran DH. Factors affecting intelligent transport systems towards a smart city: A critical review. *International Journal of Construction Management*. 2022;1–17.

[16] Francini M, Chieffallo L, Palermo A, and Viapiana MF. Systematic literature review on smart mobility: A framework for future "Quantitative" developments. *Journal of Planning Literature*. 2021;088541222199424.

[17] Visan M, Negrea SL, and Mone F. Towards intelligent public transport systems in smart cities; Collaborative decisions to be made. *Procedia Computer Science*. 2022;199:1221–8.

[18] Rey-Moreno M, Periáñez-Cristóbal R, and Calvo-Mora A. Reflections on sustainable urban mobility, Mobility as a Service (MaaS) and adoption models. *International Journal of Environmental Research and Public Health*. 2022;20(1):274.

[19] William J. Mitchell – Intelligent Cities [Internet]. www.uoc.edu. Available from: https://www.uoc.edu/uocpapers/5/dt/eng/mitchell.html

[20] Turnheim B, and Kivimaa P, *Frans Berkhout. Innovating Climate Governance*. Cambridge: Cambridge University Press; 2018.

[21] Kivimaa P, and Rogge KS. Interplay of policy experimentation and institutional change in sustainability transitions: The case of mobility as a service in Finland. *Research Policy*. 2022;51(1):104412.

[22] United Nations. World Urbanization Prospects – Population Division – United Nations [Internet]. Un.org. 2018. Available from: https://population.un.org/wup/

[23] Lu J, Zhou S, and Kwan MP. Does density promote non-motorized travel? The effects of individual space–time activity patterns. *Transportation Research Part D: Transport and Environment*. 2023;116:103621.

[24] Dargay J, and Hanly M. Volatility of Car Ownership, Commuting Mode and Time in the UK. *Transportation Research Part A: Policy and Practice* [Internet]. Dec 2007 [cited 25 Nov 2019];41(10):934–48. Available from: https://www.sciencedirect.com/science/article/pii/S0965856407000407

[25] Chatterjee K, Clark B, and Bartle C. Commute mode choice dynamics: Accounting for day-to-day variability in longer term change. *European Journal of Transport and Infrastructure Research* [Internet]. 1 Sep 2016

[cited 3 Jul 2020];16(4). Available from: https://journals.open.tudelft.nl/ejtir/article/view/3167/3355

[26] The Impact of Mobility as a Service Concept to Land Use [Internet]. [cited 17 Aug 2023]. Available from: https://aaltodoc.aalto.fi/bitstream/handle/123456789/16332/master_Rantasila_Karri_2015.pdf

[27] Richter A, Löwner Marc-O, Ebendt R, and Scholz M. Towards an integrated urban development considering novel intelligent transportation systems. *Technological Forecasting and Social Change*. 2020;155:119970.

[28] Haarstad H, Sareen S, Kandt J, Coenen L, and Cook M. Beyond automobility? Lock-in of past failures in low-carbon urban mobility innovations. *Energy Policy*. 2022;166:113002.

[29] Mobility Data Specification [Internet]. City of Los Angeles GitHub. 2023 [cited 14 Aug 2023]. Available from: https://github.com/CityOfLosAngeles/mobility-data-specification

[30] Kutty AA, Kucukvar M, Onat NC, Ayvaz B, and Abdella GM. Measuring sustainability, resilience and livability performance of European smart cities: A novel fuzzy expert-based multi-criteria decision support model. *Cities*. 2023;137:104293.

[31] Galán-García JL, Aguilera-Venegas G, and Rodríguez-Cielos P. An accelerated-time simulation for traffic flow in a smart city. *Journal of Computational and Applied Mathematics*. 2014;270:557–63.

[32] Simonyi E, Fazekas Z, and Gáspár P. Smartphone application for assessing various aspects of urban public transport. *Transportation Research Procedia*. 2014;3:185–94.

[33] Weber KM, Heller-Schuh B, Godoe H, and Roeste R. ICT-enabled system innovations in public services: Experiences from intelligent transport systems. *Telecommunications Policy*. 2014;38(5–6):539–57.

Chapter 7
MaaS in rural areas

7.1 Introduction

MaaS in rural areas is an area of increasing commercial and policy interest compared to urban areas, where solutions and policy interventions are more mature in comparison. Regardless of this, there have been a number of 'Rural MaaS' trials that have taken place, as well as examples of MaaS-style operations being delivered over areas that have been subject to a regional-style MaaS solution.

This is within the context of an increasing interest in rural accessibility issues more generally. There is a perception and in most cases an actual position, where those who live in rural areas find it more difficult to access key services, such as education and healthcare, as well as accessing employment and similar such life opportunities. Many of these challenges will be discussed later in this chapter.

It is worthwhile noting that the nature of rural areas varies significantly. For instance, across the globe rural areas are overwhelmingly poor, with extreme poverty being primarily a rural issue [1]. Places that have reduced such poverty have invested in infrastructure and public services in those areas, granted access to land and social protections, and worked closely with the most vulnerable people. Rural residents often do not travel to neighbouring urban areas regularly.

In developed countries, however, the opposite can be true. There is evidence that there is little in the way of economic lag compared to cities [2]. Car ownership is also high, though there are significant differences when it comes to income, with a mixture of high income of residents, but typically low-paid jobs [3]. Specific populations, notably young people, may also experience accessibility issues. This can also impact nearby urban areas because rural residents often drive their cars into these areas. The impact of these travel patterns on these cities can be significant.

7.2 Defining a rural area

Rural areas can be defined by a number of different criteria. For instance, the UK Department for Environment, Food, and Rural Affairs (DEFRA) has six formal classifications of a rural area [4]:

- Rural: Hamlets and Isolated Dwellings
- Rural: Hamlets and Isolated Dwellings in a sparse setting
- Rural: Village

- Rural: Village in a sparse setting
- Rural: Town and Fringe
- Rural: Town and Fringe in a sparse setting

In contrast, Canada defines rural areas as having less than 1,000 inhabitants, and less than 400 inhabitants per hectare [5]. But from a transport perspective, there are two primary ways by which rural areas are classified [6]:

- **The Geographical Perspective** – this is typically driven by population numbers and population densities. However, this can also include measures of remoteness, for example, the journey time to a nearby major population centre (which in turn is often defined by population).
- **The Accessibility Perspective** – Whilst sometimes measured in terms of journey time to major centres, accessibility is more nuanced. It can be considered in terms of journey times to major services (typically major public services, such as healthcare and education, as well as access to food stores) and the cost of transport. However, it's important to consider perspectives like digital accessibility and the requirements of various types of trips.

These concepts are important to the development of MaaS in rural areas, as they affect the type of mobility services that can be offered at a reasonable cost within rural areas. Rural transport provision is typified by a focus on improving accessibility across a wider area, typically either to the nearest service centre or to a core network of transport services operating nearby.

The value proposition of MaaS within a rural context is not necessarily in terms of providing a menu of mobility options for each trip as in urban areas, but to assist in seamlessly integrating the modes of transport that are provided within rural areas. Within Finland, a number of challenges to the effective integration of rural mobility have been identified [7] as below.

Actors
• Side effects of publicly subsidised transport • Unclear cost sharing between responsible stakeholders • No stakeholder responsible for developing and coordinating mobility services in municipalities or regions • Challenging to identify good service combinations from the point of view of society • Lack of expertise in organising and planning mobility (information technology and tendering) • Health, social services, and regional government reform and legislation create uncertainty that will complicate the development and combination of publicly subsidised services • Expense of publicly subsidised operations • Too high a service level when considering the system as an entity

(Continues)

(*Continued*)

Technology
• Lack of expertise in organising and planning mobility (information technology and tendering)
Operational environment and context
• Lack of common will and rules • Slowly changing attitudes/unwillingness towards new solutions • Irregular transport demand (e.g., school transport, seasonal variations)
Services and market
• Freight/business goes first (e.g., infrastructure investments) • Restricted market • Poor service level of public transport • Evenings and weekends have an even poorer level of service • Public transport does not fulfil accessibility requirements • Centralised services may lead to growth in transport

In practice, the boundary between a rural area and one considered to be urban is often not absolute. Urban areas with a significant population, such as a large town, can exist within a largely rural area. Additionally, major cities can have what would otherwise be considered as open countryside within their city limits. The nature of this built form plays a significant role in determining localised trip patterns that are critical to determining what types of services can be offered through the MaaS platform.

The built environment exerts possibilistic as well as probabilistic influences on human actions such as travel. Both social structures and human agency have their own properties and causal powers and built environment characteristics could be considered as a particular sub-set of social structures characterised by having a relatively high permanence (although usually being constantly subject to incremental change) [8].

7.3 MaaS and rural transport challenges

The challenge of providing transport services in rural areas is one that is well-documented through research and is a challenge that a number of MaaS trials are increasingly facing. Many aspects of rural transport are challenging, not all of which we can discuss here. But the most important aspects of rural transport for MaaS to consider are accessibility, population density, cost of transport, and inequality.

The accessibility challenge in rural areas is a multi-faceted one, and for MaaS operators is not one that they can overcome on their own but one that will determine the success or otherwise of their endeavours. The accessibility concept in a rural area has been touched on earlier in this chapter. But this has practical implications for the delivery of services in rural areas. The most practical element is the fact that access to services, access to economic opportunities, and access to personal social networks and social activities vary significantly from place to place in rural communities.

The accessibility challenges facing rural areas make guaranteeing a minimum transport service level particularly challenging. Especially when socio-economic and cultural processes are at work that reinforce the marginalisation of such communities, for example, the tendency to centralise public services [9]. The most significant impact that this has on MaaS offerings in rural areas is that the focus is not on providing a multi-modal service offering unless there are local circumstances brought about by significant public subsidy (e.g., to provide rural bike share or a regular public transport service). Instead, this is about providing transport services in support of wider policy goals and governance of public services, enabled by technology and an easy-to-use transport service offering.

With regard to population density, whilst the majority of any one country's population is likely to be concentrated in urban areas, its land mass is likely to be mostly rural and with a low comparative population. For example, in the United States, 82.87% of the population lives in urban areas [10], whereas 97% of the land mass of the country is rural [11]. For MaaS providers, this poses the same challenge that has affected many types of operation of transport services. Namely, vehicles must travel further to pick up the same number of people that they would pick up in an urban area.

It should be noted that providing a direct correlation between the viability of transport services and population density is problematic [12]. Many large cities run subsidised public transport networks, and many inter-urban public transport services serving rural areas operate purely commercially. But what is most critical for MaaS is the operator cost per passenger is higher in rural areas, necessitating a higher fare (or higher subsidy). Additionally, fare structures like zonal fares that can be found in urban areas and are attractive as a simple MaaS pricing scheme are less common in rural areas for this reason. Although this is not without precedent, for example Transport for Cornwall has adopted some flat fare zones around small towns in many areas of Cornwall.

The cost of transport issue in rural areas is simple economics. Due to the more significant distances involved in accessing and serving rural areas, as well as the greater journey times, the cost of serving rural areas with transport services is greater. These costs are attributable to drivers' wages, vehicle capital costs, and overhead costs of running the services [13]. When it comes to serving rural areas with any type of public service such as healthcare, 82% of the additional costs of doing so are transport costs [14]. The most significant challenge that this poses for MaaS providers is the question of how to provide a sufficient level of service that is attractive for people to use whilst significantly increasing vehicle utilisation.

The transport situation in rural areas creates a unique challenge for MaaS. Evidence indicates that transport networks and services are the most significant drivers of rural inequality [15]. Globally, cases of extreme poverty are more prevalent in rural areas despite significant progress in tackling the root causes of such poverty, including improving transport services and infrastructure [1].

Work in the adoption of the Smart Countryside concept (rural areas benefitting from significant improvements in technology) identifies a need for a place-based approach to tackling rural inequalities, including the delivery of MaaS solutions [16]. The development of rural needs frameworks is assisting in this regard (Table 7.1).

One of the potential concepts that could increase utilisation in rural areas is the integration of MaaS and freight. The concept of combining passenger and freight movements to serve rural areas is not a new concept. A common example is the Postbus, where postal and parcel deliveries – often from the national postal service – are combined in the same vehicle with passengers on a local bus service. Examples of this across the globe include ÖBB-Postbus GmbH in Austria (which also runs regular bus and long-distance coach services) and Swiss PostBus Ltd in Switzerland.

Conclusions from this work have indicated that there are more opportunities for involving passengers and passenger transport operators in the delivery of small freight, i.e., parcels. This could benefit some growing sectors like e-commerce, and specific segments, like e-grocery or food delivery, requiring fast deliveries. Taking advantage of the critical mass of MaaS users, it would be possible to optimise (already existing) passenger and freight trips with the minimum detour possible [17]. However, this work is currently at a conceptual stage, with little in the way of practical trials in any area.

Finally, tourism is a key part of the travel mix in some rural areas, and it can have a huge impact when it is overwhelmingly undertaken by car-based modes.

Table 7.1 An example rural needs framework [16]

Business and economy	Social and community
Green energy; Climate change risk mitigation	
Providing for an ageing population; access to health services	
Recruitment and retention of skilled workers	Social mobility and aspirations
Access to training and networking	Access to education and training
Accessible and affordable homes for employees	Affordable housing
Flexible working spaces	Home-working and flexible working arrangements
Broadband and phone coverage	Broadband and phone coverage and ICT skills
Quality of life to attract workers	Socio-cultural activities
Natural environment (especially tourism)	Green space for healthy lifestyles
Meeting spaces	Combatting isolation. 'Third places' for social interactions (e.g., pub, café, park)
Financial services (especially for cash businesses)	Essential services (e.g., PO, bank, shop, pub)
Access to/for customers	Access to other places for retail and recreation
Freight and supply chain logistics	Consumer deliveries
Premises for growth	

There has been significant work to understand the potential for replacing car trips for tourism purposes with alternative modes of transport. Buses in rural tourist areas can attract new visitors and generate spending in the area. They can also be instrumental in reducing car use in places where cars can compromise the attractiveness of the area [18].

In the United Kingdom, 89% of trips to the countryside for the purposes of tourism (often day trips to an attraction or place, short stays, and more extended stays) were made by car in 2012 [19]. It is intriguing how the matter of car-based transportation's effects on rural regions also boils down to one's point of view. For instance, in a study of the Isle of Purbeck in Dorset, UK, local residents identified a typical emphasis on local travel problems and identified tourist arrivals by car as a major contributor. Data compiled using travel diaries and a survey, however, revealed that car-based visitors found driving relatively straightforward, and were also unconcerned about their impact on the local area. Problems were seen to be minor with little or no need to modify behaviour to cope in any way [20].

7.4 The challenge of digital connectivity

Within rural areas, transport options are typically more limited, especially within more sparsely populated areas. But an equally relevant issue for solutions that involve significant elements of digital connectivity is access to a reliable, high-speed internet connection. Globally, one in ten persons living in a rural area has access to a meaningful internet connection, compared to one in seven persons living in cities [21].

Additionally, there are a number of barriers that face the deployment of digital connectivity in rural areas. In a similar way to transport networks, telecommunications networks benefit significantly from economies of scale. Similarly, the commercial deployment of telecommunications networks in many rural areas is not seen as viable now or in the future. Globally, there has been significant work undertaken by governments to address this digital connectivity barrier. Common tactics include [22]:

- Promoting and regulating infrastructure sharing between different types of infrastructure operators, for example, sharing of mobile phone masts;
- Waiving taxes and surcharges on the purchasing of devices and infrastructure to enable expansion of connectivity;
- Provide subsidies for the installation or operation of remote telecommunication infrastructure.

MaaS providers have sought to overcome the issues of digital connectivity to varying degrees, though the success in overcoming this issue is unknown in the literature. A notable example is storing the ability to pay for mobility services remotely on a device such as a mobile phone or payment card, which can provide payment for mobility services without the need for a continuous internet connection (though a connection is needed at some point to update the account).

Journey planning, in contrast, is more challenging, and requires a digital connection in order to be undertaken in real-time. When installing new mobility hubs, a WiFi connection can be provided to provide access to the internet when using the stop. Another alternative is providing a telephone connection and the capability to plan journeys over the phone. But with the more significant costs associated with this method for relatively low take up, this is not often deployed.

Globally, more and more rural areas are becoming digitally connected, with the rollout of broadband and mobile phone connectivity to more areas. But the challenge of enabling modal shift through improved information specifically is well known. It is a challenge to provide accurate, current information on arrival and departure times, particularly in rural areas. Passengers from suburban, rural, and remote areas need more reliable and sophisticated travel information compared to urban areas; because, unlike in urban areas, passengers in rural areas are provided with very limited transport facilities, generally make longer journeys, may have problems with social exclusion, and limited exposure to media and other communications [23].

7.5 Impact of rural MaaS trials and studies

The impacts of rural MaaS trials and academic studies that have taken place are still being analysed. But regardless of this, the early trials have identified some key lessons from the nature of operating MaaS in rural areas.

In assessing the impacts of rural MaaS trials in Finland, an early conclusion is that whilst there are regional variations, urban MaaS is multi-modal and based on public transport complemented with additional services, whilst rural MaaS is based on integrating different services and user groups and using mainly on-demand and sharing services [24]. This reflects the different nature of transport demand within a rural context – one characterised by low population density across significant distances.

Another key differentiator for rural MaaS is in relation to commerciality. In urban areas, achieving a commercially viable MaaS, as well as commercial viability in supporting transport services, is a more realistic public policy goal. Whereas in rural areas, where subsidised public transport is more common, the financial objective is likely to be a reduction in subsidy as opposed to a commercially viable service. This is a common criticism of rural mobility trials more generally, where time-limited funding is provided with the intention that new services become commercially viable. When this does not happen, the service is then removed as local transport authorities are unable to continue supporting the service.

Finland has undertaken trials of Rural MaaS through the research organisation VTT. The project 'Mobility as a Service Concept Promoting Service and Livelihood Development in Rural Areas' (Rural-MaaS) was co-funded by the development fund of the Ministry of Agriculture and Forestry of Finland and was conducted in 2016–2017. The project studied business, regulation, and technology aspects and proposed, e.g., rural MaaS SWOT, vision, and measures. The Rural-MaaS was based on qualitative data collected via regional and national workshops, 30 interviews, and a literature review.

This resulted in a number of practical trials. This included a Demand Responsive Transport (DRT) service called Kyläkyyti, operated by minibuses, and ran from January to May 2019. The pilot covered a narrow area from the centre of Porvoo northward to the southern parts of Askola. It was selected to include rural areas where no public transport existed in the evenings, as the city wished to offer a service targeted at young people travelling to and from leisure activities. This area has roughly 3,000–4,000 inhabitants, or 120 inhabitants per km^2 on average. The pilot included a rural area close to an urban area according to the Finnish urban–rural classification.

Another case integrated different user groups, services, and vehicles within the same mobility system, and ran from March to October 2019, starting gradually with different services. Kuru (city of Ylöjärvi) and Vammala (city of Sastamala) were selected as pilot areas because of their sparse population and long distances. There is also very limited public transport in these areas. Kuru consists mainly of sparsely populated rural areas, and Vammala includes rural heartland. Kuru has approximately 2,600 inhabitants and 50 customers, whilst Vammala has approximately 16,000 inhabitants and 300 customers.

The Open Arctic MaaS project advanced mobility services in northern Finland in 2018–19. The main goal of the project was to create new business opportunities for local companies by advancing sustainable and energy-efficient mobility. Supporting collaboration between transport service providers and tourism service providers and developing regional collaboration, especially concerning service transport is an essential part of the project. By sharing knowledge, digitising routes, schedules, and other information about destinations (attractions, services, etc.) project adds visibility to destinations in the digital world.

These trials were particularly interesting in terms of their impacts on user behaviour. In particular, the following impacts were observed:

- A shift from the use of private cars to shared and public transport;
- Improvement in the occupancy rate of vehicles in rural settings;
- Improvements in collaboration and data sharing.

When analysed in further detail, the variety of benefits across different types of areas becomes apparent. An impact analysis was undertaken for three of the Finnish trials. The Kyläkyyti and Kuru and Vammala trials have already been mentioned previously, but a further trial was also analysed called Ylläs Around, which took place in 2016–2017 in Northern Finland (a sparsely populated rural area), in the Ylläs ski resort area between the airport and railway station. The pilot included multi-modal transport services (buses, taxis, and shared taxis) through a mobile app including payment and ticketing features [24].

The impacts across a range of KPIs are shown in Table 7.2. A few interesting observations can be made. The first is that the evidence of the impact on the total number of trips is lacking, but there had been a modal shift away from the private car in all trial areas. All of the trials led to improved data sharing across all of the trial areas and encouraged an organisational change in the participants.

Table 7.2 Comparison of impacts between different trials in Finland [24]

Level	KPI	Services		
		Kyläkyyti	***Kuru and Vammala***	***Ylläs Around***
Individual/user level	Total number of trips made	↓ Parents gave fewer rides to their children.	N/A	Travelling without a car became easier
	Modal shift (from car to public transport, to sharing, to...)	↑ From private cars to DRT	↑ Increased sharing and slight change from cars to shared rides	↑ Increased use of public transport (63%) and reduction of car use (71%)
	Combining different modes of transport	The service did not offer travel chains	The service did not offer travel chains	↑ Bus and (shared) taxi on the same ticket
	Attitudes towards public transport, sharing, etc.	↑ Respondents were extremely satisfied and willing to continue the use of the service	↑ Neutral and positive attitudes mainly.	↑ Respondents were willing to continue using the service
	Perceived accessibility to transport	↑ Major improvement in accessibility	N/A	↑ Respondents found that the service simplified their mobility
	Total travel cost per individual/household	Respondents found the fare very affordable, which played a role in using the service	Respondents found the fare very affordable, which played a role in using the service	↓ Respondents (60%) found their travel expenses to be lower due to the service
Business/organisational level	Number of customers	↑ 69 customers	↑ Some new customers.	↑ 106 customers
	Customer segments	Average age 39 years. Mostly women.	Elderly people. Mostly women	Average age 43.5 years Mostly men
	Collaboration/partnerships in value chain	↑ Public-private collaboration	↑ Public-private collaboration	↑ Public-private collaboration
	Revenues/turnover	↑ New business for transport and digital service providers	New business for digital service providers is possible, but uncertain	↑ Positive economic effect for transport service providers

(Continues)

Table 7.2 (Continued)

Level	KPI	Services		
		Kyläkyyti	***Kuru and Vammala***	***Ylläs Around***
	Data sharing	↑ Data sharing between digital service providers and the client (city)	↑ Data sharing between digital service providers and the client (city)	↑ MaaS platform improved data sharing by stakeholders
	Organisational changes, changes in responsibilities	↑ Initiated larger development of the DRT service	↑ Promoted the development of regional passenger transport services	↑ Initiated larger development of mobility services in the Ylläs area
Societal level	Emissions	Reduced use of private cars and optimisation of routes	↓ Driven kilometres and CO_2 emissions reduced	↓ Reduced use of private cars
	Resource efficiency (roads, vehicles, land use, ...)	↑ Increased use of DRT and route optimisation	↑ Improved occupancy rates of vehicles, reduced driven kilometres and emissions, and cost-savings for the public sector	↑ Increased use of public transport and shared rides
	Accessibility of the public to transport services	↑ New door-to-door service (no public transport available)	↑ New door-to-door service for private customers (no public transport available)	↑ Single train/plane ticket with last mile service
	Modification of vehicle fleet	Existing fleet utilised. New driver software	Existing fleet utilised. New driver software	Existing fleet utilised. New driver software
	Legal and policy modifications	Need: interoperability of travel card systems	Need: GDPR interpretation guide. Harmonised procedures for SHST	Need: public transport data in digital format

Tayside and Central Scotland Transport Partnership (Tactran) is one of the Regional Transport Partnerships in Scotland. Tactran secured funding from Transport Scotland's 'MaaS Investment Fund' (MIF) to deliver a Mobility-as-a-Service (MaaS) software platform across the Tactran region known as the 'ENABLE MaaS Platform'.

The platform knits together transport routes, prices, and other information from across the region together into one place and is used to power three new journey planning and booking apps and websites in the region: 'Go NHS Tayside' for patients and visitor travel to NHS Facilities throughout the NHS Tayside region; the 'National Park Journey Planner' app and website which help visitors and residents get to and travel within Loch Lomond and the Trossachs National Park; and the 'My D&A Travel' app for students and staff of Dundee and Angus College.

More specifically, the project sought to deliver the following:

- Develop a Tactran region-wide 'MaaS platform' which brings together real-time journey planning data, booking/ticketing and the payment systems of public, active and shared mode transport operators where possible.
- Co-tailor (the look and function) and pilot three apps powered by the MaaS infrastructure platform with three 'service lead' organisations: NHS Tayside, Loch Lomond and the Trossachs National Park, and Dundee and Angus College.

Following the successful launch and delivery of the tool at Perth Royal Infirmary, NHS Tayside agreed to roll out the solution, promoted via appointment letters. An accessibility audit confirmed the tools are simple and easy to use and useful for enabling a virtual journey 'run-through'.

HITRANS, the regional transport partnership, commissioned a leading technology firm to produce the bespoke GO-HI branded app, to provide instant access to information on buses, trains, car hire, car clubs, bicycle hire, air travel, and ferries. This will allow users to plan their journeys and find, book and pay for all modes of transport using any iPhone or Android mobile device.

GO-HI is powered by Fleetondemand's Mobilleo Mobility-as-a-Service (MaaS) platform, which has been built with the capacity to deploy partner MaaS schemes using its white-label capabilities. Go-HI is addressing Government ambitions to reduce carbon emissions by encouraging a modal shift from sole occupancy cars to shared cars and public transport alternatives. Go-HI is also reducing the feeling of isolation among those without access to a car and creating healthier lifestyles by improving the sustainable travel choices people can make in the region, including active travel opportunities.

Go-HI put in place a seamless booking and payment integrations with a range of leading transport and travel partners including Enterprise Car Club and Car Hire, (now defunct) Bewegen bike share provider, Brompton Bike Hire, Stagecoach Bus, West Coast Motors, ScotRail, Loganair, Orkney Ferries,

Northlink Ferries, Liftango, Inverness Taxis, Leeds Institute for Transport Studies, and SkedGo.

Across all of these different Rural MaaS case studies, a relatively under-explored area is understanding the impact of these trials on rural communities themselves outside of the pure transport impacts. This started to be explored in an organisational sense by an analysis of five rural MaaS trials in Sweden:

- DalMaaS in Skattungbyn. A trial launched in November 2018 which was ongoing in 2023. The MaaS features combine a search and book functionality through an integrated app and the transport options include ridesharing (private car and special transport).
- FjällMaaS in Södra Årefjällen (February 2020 and is ongoing in 2023). The service includes a commercial bus service, home delivery, ridesharing, and public transport. The platform functionality integrates search, book, and payment functionality in an app, except for public transport, which is deep linked (i.e., a hyperlink that links to a specific piece of content on a website or app).
- Hämta in Torhamn. A short trial in March 2018 until April 2018 which included ridesharing and public transport. This combined a journey planner and booking function in a web app with a deep link to public transport.
- KomILand in Broddetorp, Timmersdala, and Lundsbrunn. This service was launched in October 2020 and is still ongoing in 2023. The modes that can be accessed via the platform are public transport, two types of car sharing, bicycle sharing, tool sharing, and taxis. The platform integrates the journey planning booking, and payment functions in the app, except for taxi and car sharing, which are deep linked.
- Mobilsamåkning in Broddetorp. A trial programme which ran for 5 years from September 2013 and September 2018. This early trial incorporated search, book, and payment functionality in a web app for ridesharing and visualised public transport schedules.

Within the context of these trials, a number of organisational challenges were identified. Most pertinent are those of local transport authorities, and the need for close engagement with users of the services. All of the trials considered that local transport authorities were the natural leaders of the trials, but this was challenging considering the limited technical capability of such authorities to deliver a trial. Furthermore, this is complicated by the fact that different local transport authorities operate their public transport networks differently in different areas.

Concerning the ability to effectively engage with local communities, this study identified a common challenge among rural studies that is pertinent here. It states that since the most marginalised areas often lack the needed capacity to engage in innovation processes, innovation strategies building on local social capital may increase spatial inequalities as villages without social capital might be left behind by areas with higher capacity to act. Hence, although spatial inequalities between urban and rural areas were a prime motive for many of the studied rural MaaS pilots, there seems to have been a lack of awareness of, or attention to, the risks of increased spatial inequalities between different kinds of rural areas among the interviewees. Measures may need to be taken to ensure equal opportunities across rural communities when developing and disseminating rural MaaS.

Analysis of the nature of MaaS in rural areas and the trials to date point to a question of the required degree of integration with existing transport systems that is considered desirable, primarily from a public policy outlook. Krauss *et al.* (2022) define MaaS in the context of a rural area in terms of different levels of morphologies. Doing so enables potential MaaS operators to consider effective bundles to offer in a low density, and consequently lower trip, environment.

This particularly holds for the customer segments as it might be worthwhile to first focus on specific groups such as commuters or tourists [24]. Depending on the specific group(s) selected, the number of modes required can be limited to keep costs low and bundling might be adapted to the customer types and styles (e.g., specific commuter bundles or tourist bundles with access to all modes for just 1 week) (Figure 7.1) [25].

There have also been some limited trials of MaaS specifically focussing on tackling issues associated with tourism in rural areas. Many rural areas have been especially proactive in the development and delivery of sustainable transport services as a means of mitigating the impacts of car use by tourists. And MaaS is being trialled as part of this toolset.

In a study of the applicability of regional MaaS solutions for tourists in Queensland, Australia [25], several strengths and weaknesses of MaaS offerings were identified. This work demonstrated the challenges associated with the seasonality of tourism demand. A future MaaS pilot for the area is most suitably recommended in Townsville first because of its relatively higher population base, diverse economy, proactive government and business lobby, and more transport and tourism services ready for integration. It should be acknowledged that Cairns, with its larger tourist base, might also be a good candidate for MaaS trials, however, the feasibility necessitates further investigation as the region is subject to seasonal tourist demand, a narrower economic base, and more dispersed settlement patterns.

Meanwhile, a stated preference study for a tour-based MaaS on Jeju Island in South Korea found that individuals' travel patterns whilst visiting an area have a significant effect on preferences for different modal bundles. A shared van is the most preferred for both a trip and a tour. Persistence to using the same mode and synergy of using different modes are captured through significant cross effects of sequence-specific travel modes in a tour mode bundle [26]. These findings indicate that more work is required to specify specific service offerings for rural areas.

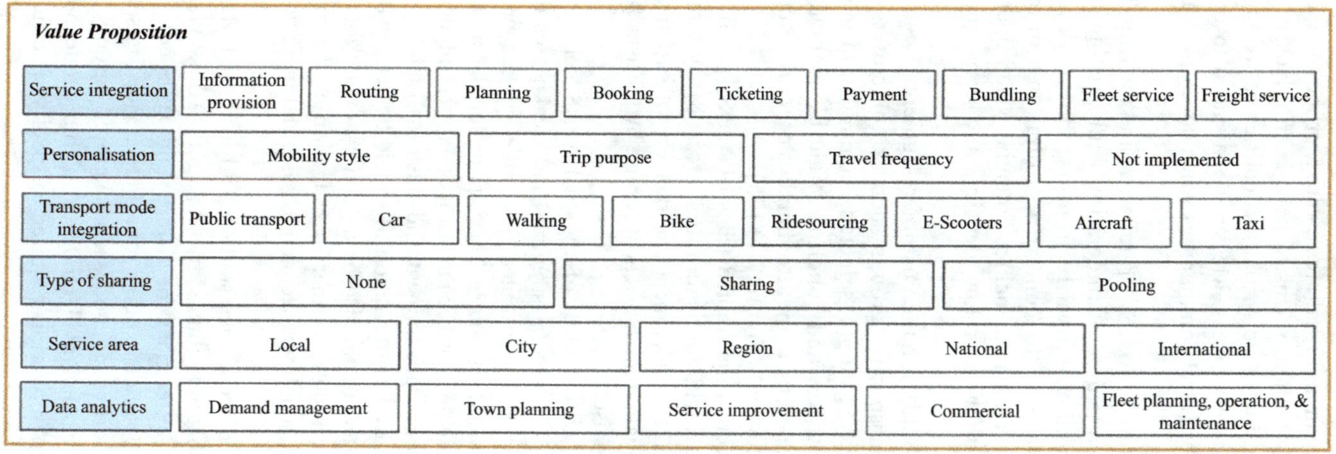

Figure 7.1 Morphologies of MaaS, with particular reference to rural areas [25]

7.6 Summary of MaaS in rural areas

The characteristics of rural areas – low population density and smaller towns and villages in rural areas contribute to people travelling longer distances on average. The demographics and topography create unique challenges for transport in general and MaaS in particular.

Transport is more expensive to provide because trips are on average longer. In addition, with fewer people to make them, the average cost per person per trip kilometre is likely to be higher than in urban areas.

The value proposition for MaaS in rural areas is different and tends to focus on meeting those challenges whilst achieving the best vehicle utilisation possible. Many examples of rural MaaS in Scandinavia include high levels of shared mobility – car clubs and ride-sharing. On-demand transport is also regarded as a key component, thus avoiding empty vehicle miles.

Tourism transport is another potential area in which MaaS could switch visitors away from car travel in fragile environments and where infrastructure is easily overwhelmed. There are some limited trials that are ongoing at the time of writing.

The nature of rural communities means that a high level of engagement is needed to ensure that people understand new services and technologies – and that as high a proportion of people as possible use them. The service information needs to be the gold standard, ensuring that people are confident they will not be stranded in a remote area.

Besides patchy services, long distances, and sparse populations, there is an additional discontinuity between areas with different levels of social capital.

Taken together this means there is unlikely to be a universal rural MaaS solution. Rural MaaS needs to be tailored to make the best use of the resources and community skills on the ground.

References

[1] United Nations. Reducing Poverty and Inequality in Rural Areas: Key to Inclusive Development | DISD [Internet]. www.un.org. 2021. Available from: https://www.un.org/development/desa/dspd/2021/05/reducing-poverty/

[2] Bertolini P. Expert Group Meeting on Eradicating Rural Poverty to Implement the 2030 Agenda for Sustainable Development United Nations Economic Commission for Africa ▪ Addis Ababa Overview of Income and Non-Income Rural Poverty in Developed Countries [Internet]. 2019. Available from: https://www.un.org/development/desa/dspd/wp-content/uploads/sites/22/2019/03/bertolini-Overview-rural-poverty-developed-countries-1.pdf

[3] Reimagining the Rural: What's Missing in UK Rural Policy? [Internet]. Available from: https://www.ncl.ac.uk/media/wwwnclacuk/centreforruraleconomy/files/reimagining-rural.pdf

[4] Defining Rural Areas [Internet]. Available from: https://assets.publishing.service.gov.uk/government/uploads/system/uploads/attachment_data/file/597751/Defining_rural_areas__Mar_2017_.pdf
[5] Government of Canada SC. Population Centre and Rural Area Classification 2016 [Internet]. www.statcan.gc.ca. 2017. Available from: https://www.statcan.gc.ca/en/subjects/standard/pcrac/2016/introduction
[6] Innovations for Better Rural Mobility [Internet]. Available from: https://www.itf-oecd.org/sites/default/files/docs/innovation-rural-mobility.pdf
[7] Eckhardt J, Nykänen L, Aapaoja A, and Niemi P. MaaS in rural areas – case Finland. *Research in Transportation Business & Management*. 2018;27:75–83.
[8] Næss P. Built environment, causality and travel. *Transport Reviews*. 16 Mar 2015;35(3):275–91.
[9] Vitale Brovarone E, and Cotella G. Improving rural accessibility: A multilayer approach. *Sustainability*. 2020;12(7):2876.
[10] Share of People Living in Urban Areas [Internet]. Our World in Data. Available from: https://ourworldindata.org/grapher/share-of-population-urban
[11] Haya El Nasser. What Is Rural America? [Internet]. The United States Census Bureau. 2018. Available from: https://www.census.gov/library/stories/2017/08/rural-america.html
[12] Cooke S, and Behrens R. Correlation or cause? The limitations of population density as an indicator for public transport viability in the context of a rapidly growing developing city. *Transportation Research Procedia*. 2017;25:3003–16.
[13] Ceglowski K, Lago A, and Burkhardt J. Rural Transportation Costs [Internet]. [cited 17 Aug 2023]. Available from: https://onlinepubs.trb.org/Onlinepubs/trr/1978/661/661-003.pdf
[14] Smith N, Hirsch D, and Davis A. Accessibility and capability: The minimum transport needs and costs of rural households. *Journal of Transport Geography*. 2012;21:93–101.
[15] Department for Environment, Food & Rural Affairs. Rural Poverty Statistics [Internet]. GOV.UK. 2013. Available from: https://www.gov.uk/government/statistics/rural-poverty
[16] Bosworth G, Price L, Collison M, and Fox C. Unequal futures of rural mobility: challenges for a "Smart Countryside." *Local Economy: The Journal of the Local Economy Policy Unit*. 2020;35(6):586–608.
[17] Pira ML, Tavasszy L, Correia G, Ignaccolo M, and Inturri G. Opportunities for integration between Mobility as a Service (MaaS) and freight transport: A conceptual model. *Sustainable Cities and Society*. 2021;103212.
[18] Providing Public Transport for Tourists in Rural Areas [Internet]. Available from: https://clok.uclan.ac.uk/7606/5/7606_Guiver.pdf
[19] England's Countryside: What Are the Opportunities? [Internet]. [cited 17 Aug 2023]. Available from: https://www.visitbritain.org/sites/default/files/vb-corporate/Documents-Library/documents/England-documents/e_countryside.pdf

[20] Dickinson JE, and Robbins D. Representations of tourism transport problems in a rural destination. *Tourism Management*. 2008;29(6):1110–21.

[21] Meaningful Connectivity for Rural Communities Geographic Barriers & Policy Strategies for Digital Inclusion [Internet]. [cited 23 Jan 2023]. Available from: https://webfoundation.org/docs/2022/04/MC-Rural-Report-English.pdf

[22] Global Connectivity Report 2022 – ITU Publication [Internet]. ITU/UN Tech Agency. 2022. Available from: https://www.itu.int/hub/publication/d-ind-global-01-2022

[23] Velaga NR, Beecroft M, Nelson JD, Corsar D, and Edwards P. Transport poverty meets the digital divide: Accessibility and connectivity in rural communities. *Journal of Transport Geography [Internet]*. 2012;21: 102–12. Available from: https://www.sciencedirect.com/science/article/pii/S0966692312000026

[24] Eckhardt J, Lauhkonen A, and Aapaoja A. Impact assessment of rural PPP MaaS pilots. *European Transport Research Review*. 2020;12(1).

[25] Krauss K, Moll C, Köhler J, and Axhausen KW. Designing Mobility-as-a-Service business models using morphological analysis. *Research in Transportation Business & Management*. 2022;100857.

[26] Kim EJ, Kim Y, Jang S, and Kim DK. Tourists' preference on the combination of travel modes under Mobility-as-a-Service environment. *Transportation Research Part A: Policy and Practice*. 2021;150:236–55.

Chapter 8
MaaS in regions

8.1 Introduction

The use of transport networks and services can vary significantly between individuals and within countries. Evidence from numerous countries indicates that the majority of trips undertaken by individuals are local in nature. For example, in the United Kingdom, the majority of trips undertaken by individuals are less than 5 miles in length [1]. Regardless of this, trips that are regional and even national in nature are of interest to policymakers and planners alike.

Regardless of how a region is conceptualised (more on this in this chapter), delivering MaaS at a regional level is becoming of more interest generally. This is for a number of reasons, notably that transport operations and planning can be and sometimes are undertaken at a regional scale. But also because whilst urban and rural areas pose their own unique challenges, in some areas, the interface between urban and rural communities is much closer, and the movement of people from city to countryside (and vice versa) has a measurable impact. This necessitates closer planning of transport services and infrastructure at a regional scale.

It should be noted that this planning should not be mistaken for planning rural services and rural MaaS across an extensive area. Whilst geography plays a role in defining a region, as we will explore this is not the only way by which a region is defined. It may be a case that a MaaS provider covers an extensive area, but is not a regional operator.

8.2 Concept of regions

Between different countries, the concept of a region significantly varies, with consequences for how MaaS serves a region. Regions are typically administrative concepts related to the development and establishment of transport powers and decision-making through varying degrees of regional governance. Regional governance often refers to the rules, procedures, and practices used by institutions at a regional level [2]. Within transport, such regions typically manifest themselves in one of two ways:

- The **City Region** is typified by usually one city acting as the social, economic, and cultural centre of the region. Accordingly, the transport network and services are not only the most extensive in that city, but such services and networks elsewhere within the region typically serve that city or reflect its impacts on the wider region. Examples of city regions include the Île-de-France around Paris, and the Combined Authorities found in the United Kingdom.

- The **Geographic Region** is typified by a regional government holding powers over an extensive area, constituting a mix of rural and urban areas, and whilst there may be large towns or even cities within it, none have overwhelming social, economic, or cultural influence over the region. Transport networks and services tend to be more dispersed than in City Region, although still concentrated around major towns and cities. Examples include the Cantons in Switzerland.

A significant challenge facing the deployment of MaaS at a regional level is the blending of the characteristics of different area types (e.g., urban and rural) to produce a viable MaaS solution that integrates with regional governance frameworks. Within a city region, Transport for Greater Manchester stated that there are a number of alternative MaaS models that could be deployed and be feasible, and consequently any procurement of MaaS should hold the capability of delivering such models [3]:

- Model A (direct): Public Sector is the MaaS operator and uses in-house resources
- Model B (external provision of services): Public Sector is the MaaS operator but outsources all of its responsibilities (becomes like a commissioning authority)
- Model C (operational commissioning): Public Sector is the MaaS operator but outsources all of its responsibilities except financial transactions
- Model D (joint provision, e.g., partnership): Public Sector is the MaaS operator but brings in a partner to manage and operate the system
- Model E (Spin-out, mutual): Public Sector is the MaaS operator but shares platform/resources with other providers to make financial savings and bring efficiency
- Model F (private sector operation): Private sector is the MaaS operator and has full control of its operation.

8.3 The role of regional governance in establishing MaaS

Regional authorities can play a role in establishing common rules and regulations for MaaS across a variety of areas. This is not to prevent the establishment of MaaS solutions at a more local level, but to provide the opportunity for such solutions to scale should they become successful, in the knowledge that they can apply their solution across a set geography. There is a desire on behalf of MaaS operators for regional authorities to play this more proactive role in establishing common market rules and enabling this scale of deployment [4].

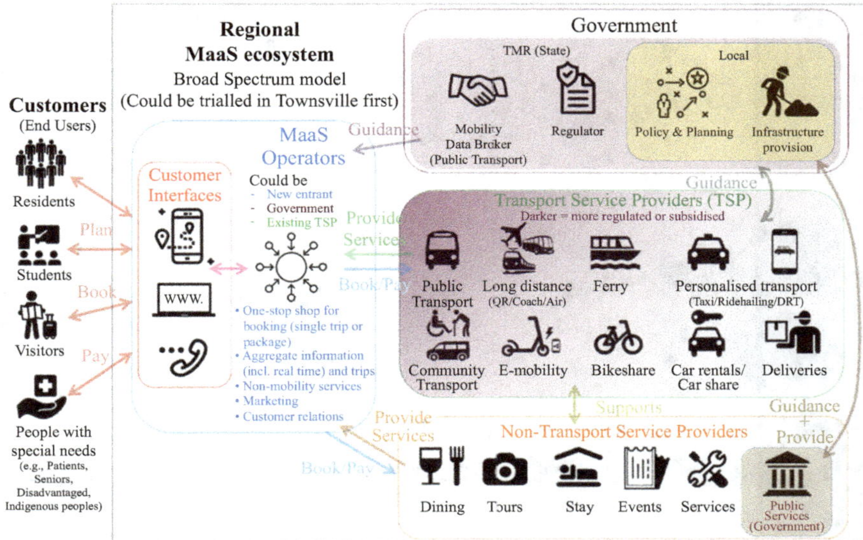

Figure 8.1 Components of a regional MaaS ecosystem [5]

The reality is that the role of regional governance in the establishment of MaaS trials is mixed and highly dependent on how regional governance works in different areas. In some countries, authorities with the necessary transport powers to facilitate the development of MaaS are regional in scale and consequently can play a much more direct role in deployment.

In addition to this, regional authorities can facilitate joint planning and procurement of MaaS, and act in a super-regional manner. This may be most attractive where there is a lack of capacity in regional governance to facilitate the deployment of MaaS, or where there is a lack of technical capability. This approach can also be beneficial for the latter in terms of ensuring that technologies and integration with IT systems are undertaken in a consistent manner (Figure 8.1).

Within the Netherlands, a number of regional trials were established within existing governance arrangements in the country. A national MaaS programme was established, with cities and city regions responsible for the regulation of new services, that were procured through a common procurement framework overseen by the Ministry of Infrastructure and Water Management (Figure 8.2).

Since mid-2016, the Ministry of Infrastructure and Water Management (Ministry of IenW) has been discussing MaaS with regional authorities and private companies. In 2017, the Ministry of IenW published a white paper discussing definitions and models of MaaS. The same year, they began the collaboration with regional authorities to set up four to seven MaaS pilots. At the end of 2017, the Ministry of IenW presented the seven pilots and started the market consultation in preparation for the tender process of the Framework Agreement for seven national scalable MaaS pilots resulting in eight apps. In 2018, agreements were made on the

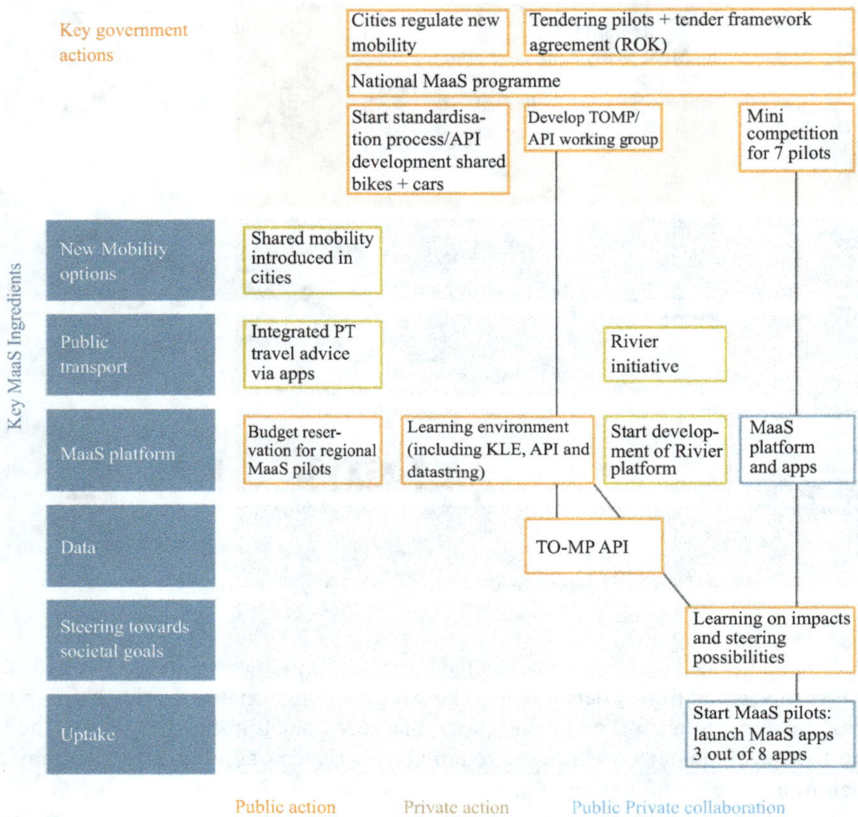

Figure 8.2 Components of the regional approach to MaaS in the Netherlands [6]

budgets. The MaaS pilots are co-financed by the regional authorities and the Ministry of IenW up to a total of €20 million (for seven pilots). The financing is meant to cover the start-up costs of a MaaS. The private companies executing the pilots are largely funded by public authorities. The idea is that after the pilots, they will have established a viable business case [6].

Another strength of regional governance is that trials can be undertaken at a scale that enables a MaaS solution to be trialled in a variety of types of areas that, whilst their geography may differ, can be similar in terms of social and economic factors. This can give confidence to regional authorities and MaaS providers that MaaS solutions can be scalable within the region should the pilot prove successful.

An example of this is the Île-de-France. Île-de-France Mobilités (IDFM), the regional transport authority, has its own MaaS platform and app (ViaNavigo) which provides planning, booking, and ticketing. This has been enabled through IDFM establishing a formal collaboration between public and private sector parties.

MaaS is seen by IDFM as a key lever to achieving the regional (and local to Paris) goals and strategic objectives, and they have had a Smart Mobility programme in place since 2016. Regarding MaaS, IDFM's policy is based on three main pillars:

1. playing its full role as a mobility authority (by clarifying relations between actors and disseminating good practices),
2. being a MaaS app operator (offering digital services to travellers), and
3. being the owner of a MaaS platform (for businesses and communities offering data and services for re-users).

A fourth element that is essential in achieving these regional goals is regulation. Providing the ecosystem with a platform that is optional and non-mandatory to be used by all operators and building a framework on how all of these elements work together is seen as an important task by IDFM. This calls for formalisation regarding how to collaborate, which has been taken up in the 'Guide de reference', a reference guide on how to collaborate with all ecosystem parties: travellers, MaaS actors (both IDFM and other MaaS providers, including public and private platforms), and data providers (public transport operators, New Mobility operators, and local community operators). The reference guide includes agreements, permits and conventions with local authorities, details on data and service provision, and contracts and conventions with IDFM [6].

An additional complexity at the regional level is the interface between national and regional governance on transport policy matters. In many nations, at a regional level national transport organisations and government departments can play a significant role in the governance of regional services, and thus significantly influence the degree to which regional bodies can establish MaaS ecosystems.

A good example of this is Switzerland, which has a strong degree of delegation of powers from the Federal Government to the government at the level of the canton. One of the seven departments of the Federal Government is the Federal Department of Environment Transport, Energy, and Communications (DETEC). The Federal Government has set DETEC the primary task to 'assure the sustainable provision of primary services in Switzerland ... to meet present requirements for infrastructures and at the same time to secure for future generations the chances of an intact environment'. Whilst cantons have most of the legal power in transport, DETEC, by its decisions and support to natural monopolies in Switzerland (e.g., Swiss Federal Railways and PostBus), has had an influential role in the development of the Swiss public transport network.

Thus, despite largely independent cantons in Switzerland, public transportation is primarily developed under a strong national planning paradigm, with national transportation monopolies setting the pace and more dynamic last-mile solutions serving regional and local needs. There is considerable path dependency in the development of Swiss public transportation, such as the timetable approach mentioned above, and different bottom-up social innovations that have been primarily witnessed within single cantons.

In the Basel region, this has posed some challenges in terms of the deployment of innovations such as MaaS. As a positive result of collaboration, the respondents state the increase in the quality of life, and especially appreciated is the promotion of active modes of transport (e.g., riding a bike or walking) [7]. A further positive experience mentioned was the fact that stakeholders and companies from different modes of transport want to cooperate, share knowledge, and learn from each other. In Switzerland, even the most significant natural monopoly companies, such as Swiss Federal Railways (SBB), have also demonstrated their commitment to knowledge sharing and cooperation. In contrast, some influential lobbyists (e.g., automotive associations) are very much focused on sustaining their power and are somewhat conservative regarding new mobility innovations [7].

8.4 Tackling the urban-rural dynamic

Regional authorities can also experience the two sides of urban and rural population issues. Whilst urban populations globally tend to be expanding, rural populations tend to be shrinking [8]. This poses a number of challenges to regional transport operators, notably how traditional public transport services in rural areas are becoming increasingly unviable, whereas there is more significant pressure placed on the transport networks of urban areas.

Some regional transport authorities, with the support of the national government, have been utilising MaaS as a means of introducing more innovative transport services. In Japan, for instance, there is a twin issue. Not only is the rural population decreasing, but it is also rapidly ageing [9]. To partly tackle this issue, the Japanese Government established the National MaaS Framework. It is envisaged that MaaS in the major cities (e.g., Tokyo, Osaka) would be facilitated by the existing transport companies. However, in the regions, where public transport is less profitable, and it is envisaged that the public sector will take on a more active role. This work would be led by the government at the city, region, or prefecture level (Figure 8.3).

Under the 2020 Smart Mobility Challenge Project, 52 regions were selected to demonstrate smart mobility solutions, including MaaS at a regional level [11]. In considering these location types, the Japanese government has developed a typology that can be utilised to develop regional MaaS initiatives. Underpinning this work is a national standardised API that is currently in development (Figure 8.4 and Table 8.1).

8.4.1 Case study: not for profit uberisation in rural Japan – Sasaeai Kotsu a case of community transport in Tango Peninsula, Kyoto Prefecture

In Japan, in 2016, Uber entered the Tango area where the population is ageing, along with depopulation [14]. Conventional taxi operators abandoned operation. Uber started operations with a local non-governmental organisation (NGO),

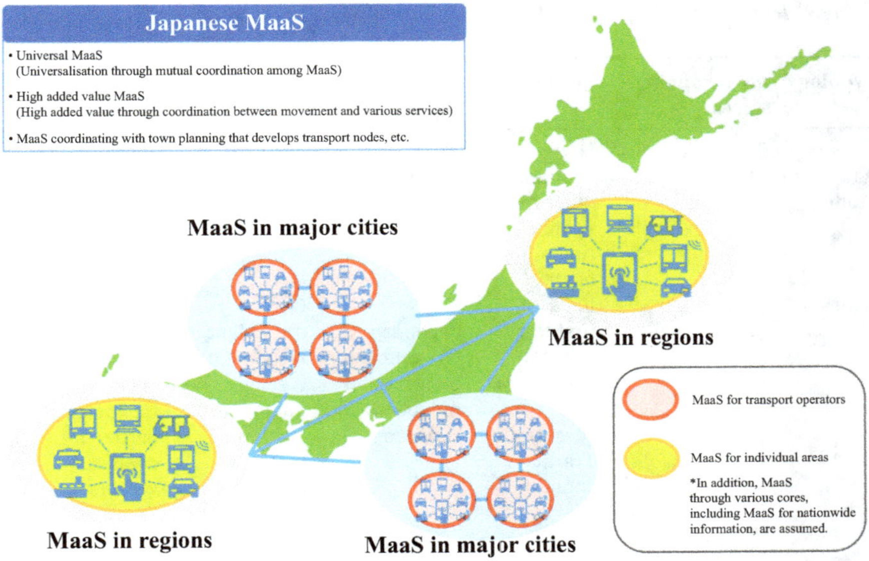

Figure 8.3 The Japanese national MaaS framework [10]

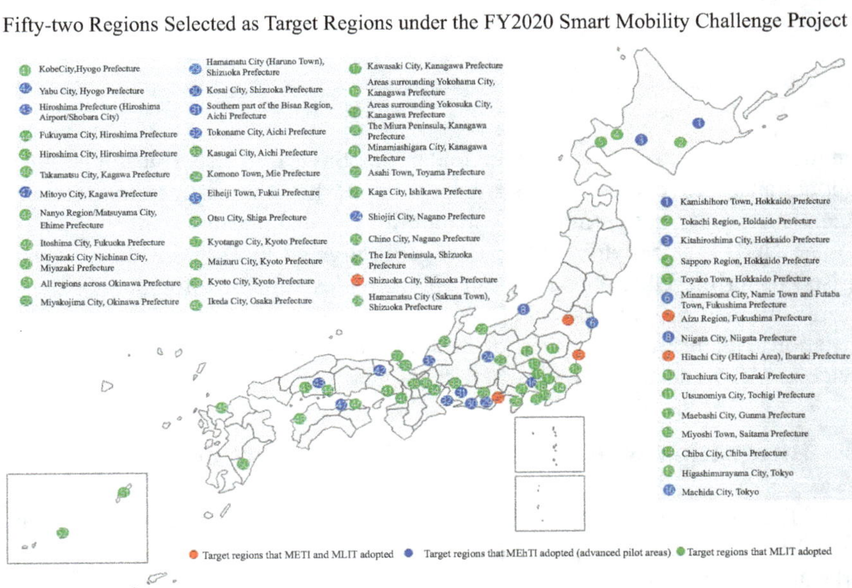

Figure 8.4 Location of MaaS projects in Japan [12]

Table 8.1 The typologies of MaaS in Japan [13]

Typology	Metropolitan area	Suburban	Regional town	Rural	Tourism
Density	Very high	High	Medium	Low	Variable
Dominant Mode	Rail (Heavy/ Metro)	Rail and other motor vehicles (including bus)	Motor vehicles	Motor vehicles	Variable
Issues	Peak hour overcrowding	Lack of first/last mile connection and accessibility Traffic congestion	Car dependency Government budgets and reduction of services	Car dependence Depopulation Ageing population Declining transport options Transport exclusion for non-car owners	Seasonal demand Variable needs Language barrier for overseas visitors
MaaS goals	Divert peak hour rail to other modes Door-to-door services with small vehicles	Provide first/last mile connection Taxi-like services Reduce car use	Provide non-car travel options	Make rural areas more livable Encourage retention of young people	Connect to airport, accommodation, and tourist attractions

supported by the local government under the Sasaeai Kotsu (meaning mutual support transport) scheme. This provides a somewhat unique operation for Uber (which is normally not permitted to operate for ride-hailing in Japan).

The NGO matches drivers to passengers (targeted towards local residents or tourists) using the Uber Japan system and provides support through subsidies to drivers with their own private cars [15]. All operations are regulated under national legislation. Against a background of declining taxis and buses in rural areas, this service model could be used elsewhere to support rural communities [16].

8.5 Integration with regional public transport networks

One of the most common types of transport service operation that operates at a regional level is that of the public transport network. In many countries, such networks are operated, whether as public or private systems as an integrated system. Accordingly, there is often a demand for MaaS providers to be deployed at a scale consummate to the scale of the public transport network.

The nature of this agreement and integration at a regional scale depends on the type of operation of the public transport service. Early recommendations for the development of MaaS gave guidance that in a MaaS system with commercial MaaS providers, the providers will charge a commission on ticket resale. Under this approach, the public transport operator would be free to contract and negotiate with the MaaS providers. They could award a discount or commission based on the value they perceive the MaaS provider bringing as a seller of public transport services to a potentially new or wider audience. MaaS providers would also need to identify and demonstrate the value they bring to progress from initially short-term contracts to longer-term agreements. They would need to show that their involvement helped realise policy outcomes, such as improved accessibility and increased public transport ridership.

The alternative is the 'Public MaaS' model that bears the attendant financial and reputational risks. They could also potentially subsidise private aggregators in the interests of reaping the social benefits of a modal shift towards sustainable mobility or improving accessibility in poorly served regions [17].

8.5.1 Case study of regional MaaS connecting into regional rail services: Mobility on Demand (MOD) Sandbox Demonstration: DART First and Last Mile Solution: Dallas, USA [18]

In October 2016, Dallas Area Rapid Transit (DART) was selected as 1 of 11 MOD Sandbox Demonstration projects; the DART First and Last Mile Solution project received $1.5 million, most of which was ultimately used for technology, with DART providing operational funding.

The project goals were to improve first mile/last mile (FMLM) access to DART transit for all people including individuals with disabilities, increase transit ridership on DART, improve the experience of transit, provide alternative transportation/multimodal travel options, expand service within certain areas and improve access to jobs, replace ineffective and costly fixed-route transit with MOD services, and improve customer satisfaction. The project also intended to provide same-day service for riders with disabilities who use wheelchair-accessible vehicles (WAVs) instead of next-day, demand-responsive service and to comply with the American with Disabilities Act (ADA) through meeting an equivalent level of service requirement.

Through its MOD development process, DART began to collaborate with technology providers; in October 2017, approximately 1 year after the Federal Transportation Agency (FTA) selection announcement, a pilot MOD test was conducted that proved to be an operational success using DART's GoPass applications (app) and vehicles supplied by a DART contractor.

The evaluation of this project showed that DART improved transit access for all people by implementing MOD using GoLink and Uber. Response and travel times to access the system improved over time and were enhanced through the inclusion of Transportation Network Companies (TNCs). Moreover, MOD and Uber services were added without additional expense to riders. Results of a MOD customer survey showed major improvements in overall customer satisfaction.

MOD extended coverage to areas with little or no transit, which increased overall transit ridership and reduced costs to DART. The performance of TNCs was critical with the furthering of project objectives.

8.6 Summary of MaaS in regions

Regions incorporating cities and their outlying areas often have good reasons to deploy MaaS. They are of sufficient diversity and scale to incorporate urban and regional rapid transport services with local shared solutions in periurban and rural areas.

Many of the issues regions face arise because of the different population patterns and distributions which make 'one-size-fits-all' solutions inappropriate. They often see significant movements of people between areas (for instance commuters converging on urban centres from outlying areas) and with a broader view of these movements can (in theory) take action to plan networks and services appropriately.

Examples of MaaS solving regional problems include improving the links between Paris and the greater Ile-de-France area; and enabling tailored solutions for the very different needs of rural populations (typified by ageing and depopulating demographics) adjacent to urban areas (which are typically growing and have a younger population) in Japan.

Regions have the advantage of oversight and network planning whilst MaaS can combine rapid transit and on-demand or shared services, making a potentially powerful combination.

References

[1] Average Number of Trips Made and Distance Travelled [Internet]. GOV.UK. Available from: https://www.gov.uk/government/statistical-data-sets/nts01-average-number-of-trips-made-and-distance-travelled

[2] Widuto A. Briefing EPRS | European Parliamentary Research Service [Internet]. 2018. Available from: https://www.europarl.europa.eu/RegData/etudes/BRIE/2018/628244/EPRS_BRI(2018)628244_EN.pdf

[3] MaaS Movement? Issues and Options on Mobility as a Service for City Region Transport Authorities | Urban Transport Group [Internet]. www.urbantransportgroup.org. [cited 17 Aug 2023]. Available from: https://www.urbantransportgroup.org/resources/types/report/maas-movement-issues-and-options-mobility-service-city-region-transport

[4] Maas B. Literature review of Mobility as a Service. *Sustainability*. 2022;14(14):8962.

[5] Mobility as a Service – Regional Research [Internet]. www.griffith.edu.au. [cited 17 Aug 2023]. Available from: https://www.griffith.edu.au/cities-research-institute/research/transport-group/mobility-service

[6] Case Study Analysis: Governance Models for Mobility as a Service MaaS [Internet]. tno.nl/en. [cited 17 Aug 2023]. Available from: https://www.tno.nl/en/newsroom/2021/01/case-study-analysis-governance-models/

[7] Surakka T, Härri F, Haahtela T, Horila A, and Michl T. Regulation and governance supporting systemic MaaS innovations. *Research in Transportation Business & Management*. 2018;27:56–66.

[8] United Nations Population Fund. Urbanization [Internet]. Unfpa.org. 2015. Available from: https://www.unfpa.org/urbanization

[9] Semuels S. A. Can Anything Stop Rural Decline? [Internet]. *The Atlantic*. 2017. Available from: https://www.theatlantic.com/business/archive/2017/08/japan-rural-decline/537375/

[10] Leung A, Prof A, Burke M, Akbar D, and Kaufman B. Mobility as a Service Regional Research [Internet]. 2021 [cited 17 Aug 2023]. Available from: https://www.griffith.edu.au/__data/assets/pdf_file/0040/1379947/MaaS_Regional_Research_Report_FINALREPORT_Final1.pdf

[11] Creation of New MaaS in Various Regions to be Promoted [Internet]. www.meti.go.jp. [cited 2023 Aug 17]. Available from: https://www.meti.go.jp/english/press/2020/0731_001.html

[12] Creation of New MaaS in Various Regions to be Promoted [Internet]. www.meti.go.jp. Available from: https://www.meti.go.jp/english/press/2020/0731_001.html

[13] *Recent Trends in Transportation Policy Toward the Realisation of Japanese Version of MaaS*. Tokyo: Ministry of Land, Infrastructure, Transport and Tourism. 2019.

[14] Nomura M, and Takahashi Y. Uberisation in rural Japan: A case of community transport in Tango Peninsula, Kyoto Prefecture. In: *International Conference Series on Competition and Ownership in Land Passenger Transport*. 2017.

[15] 気張る！ふるさと丹後町 | 公ホーム気ージ [Internet]. [cited 17 Aug 2023]. Available from: http://kibaru-furusato-tango.org/

[16] Mulley C, and Kronsell A. Workshop 7 report: The “uberisation” of public transport and mobility as a service (MaaS): Implications for future mainstream public transport. *Research in Transportation Economics*. 2018;69: 568–72.

[17] Integrating Public Transport into Mobility as a Service [Internet]. ITF. 2020. Available from: https://www.itf-oecd.org/public-transport-mobility-service

[18] Mobility on Demand (MOD) Sandbox Program [Internet]. Federal Transit Administration. 2016 [cited 31 May 2019]. Available from: https://www.transit.dot.gov/research-innovation/mobility-demand-mod-sandbox-program

Chapter 9
Specialist use cases for MaaS

9.1 Introduction

Despite the significant variety of trials of MaaS, and their significance in number, these trials have often coalesced around a series of common use cases and contexts. Notably, the delivery of MaaS trials primarily on or focuses on urban areas, seeking to establish commercially viable service operations for the MaaS operator, as well as some trials in rural areas. This often aligns with the original vision of MaaS being a multi-modal service utilising new technologies to deliver an enhanced service offering.

Despite this, there have been some trials in more specialist use cases focused on providing services and infrastructure to support universal mobility. To take a common example, 1.3 billion people globally – 16% of the population – are estimated to have some form of disability [1]. However, within this population exists a variety of conditions that significantly affect people's use and experience of transport services. These can include physical mobility issues, sensory deficits, cognitive and neurological differences, and mental health issues, both permanent and temporary. Each one of these presents different use cases for MaaS.

As MaaS is still an emerging concept, the number of trials in specialist use cases is still relatively limited. But this chapter highlights some of the more interesting of such use cases for which there have been trials, deployment of solutions, or even theoretical discussions.

9.2 Accessible MaaS

The issue of ensuring that MaaS are accessible has been discussed since the emergence of the concept. Being able to customise trips to the needs of people making them is, prima facie, very attractive.

The accessibility implications of MaaS are considered in two aspects. The first is the accessibility of the transport services themselves, the second is the accessibility of the MaaS. The former has been a long-standing issue across transport networks across the globe and has direct implications in terms of the number of trips that people with disabilities and diverse mobility needs face.

It is important to consider at this stage that people with physical or mental disabilities are a far from homogenous group, and the breadth and scale of which cannot adequately be covered in this small section of this book. Whilst an estimated 16% of the global population experiences some form of disability [1], experiences of disability vary depending on the social and legal context and the social and medical models underpinning those definitions. Whereas the medical model focuses specifically on the impairment, for example 'that person cannot access local buses because they are in a wheelchair', the social model focuses on the physical and social world that is organised in a way that excludes the person with the impairment. For instance, should a person with limited sight find it difficult to use a MaaS app, the social model would state that this is an issue with poor app design, as opposed to an issue with the person having limited sight.

People with disabilities also experience more general issues associated with digital inclusion. Across the world, cities have been rapidly expanding digital services and smart solutions as a way of serving the needs of their communities more efficiently. Whilst digitally driven urban development has become a new paradigm for cities, the digital divide is emerging as a serious concern when it comes to the goal of making cities accessible and inclusive for everyone, especially the most vulnerable. Issues such as digital inclusion and access to key urban services, information, and experiences by people with disabilities are critical areas in which cities face important challenges related to ethical and responsible governance.

For this purpose, the barrier-free urban logic is defined based on four key pillars. Digital Inclusion and Equity Leadership focuses on institutional and leadership factors within public services that are necessary to deliver an inclusion-first approach. The Culture of Universal Accessibility focuses on staff training and development. The Standardisation of Digital Access seeks to establish common best practice frameworks and principles of universal design and accessibility in the development and delivery of services. Meanwhile, a Shared Commitment focuses on developing partnerships with businesses and the community to deliver.

Minimum standards for the accessibility of digital services are critical to ensuring that services that are offered to the public can be used by a wide variety of people as possible. The Web Accessibility Directive (Directive (EU) 2016/2102) has been in force since 22 December 2016 and provides people with disabilities with better access to websites and mobile apps of public services. This states that EU Member States must ensure that websites and mobile applications of public sector bodies are 'more accessible', particularly for people with disabilities, by making them 'perceivable, operable, understandable and robust'. The accessibility standard is set out in the harmonised European standard EN 301 549 v3.2.1. All websites and applications provided by public sector organisations must be accessible to these standards from June 2021.

At a global level, the Web Accessibility Initiative (WAI) by the World Wide Web Consortium (W3C) has developed standards for making web and web-based content more accessible. The Web Content Accessibility Guidelines 2 [2] explain how to make web content more accessible to people with disabilities. Web 'content' generally refers to the information in a web page or web application, including natural information such as text, images, and sounds, as well as code or markup that defines structure or presentation.

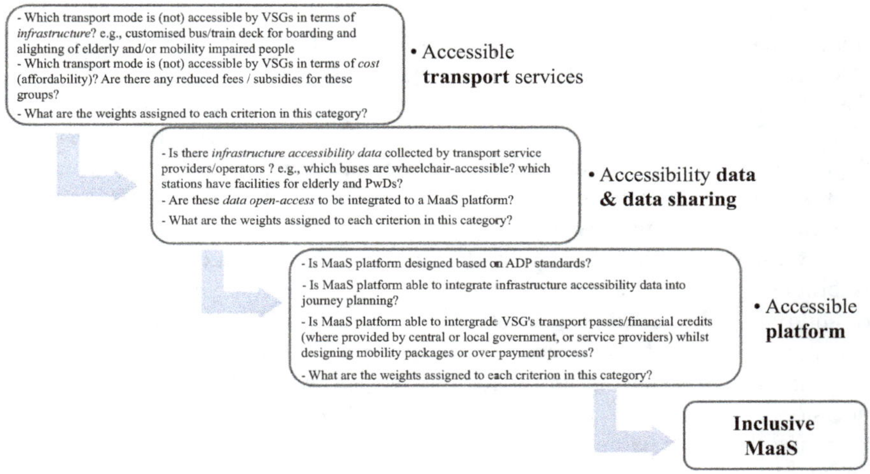

Figure 9.1 Inclusive MaaS framework [3]

Whilst the WCAG is a well-established standard, its use and adoption is voluntary. Furthermore, unless there are specific accessibility requirements written in law (e.g., the Web Accessibility Directive), there is no punishment for not meeting these standards. Within this context, it is not surprising that there has been no audit of the adoption and use of accessibility standards by MaaS providers and operators.

More generally, the academic work in relation to MaaS and accessibility has focused on accessibility in its broadest terms as a matter not related to physical and mental impairments, but of access to services. A notable exception to this is a study that developed a MaaS Inclusion Index (MaaSINI) against which to assess the inclusivity of MaaS offerings. This proposes to include accessible transport services, accessibility data and data sharing, and an accessible platform [3]. The suggested framework consists of the following three main categories: (i) Accessible transport services (ii) Accessibility data and data sharing, and (iii) Accessible platforms. This framework is yet to be tested on any MaaS applications or services (Figure 9.1).

An interesting use case for meeting accessibility needs, for which there is some research in relation to MaaS, is that of community transport. Sometimes referred to as paratransit in other areas, within the UK community transport is broadly defined as providing flexible and accessible community-led solutions in response to unmet local transport needs. It often represents the only means of transport for many vulnerable and isolated people, often older people or people with disabilities [4]. Though within legislation its operational characteristics are more clearly defined. In Australia, by contrast, in general, there is no accepted definition but defining community transport as not-for-profit, being based on community need and consisting of targeted and flexible services that would cover the type of service provided in each State [5].

A study of five community transport operators posed the concept of five different MaaS-style operations as potential service options for community transport customers. The findings indicate a strong enthusiasm among CT operators for offering packages of mobility services to their current users and investigating the delivery of packages to non-users/new users. The packages were perceived as advantageous for organisations due to the potential to retain customers, as there was a significant concern, and quite possibly a valid one, that operators might be left primarily offering costly trips.

A clear theme that emerges is that transport for the aged and people with disabilities is treated as a welfare issue in Australia. The MaaS approach offers the possibility to take mobility to another level where transport or mobility is not simply regarded as an entitlement (under threat by Government cuts) but an essential part of engaging in society. However, whilst the paradigm shift is understood and recognised by CT operators, being more entrepreneurial and expanding the client base, will only ever be of limited success until the full social benefit of mobility services is recognised.

Regardless of this work, providing dedicated MaaS-style solutions for people with disabilities is still a work in progress. Several applications improve the travelling experience for people with physical and mental impairments. A notable example is Passenger Assistance, which enables people with impairments to book assisted travel on the UK rail network through either an app or a booking website. At the time of writing this book, there have been no formal reviews of the effectiveness of this service in enabling travel by those with impairments, though the application has an average score of 4.6 stars out of 5 on Google Play.

Another example that focuses on navigation is that of Briometrix Navability. This provides a 'Google Maps for people with wheelchairs'. Taking account of personal travel and fitness data, in tandem with the paths and environment ahead, it provides individual journey analysis. Every Navability map and/or path reflects the persona's travel ability and effort as a distinct colour. That way, the user will know how a path will affect them so they can make an informed decision about where they will travel.

Navability will select paths to keep the user within their ideal effort zone, which aims to preserve physical energy or power chair battery life. Whatever path the user chooses, they can continually assess throughout the day how travel is impacting the user. An interesting aspect of Navability is the integration with other healthcare technology, especially wearables and a smart wheelchair. Should people with physical or mental impairments make use of such technology, this can be integrated with their navigation and transport planning applications to provide a specific benefit to individual users.

9.3 MaaS on islands

Islands represent a unique transport use case, and consequently a unique use case for the adoption of MaaS. Much like rural areas, to which they are often compared, islands vary in their challenges and nature. As an example, consider an island that boasts robust connections and seamless integration with its public transportation system. In contrast, islands like those found within the city of Stockholm, which enjoy different transportation and Mobility-as-a Service (MaaS) scenarios, differ significantly from remote islands with sparser populations. It is within the latter context that we mainly focus our attention here.

On the surface, their transport challenges are somewhat similar to rural areas – namely around accessibility and a lack of scale of population to support commercially viable services. But this is to misunderstand the nature of the severity of the challenges facing island communities. Often, public transport networks on islands are few outside of connecting the main settlements on the island, should they actually operate at all.

The second element is connecting the island to either other islands or to a nearby mainland. This means that MaaS will be required to integrate either with passenger ferry services or with passenger air services that provide this connectivity. Within both, practice at present is relatively limited, though some research has sought to analyse the potential of MaaS to add value to such services.

One such study has focused on the case study of ferry services between the Greek Islands in the Aegean Sea. All operators that participated in the study were keen to participate in a MaaS scheme. An important consideration came from operators who stated that they want their firm's name to be visible within MaaS (i.e., the end-user should be aware of the ferry operator who offers the service), whereas other MaaS actors do not set this as a prerequisite. Furthermore, ferry operators would like passengers to be able to use the loyalty programmes they currently offer, even if the passenger has purchased his/her ticket through a MaaS platform [6].

Operators also expressed their concern regarding the availability of resources (e.g., adequate human and economic assets) for the adoption of MaaS. All ferry terminal operators denoted that they have neither the appropriate human assets nor the economic capability to develop a MaaS on their own. The same applies also to all small transport operators with few employees.

One of the examples of a real-life deployment of a MaaS that incorporates island connectivity is the Go-HI app coordinated by HI-TRANS in Scotland (see Section 7.5). This offers seamless journey planning and integrated payment for transport services across the HI-TRANS region. This includes integrated journey planning and payment with ferry and air service operators, notably Northlink Ferries (that sail from Aberdeen to Shetland via Orkney), Orkney Ferries (that sail between Orkney's mainland and 13 other islands), and Loganair (who operate several flights from Inverness Airport to Orkney, Stornoway, and Shetland). The impact of this MaaS in terms of changes to the connectivity of islands has not yet been assessed.

9.4 Travel to education

The transport of school-aged children to and from their education is a long-running transport challenge. With the right to an education, as well as attending education a legal requirement in many countries, it has often been a requirement of the public sector – whether the education or transport authority – to provide transport to and from school. In the United Kingdom, for example where certain minimum criteria are met, every child is entitled to free travel to and from school up until the age of 16, with over £1 billion spent every year to meet this requirement [7].

In practice, school travel is provided by several means. This ranges from demand-responsive services such as taxis to buses specifically for school trips for which there are dedicated vehicles, for example, the well-known yellow school buses in America. In some places, free travel for school-aged children is provided on public transport. For example, the Zip Oystercard provided for 11–15-year-olds entitling them to free use of public transport throughout London [8].

The consideration of school travel within a more broad MaaS concept has only recently started in earnest. A notable paper on this by David Hensher [9] considers future bus transport contracts more broadly in a MaaS age, which includes school travel as one element. This paper notes that where contracted school services are provided, the constraints on them would make it difficult for a MaaS-type solution to be implemented. The observation is that the inclusion of school transport services with a MaaS offering is subject to a number of specific matters related to the operation of those services. He states that for school bus contracts, it is unlikely, but not impossible, that the smart point-to-point service model would apply, and since in most contexts in many countries, including Australia and the United States, school children are the backbone of their patronage, the future of point-to-point services is likely to be focussed primarily on the very small segments of the non-school children market. However, as a counterargument, some social aspects may push parents into using smaller bus-like vehicles with point-to-point travel provisions (a facility in the 2016 legislation that is now possible in New South Wales provided that the bus has a carrying capacity not exceeding 12 seats). Parents are also increasingly strongly driven by safety associated with the notion of 'stranger danger', including who they might meet on a bus (although this tends not to be an issue with exclusive school buses). Combined with the growing danger of getting to and from the bus stop, more specific service transport might be tempting, which reinforces a growing preference of parents to have their children only mix with other children that the parents know.

A similar issue plays out in the United Kingdom. Under the current legislation where Local Education Authorities are required to provide free home-to-school transport to eligible children, Local Education Authorities can simply provide free passes to use on local public transport services that are accessible and are used by the public every day, so long as the respective children are provided with a reasonable and safe journey to school. This enables them to meet their legal obligations without providing a dedicated service. However, where transport services are provided under contract for the specific purposes of transporting children to and from school, these are often subject to additional requirements (notably around safeguarding of children such as background checks on drivers, and specially adapted vehicles for those with special needs) which make them unsuitable for wider use, and consequently unable to be part of a MaaS-style service. To further complicate matters, bus operators can publicly register school services so that any person can board, though in practice as these services typically run once a day in each direction and only on a specific route, they are rarely used.

However, some examples in the United Kingdom demonstrate how school transport may be incorporated into MaaS. In many regions, whilst not MaaS in the strictest sense, there is increasingly a requirement for users of school transport services to do so with a smart card or application. Nominally this is for two reasons. The first is revenue protection. Many local education authorities are shifting their payment model away from purely providing a subsidy to run a service, to that based on the actual utilisation of the service, with basic operational costs also covered. A smartcard operation makes this a more practical proposition.

The second depends on the technological capability of the ticketing provider, and that is child safeguarding. In the event of a child not turning up for school, tracking movements via a ticketing system can verify that they boarded a school transport service (or did not). Some ticketing providers can go further, providing parents with live details of their child's boarding and alighting services, and tracking the vehicle in real-time.

An example is East Lothian, Argyll & Bute, and the Western Isles in Scotland. In 2014, a Home to School Transport Initiative provided pupils with either a Kidz or a Young Scot card, with the aim of being able to improve planning and ticket offerings for school pupils across the area. Though there is no evidence provided of its impacts on the experience of school children or the local education authorities, this case study provides an interesting example of integration with other products and the challenges associated with it.

On 31 January 2022, Young Scot was significantly expanded for non-school travel. The expansion meant that everyone aged under 22 years old could travel for free on buses across Scotland [10]. Furthermore, people aged 16, 17, or 18 years old could benefit from discounted travel on rail services, as well as up to four free trips per year on ferry services. This posed a significant operational challenge for home-to-school transport providers, as whilst home-to-school transport services are not public buses and are therefore not subject to the same rules regarding acceptance of Young Scot tickets, they are required to accept Young Scot cardholders who are eligible for free home to school transport services. A similar, though not as significant in terms of scale, issue emerged on ferry services.

A lack of an integrated ticketing system has necessitated local education authorities to provide technical (and sometimes manual) workarounds for home-to-school transport services. For instance, West Lothian Council provides a separate smart card ticketing offer to secondary school children who are eligible for home-to-school travel. But this is integrated with the Young Scot card, and parents have to provide the ITSO card number as part of their application for free home-to-school transport. The entitlement to free home-to-school transport is then validated on the vehicle [11].

The overarching view currently is that school transport services are likely to remain in their current form as the matters associated with their integration with MaaS are contractual, and they are highly specialised in their operations. However, this has not meant that MaaS-like services, especially in relation to ticketing, have not been launched or trialled.

One example in Australia was Rydhero which emerged as a provider of on-demand home-to-school transport services through an app. The founders of Rydhero intended that the company would [12]:

1. Create an efficient and nicely designed app to help customers find a safe driver;
2. Broaden our target customers to include the elderly, whilst ensuring a secure and reliable service;
3. Build partnerships with companies offering services for kids and the elderly to provide plus-value to the customers;
4. Start an Open-Source, Non-For-Profit Self-Driving-Car project to bring even more security and efficiency to the already existing services.

The cost was $15 (Australian) for the first child for a distance of up to 5 km ($5 per extra child) and then a cost based on distance ($1/km) and time. This company is no longer operational.

9.5 Youth travel

The travel patterns of young people have been considered by many practitioners to be a significant potential market for MaaS. With perceptions about young people seeking to defer car ownership, and instead preferring to own and use a smartphone over wishing to buy a car, as well as perceptions about them wishing to move to cities and live throughout their lives, they consequently made for an attractive potential market for the adoption of MaaS.

In reality, young people are every bit as complex as other demographics. Across many Western societies, there has been a sustained decline in both car use and car ownership of people aged 25 years old or younger. Though this has been more a factor of rising motoring costs (e.g., insurance) as opposed to there being less of a wish to drive, as well as wider social changes over the last 30 years, notably declines in home ownership, change in educational attainment and a more volatile employment situation [13]. Such observations contrast with the common perception of young people willingly giving up their cars and have posed a challenge in the adoption of MaaS solutions specifically for young people.

9.5.1 Case study: NaviGoGo

NaviGoGo was Scotland's first MaaS implementation, developed with Innovate UK funding (project title: 'Pick & Mix'). It aimed to improve how young people relate to, use and combine travel modes and to this end was co-designed by young people for young people (aged 16–25). An initial 6-month pilot ran from October 2017 to March 2018 and involved 98 participants. Participants were given access to streamlined and personalised information and payment (where available) for trains, taxis, bike schemes, buses, car clubs, and walking – all in one single hub. The Pick&Mix project was a partnership between ESP Group (lead partner), SYSTRA,

Young Scot, Mudlark, RouteMonkey, and the Scottish National Entitlement Programme Office.

By choosing a very narrow demographic it was expected that more value could be provided, rather than designing one service to appeal to everyone. Furthermore, the project called 'Pick&Mix', set out to co-design the solution with young people themselves. Co-design is a process whereby future users are engaged and empowered within the design process. It can allow for a deeper understanding of the issues to be tackled and the needs of the users as well as the participation of users in the design, prioritisation, and testing of potential solutions.

There was already some understanding from existing research that this demographic is generally more pragmatic about their travel choices – i.e., they are more likely to be multi-modal – and may have a high willingness to try different modes and travel services. As such, a MaaS approach appeared to be a strong fit. Indeed, once the co-design process began, it quickly became clear that young people did face many challenges that a MaaS solution would be able to assist with. Participants guided the development of each feature of the service whilst also contributing to the branding and user experience and testing of the service. Certain features, such as the Taxi Splitter Tool, would not have been developed without this level of participation of users from the outset.

The Pick&Mix consortium included Young Scot (Scotland's national youth information and citizenship charity) and the National Entitlement Card Programme Office (NECPO) which is run by Dundee City Council on behalf of the Scottish Government. The inclusion of these two partners was seen as important to ensure a strong alignment with local and national policy. At the start of the project, the consortium partners agreed on several objectives to guide the development of NaviGoGo. These were chosen to align with local and national policy objectives relating to travel/transport and young people.

Primary: key policy objectives for the project

- Support and help young people to make the most of the transport services available to them in their area
- Make transport easier within local areas (urban and rural)
- Increase participation in education, volunteering, employment and leisure activities by helping to remove the transport-related barriers to access
- Create understanding and data to inform general policy

Secondary: supporting policy objectives for the project

- Make transport easier between areas (long distance)
- Influence future transport services
- Help improve health
- Create lifelong sustainable transport users

The success of the approach and the resultant NaviGoGo solution can be seen through the behavioural and attitudinal impacts:

- 50% of participants agreed or strongly agreed that NaviGoGo made public transport more attractive

- 54% of participants agreed or strongly agreed that NaviGoGo made their travel easier
- 46% of participants agreed or strongly agreed that NaviGoGo allowed them to travel with confidence

NaviGoGo operated as a 6-month pilot within a 2-year funded research project. This context placed certain boundaries on the scale of service development possible and, therefore, the subsequent impact that could be achieved. The young co-designers uncovered many more potentially valuable features and functionalities than the team was able to develop in the relatively short timeframes and limited budgets. Furthermore, the scope of the pilot meant that user numbers were limited.

Within the project, it was not possible to fully integrate bus use. Weekly bus tickets could be loaded onto the users' existing Young Scots cards, but no further ticketing or payment options were available for the bus. Participants anticipated that when the bus was fully integrated, public transport would have been even more attractive and travel would have been even easier. Additionally, there was a clear demand for NaviGoGo to include longer-distance transport options, which would create further value for users.

During the NaviGoGo pilot, there was a substantial increase in reported taxi use, although these were predominantly shared taxi trips. Whilst this is less environmentally sustainable, it still has value as a shared journey and supports many of the key policy objectives for the project. Furthermore, it appears that the increase in taxi use was largely due to participants having access to incentive credit on their NaviGoGo accounts. As such, if the incentive credit is removed in future iterations of the service, taxi use is likely to decline.

Having the ability to pay for all forms of transport (whether through a subscription, a stored account/wallet, or a pay-as-you-go model) is seen as a central piece of any MaaS implementation. Within this iteration of NaviGoGo, payment through the NaviGoGo account was limited to certain modes (where technical integration was simple) and was not always as straightforward as being able to 'tap' your NaviGoGo card (i.e., the user's Young Scot card). As the case for MaaS – specifically the generation of bus travel – is more clearly proven, the integration of these modes will become more commercially viable.

Whilst the project was initially funded through innovation funding, the creation of a financially viable and sustainable business was always a central objective. The 6-month NaviGoGo pilot produced valuable insights for refinement of the Business Case for such a service, but more work is required to prove the case at scale and to test the refined business case assumptions made possible by the insights generated by the NaviGoGo pilot.

Some of the key learning points were that it is unlikely that a single solution will be valuable for all potential users. Future trials should give clear consideration to who their desired market is, what needs are being supported, and what behaviours are being targeted. In addition, the co-design process was crucial. Without participation from potential future users, it was clear that the real needs of 16–25-year-olds would not have been understood. The final service would have looked very different and would not have provided such high value for users. However, it is recognised that pursuing a co-design approach requires ongoing commitment and resources – for example, there is an imperative to include a range of people from the target segment in the co-design activities, some of whom may be more difficult to recruit or involve. Once a strong basis of co-design has been undertaken, there is the opportunity to tailor the solution, building on the existing, proven offering.

The team found that users' circumstances or needs will change over time. In response, the MaaS offering could also change. Users should be able to easily transition to a different MaaS solution, or changes to their profile should trigger newly personalised experiences.

However, one of the barriers to the success of this implementation was the lack of confidence among some transport operators regarding their involvement in MaaS implementations. Many are still cautious that such schemes will erode their markets and direct contact with customers. As such, future trials must gather additional data to demonstrate the direct and indirect benefits for operators. In turn, this should encourage operators to release more of their data to MaaS operators thereby allowing for more sophisticated MaaS offerings.

Whilst this was a specialised approach to MaaS, it is clear that this methodology could be used to create tailored MaaS solutions.

9.6 Travel to healthcare services

In many countries, a variety of healthcare services are often provided by a mixture of national government, regional government, community enterprises, and even the private sector. Being able to successfully access a variety of healthcare services is essential to living a long and healthy life, and the significant variety of healthcare services that are available makes this a challenge. Taking the UK's National Health Service as an example of a comprehensive offering of healthcare services, these can include (but are not limited to):

- General Practitioners (Doctors)
- Dental
- Physiotherapy
- Pharmacies
- Hospitals (without Emergency Departments)
- Hospitals (with Emergency Departments)
- Pre-natal and Ante-natal Services
- Mental Health Services
- Sexual Health Services
- Opticians
- Alcohol addiction services
- Cancer screening and surgery
- Cardiology
- Carer Support Services

- Disability Support
- Drug addiction support
- Ear, Nose, and Throat

Research work undertaken in the early 21st century revealed the potential of alignment with information technology services and transport to healthcare and the opportunities that this would bring. For example, in studying access to the John Radcliffe Hospital in Oxford, the opportunity to utilise a demand-responsive transport service (booked online or via the telephone) indicated substantial accessibility gains through the deployment of such demand-responsive services. Overall, about 40% of the hospital orders were assigned to a service route within a 20-min time window for this initial set up. Note the same 'success rate' applies to the GP surgery services developed separately from the hospital ones. For non-car drivers and/or the elderly, the DRT service routes have significant accessibility benefits. This is particularly true where access to public transport is poor [14].

There is extensive evidence that indicates that improving access to healthcare services by non-car modes plays a significant role in improving the overall health of the population. For example, an economic assessment of changes to the public transport network in Melbourne, Australia, multi-directional, high-frequency improvements to the public transport network resulted in significantly greater health and economic gains than current policy directions in relation to physical activity (mean 6.4 more MET minutes/week), body weight (mean 0.05 kg differential), health-adjusted life years gained (absolute difference of 4,878 HALYs gained) and healthcare cost-savings (absolute difference of AUD 43 million), as compared to business as usual under both scenarios ($n = 2,832,241$ adults, over the life course). Improving accessibility to public transport will improve population health by facilitating physical activity and lead to healthcare cost savings compared with business-as-usual [15].

Solutions that are often part of MaaS offerings, especially in cities, also have healthcare benefits. A notable example of this is bike share systems that facilitate cycling in some cities across the world. Primarily due to the health benefits associated with cycling, it is estimated that for every £1 invested in cycling initiatives, £5.62 in wider economic benefits would accrue (in 2014 values) [16].

The role of bike sharing in MaaS is discussed in greater detail later in this book. But for now, it is sufficient to state that early evidence indicated that people seeking to use MaaS-style services are willing to consider subscribing to products that have shared bikes as part of the package offering [17].

Work to date on understanding the potential of MaaS in relation to healthcare services has primarily been about improving access to healthcare services alongside the introduction of new transport services. This is especially the case in rural areas, for which wider evidence suggests that there is poorer accessibility to healthcare services on average in these areas [18].

In a study of Shobara, Japan, a largely rural area of Japan with a significant elderly population, the potential of delivering a MaaS offering alongside an AI-powered demand-responsive transport service was considered from an equity perspective. The provision of the AI on-demand shuttle bus resulted in a remarkable decrease in the accessibility inequity of two out-of-home activities compared to a regular bus and seemed to benefit almost all Shobara areas.

This study focused on equity outcomes of accessibility in certain regions. The place-based accessibility measure successfully showed where accessibility levels are high and low on a map and provided an overview of accessibility levels. In contrast, the person-based accessibility measure highlighted the relationships between accessibility and multiple sociodemographic characteristics. Furthermore, the results highlighted the importance of implementing MaaS, such as the AI on-demand shuttle bus, in rural areas to provide a more equal distribution of transportation benefits [19].

The healthcare implications of MaaS can also be understood in terms of its component parts, and what the overall package intends to achieve in terms of transport impacts, for example reducing levels of congestion (Table 9.1). The component parts of MaaS can have positive healthcare impacts in the following ways.

A significant study of the practice of innovative healthcare mobility services in the United States identified a number of practices that could be broadly classified into the following terminologies and groupings [21]:

- **Health care provider leverages ride-sourcing technology**. The first type of innovation is when a healthcare provider leverages ride-sourcing technology to book patient trips. This was the most common type of innovation found and it primarily involves transportation companies tailoring the ride-sourcing experience to the healthcare industry. The critical feature of this innovation is the added HIPAA (Health Insurance Portability and Accountability Act) compliance of the booking process.
- **Insurer partners with Transportation Network Company (TNC)**. When an insurer partners with a ride-sourcing company. This is when a health plan or insurance company formally partners with an existing ride-sourcing service(s) to expand transportation services available to beneficiaries or offer these transportation benefits for the first time.
- **Paratransit provider partners with TNC**. When a paratransit provider partners with a ride-sourcing company. Due to the demand-responsive nature of paratransit provisions (e.g., services do not operate over a fixed schedule like a standard public bus; rather, vehicles are dispatched on request and operate door-to-door), paratransit services have been said to be a sort of progenitor of mobile app-based TNCs. Whilst not all paratransit trips are for health-related purposes, paratransit is especially important for people with disabilities who may have no other mode of reaching healthcare facilities

Table 9.1 Health impacts of different transport policy interventions [20]

Transport policy measure that can potentially be part of a MaaS offering	Pathway of action	Health impact
Intelligent transport systems	Cover a wide range of applications of information and communications technologies to transport. Can lead to increases in effective capacity	Impacts unclear and depend on changes in road capacity and traffic flow parameters. Impacts will vary considerably based on the application
Demand responsive transport	Provide a service for those who otherwise have limited or no public transport service. May cause modal shifts from car	Possible positive impacts through some increases in active travel and physical activity, reductions in traffic, air pollution, heat island effect, noise and climate change effects. Possible positive impacts through reduced inequalities
Promotional activities	Varied by type of promotional activity. More effective if they are combined with 'hard measures' like improvements in the infrastructure	Depends on the message conveyed with possible positive impacts through increased active travel and physical activity, reduction in air pollution, heat island effect, noise and climate change effects, and possibly motor vehicle crashes
Personalised journey planning	Reductions in car use through providing targeted information on alternatives to the car for particular trips and encouraging the use of alternatives	Possible positive impacts through increased active travel and physical activity, reduction in air pollution, heat island effect, noise and climate change effects and possibly motor vehicle crashes
Ridesharing	Reduction of the number of cars on the road	Positive impacts through some reduction in air pollution, heat island effect noise, and climate change effects. Further positive impacts are possible through reduced inequalities

(Continues)

Bike sharing	Reduction in car use and increase in cycling and transit usage	Positive impacts through the increase in active travel and physical activity, reduction in noise. Small negative impacts through some increase in personal air pollution exposure in cyclists and increased risk for motor vehicle crashes in those switching to cycling. Positive impacts on the general population through reduction of air pollution, noise, climate change effects, and motor vehicle crashes
Car clubs	Reduction in car travel and usage and the need for car ownership	Positive impacts through some reduction in air pollution, heat island effect, noise and climate change effects, and an increase in active travel and physical activity. Reduced parking needs allow more public space. Further positive impacts are possible if there is an increase in exposure to green space and a reduction in inequalities
Real-time passenger information	Can reduce the psychological anxiety associated with waiting for public transport as well as uncertainty and frustration. May prompt behaviour change towards increasing use of public transport and modal shift from car	Possible positive impacts through increased active travel and physical activity, reduction in air pollution, heat island effect, noise and climate change effects if public transport vehicles are not high emitter noisy vehicle. Possible reductions in motor vehicle crashes and reduced anxiety
Fare structures	Depending on direction: profit or welfare maximisation	Depends on the magnitude and the direction of the fare structure (changes). A reduction in fares could lead to positive impacts through some increases in active travel and physical activity and possible reduction in traffic and air pollution, heat island effect, noise, climate change effects, and motor vehicle crashes. Possible positive impacts can occur through reduced inequalities

(Continues)

Table 9.1 *(Continued)*

Transport policy measure that can potentially be part of a MaaS offering	**Pathway of action**	**Health impact**
Fare levels	Changes in the monetary charge for making a trip by public transport may impact the level of demand for public transport and private cars	Depends on the magnitude and the direction of the fare level (changes). A reduction in fares could lead to positive impacts through some increases in active travel and physical activity and possible reduction in traffic and air pollution, heat island effect, noise and climate change effects, and motor vehicle crashes. Further possible positive impacts through reduced inequalities
Integrated ticketing	Increasing the convenience of public transport use. Possible shift from car use to public transport and reduction in congestion	Positive impacts through slightly increased active travel and physical activity, reduction in air pollution, heat island effect, noise and climate change effects, and possibly motor vehicle crashes. Further possible positive impacts through reduced inequalities

9.7 MaaS and COVID-19

The impact of the Coronavirus (COVID-19) pandemic since 2020, and the associated public health interventions, have had a significant impact on transport systems across the world. As of August 2023, there have been nearly 769 million cases of COVID-19 across the world [22]. Consistent global data on the impact of COVID-19 on travel is hard to come by, but it is understood that the impact of travel restrictions and lockdowns has had a significant impact on how people travel. A few examples are demonstrated below.

In the United Kingdom, the Department for Transport has published regular daily usage statistics of the transport network since 1 March 2020, with the use of different modes of transport compared to the equivalent week in 2019. The usage of the network changed as so for the following modes of transport [23]:

- Car use declined as low as 22% compared to pre-COVID levels (12 April 2020). As of the week commencing 10 July 2023, use by motor vehicles generally – including goods vehicles – is 103% of pre-COVID levels;
- Rail use declined as low as 5% compared to pre-COVID levels (5 May 2020). As of the week ending Sunday 2 July 2023, use is 104% of pre-COVID levels;
- Bus use declined as low as 10% compared to pre-COVID levels for buses outside of London (15 April 2020), and 14% for buses in London (12 April). This usage level has recovered to 81% for buses outside of London (10 July 2023) and 82% for buses in London (7 July 2023);
- Use of the London Underground declined to as low as 3% compared to pre-COVID levels (11 April 2020). As of Friday 7 July 2023, use is 81% of pre-COVID levels.

In the United States, the Bureau of Transportation Statistics described the impact of COVID-19 on travel trends in its Transportation Annual Statistics Report 2022 [24]:

> The long-term expansion in travel with population growth was severely disrupted by the COVID-19 pandemic starting in March 2020. Figure 2-2 summarizes the disruptions following the last pre-pandemic 'normal' year of 2019. Travel by all passenger transportation modes collapsed in early 2020, followed by some recovery in 2021, and a return to near normal levels in 2022 for some modes (primarily personal vehicles and commercial aviation) but a weaker recovery rate for other modes (especially transit and intercity bus). Even where passenger volumes are close to the 2019 'normal', their characteristics have changed appreciably. Personal vehicle travel has shifted in terms of usage times and directions and air travel has seen an increase, although airlines report that most returning customers are not business travelers who typically pay higher ticket prices. The impacts of these pandemic-induced pattern shifts go well beyond transportation, affecting downtown offices and services, home arrangements and locations, and the use of technology and

communications as substitutes for travel, although the duration of these shifts is uncertain.

Research by Union Internationale des Transports Publics (UITP) in 2020 on the impact of COVID-19 on public transport use globally found that the drop has been as much as 80%–90% in major cities from China to North America. Shared mobility service usage has decreased dramatically during and after the lockdown, whilst citizens have chosen individual modes such as cars, bicycles, and walking [25].

The COVID-19 pandemic, and the associated healthcare emergency, led to a significant decrease in the use of MaaS [26] across the globe, primarily associated with travel restrictions associated with the healthcare emergency. In the immediate aftermath of such shutdowns, many operators of MaaS and shared vehicle services adapted their services to the circumstances. For instance, many offered assistance to essential workers who were required to continue travelling despite restrictions on travel for others.

Whilst passenger numbers on fixed routes were often a small fraction of their normal capacity, leading to reduced services, on-demand transport was used in several places to fill in gaps and provide solutions for people who had to travel.

In Kent, UK, the local bus operator, Go Coach suspended seven fixed-route lines due to the reduction of ridership and replaced them with a single area-wide on-demand service. This was a trial temporary solution to providing transport for the essential workers who were required to work whilst most people were required to stay at home. The managing director, Austin Blackburn, was fascinated by the new travel patterns that emerged as people booked the rides directly between their start and end points rather than fitting the journeys to fixed bus routes.

A different example of this was Uber Medics. Uber Medics is a ride option designed exclusively for NHS and care home staff who continued working through the COVID-19 pandemic [27]. This was offered across many geographies, but the offer in the United Kingdom consisted of the following. There is no data available on its use:

- 25% cheaper rides compared to regular Uber fares for NHS and UK care home staff (capped at £30 saving per ride)
- More flexibility on waiting time, to ensure staff can get to their pick-up point
- Uber drivers receive their normal fee and Uber does not make a profit from Uber Medics rides
- Available all day, every day until at least 28 February 2021

In a similar vein, many operators of new mobility services offered free or discounted travel for people wishing to access vaccination centres in order to receive their initial and booster COVID-19 vaccinations. Uber, for example, sent reminders to all users in August 2021, encouraging them to get the vaccine, and offered discounted Uber rides and meals on Uber Eats for young adults who get the jab. This followed Uber campaigns offering free trips to vaccination centres earlier in 2021 and for NHS staff during the height of the pandemic. Bolt also offered free ride credit to vaccination centres. A similar scheme by Bolt included offering £250,000 in free ride credit to London vaccination facilities earlier this year.

A unique factor that had a significant impact on shared transport services and public transport, and consequently the use of MaaS, was that of hygiene. Evidence from the pandemic indicated that the need to maintain effective social distancing, and concerns about hygiene on shared transport services provided a deterrent to the use of such services. A notable example is the experience of the SEAT MÓ motorbike sharing service in Barcelona.

An online survey of users of the scheme identified that during COVID-19, users of the scheme placed a high value on the application of a cleaning regime on the motorcycle fleet. Survey participants were asked to choose between 'Have hygienic measures such as hydro-alcoholic towels' and 'That the vehicle and accessories are protected against viruses and bacteria'. The comparison indicates a relatively even distribution of preferences, however, shows a slight tendency towards 'Vehicle and accessories are protected against viruses and bacteria' (38%). 27% solely prefer to have hygienic measures such as hydro-alcoholic towels and the remaining 35% prefer the implementation of both measures [28].

These results should be considered in context, however. Whilst users indicated a high preference for hygiene regimes to be implemented, in the case of Barcelona only 28% of survey participants were aware that this regime was even in place. Additionally, survey participants indicated a high valuation of availability, easy use of the application, the proximity of the vehicle, and the price. With many MaaS operators providing health advice related to social distancing and other localised healthcare advice during COVID-19, the effectiveness of doing so can consequently be questioned.

Combined, this has profound effects on the business model, and in some cases the financial viability of MaaS. These are discussed in more detail in the MaaS Business Models chapter of this book. But for the purposes of this chapter, a significant drop in the usage of transport services posed a significant risk to MaaS, which in turn are reliant on the use of transport services to re-sell tickets and to purchase passes – their primary revenue streams.

COVID-19 also led to much in the way of reconsideration of the future of MaaS and its potential role in the future of mobility. In an admittedly short and speculative paper in July 2020 [29], two different scenarios for the future of MaaS in Australia were speculated:

> Scenario 1: Travel will return to the pre-Covid-19 normal within a few months, with public transport, ride share and private car use showing very similar levels of use, crowding and congestion as before. Working from home will have a limited impact. The rationale is that, although the situation is somewhat fluid and the likely response is very uncertain, with Australia's success compared to other countries in minimising exposure and transmission ('flattening the curve'), there is a real possibility that normality might return quicker, with perceptions of risk dissipating at a fast rate. Habit persistence is also a significant trait of human beings.

Crowds, described as heaving, at shopping precincts on Mother's day (10 May 2020) in Melbourne and Sydney, despite social distancing requirements in place, highlights this outcome.

Scenario 2: One of the most important policy levers now available, in contrast to pre-Covid 19, is the effectiveness and growing acceptance of WFH. We have never had a real experiment of what might happen to the transport network in the presence of a growing interest in WFH. One of Australia's leading banks, the National Australia Bank (NAB), for example, is reporting a 15% increase in productivity (associated with WFH) since travel restrictions. This evidence, and growing anecdotal evidence together with the findings from a National Survey, suggest the possibility of a noticeable shift to WFH and consequent changes in commuting (and non-commuting) travel demand. WFH will be encouraged all the while offices are required to practice social distancing and hence have to stagger working hours for staff, including the possibility of less days in the main office and the balance as WFH. Firms will be interested, as they can save on office space in the longer term (although NAB have just invested in a huge building or buildings at Redfern near the City of Sydney).

The significant challenge facing MaaS is the recovery of pre-pandemic trip patterns and its subsequent implications for the MaaS business model. In a study of returning ridership on public transport in Chicago to pre-pandemic levels, Soria *et al.* [30] state that the role of MaaS is that of fare integration as a means of lowering barriers to groups who have discontinued public transit ridership, but who have stated an intent to return. Regardless, the research states that a number of more pressing factors are bigger barriers to people returning to public transport, and ultimately to the use of MaaS.

The results indicate that teleworking, unemployment, and vehicle access are the major factors behind discontinued transit ridership. Analysis of race, ethnicity, and gender effects reveals that vulnerable users often have a higher risk of abandonment coupled with a lower likelihood of returning. These results point to the need for transit agencies to consider the specific concerns of ethnic/racial minorities and women. Encouragingly, there is an opportunity for agencies to attract more ridership with fare integration.

This evidence is further supported by a study of commuters in Padua, Italy [31]. Surveys were undertaken between October 2020 and January 2021, the results of which informed the calibration of ordered logit and mixed logit models to estimate the demand for hypothetical MaaS offerings. The conclusions strongly indicated that the choice of using MaaS correlated closely with modal choices.

As regards the probability of acceptance of the proposed MaaS bundles, the interviewees declared a lack of interest in some services offered within the packages, probably due to the general characteristics of the population of respondents (e.g., car sharing is not a useful service for municipal employees probably because most of them, according to reports, own a car). The general perception of the pandemic situation does not influence the probability of joining MaaS; however, the perception of the health risk linked to the use of some means (e.g., car passengers), has an impact on the choice. Finally, and most importantly, as the desired weekly telecommuting frequency increases, the probability of accepting MaaS decreases; in other words, the diffusion of telework induced by the pandemic can negatively impact the choice to adopt MaaS. This could be due to the structure of the bundles themselves, since the services were offered as subscriptions and not as pay-as-you-go; moreover, teleworkers tend to travel less than commuters and their interest in mobility services (e.g., MaaS) is lower.

However, a study of Shotl on-demand transport in the suburbs of Barcelona by the Mobility Institute Berlin showed that when traditional public transport was suspended, users shifted to on-demand services. The on-demand services recovered better than fixed-line services – reaching 87% of its previous demand within 4 months of the start of the crisis whilst fixed services only reached 45% over the same period [32].

9.8 Summary of specialist uses cases for MaaS

MaaS has the potential to provide people with options that are tailored to their particular needs or circumstances, and also to provide tailored transport in particular use cases. Whilst it is not true the MaaS is a solution in every situation, it has a potential role in many transport scenarios.

MaaS that takes account of people's mobility needs – for instance wheelchair use or sensory impairments – has the potential to make trips easier by only suggesting step-free routes and providing turn-by-turn navigation (either audio or visually as appropriate).

Common transport needs where people require really good transport information, and where flexible and on-demand services play a role are also strong use cases. Integrated transport information on islands, which must be integrated with how to get on or off the island, is one such. Healthcare transport is another, particularly as people travelling to healthcare services are often vulnerable.

In these cases too, the goal of improving access to publicly accessible transport has wider benefits. There are benefits for individuals in terms of reducing the cost of car use and providing access for non-drivers. There are similar benefits for sites which can reduce the pressure on car parking provision.

Education involves very circumscribed and regular journeys, usually by bus, which makes MaaS as a solution less likely to show net benefits for operators or users. However, where more tailored services are necessary, or where travel information needs to be collected for safeguarding or parental peace of mind, MaaS has some potential.

The impacts of the COVID-19 pandemic on transport, in general, have been profound and wide-ranging. However, they have also fostered innovation and diversity of services which is unlikely to be reversed.

References

[1] World Health Organisation. Disability [Internet]. www.who.int. Available from: https://www.who.int/health-topics/disability

[2] W3C. Web Content Accessibility Guidelines (WCAG) Overview [Internet]. Web Accessibility Initiative (WAI). 2018. Available from: https://www.w3.org/WAI/standards-guidelines/wcag/

[3] Dadashzadeh N, Woods L, Ouelhadj D, Thomopoulos N, Kamargianni M, and Antoniou C. Mobility as a Service Inclusion Index (MaaSINI): Evaluation of inclusivity in MaaS systems and policy recommendations. *Transport Policy*. 2022;127:191–202.

[4] Community Transport Association [Internet]. ctauk.org. Available from: https://ctauk.org/about-cta/what-is-community-transport/

[5] Mulley C, Nelson J, Reedy L, and Wright S. Community Transport Meets Mobility as a Service: On the Road to a New a Flexible Future [Internet]. [cited 17 Aug 2023]. Available from: https://aura.abdn.ac.uk/bitstream/handle/2164/12033/Accepted_Manuscript_Community_Transport_meets_Mobility_as_a_Service.pdf

[6] Papaioannou G, Polydoropoulou A, Tsirimpa A, and Pagoni I. Assessing the potential of "Mobility as a Service" in passenger maritime transport. *Frontiers in Future Transportation*. 2022;2.

[7] Home-to-School Travel and Transport [Internet]. GOV.UK. Available from: https://www.gov.uk/government/publications/home-to-school-travel-and-transport-guidance

[8] Free and Discounted Travel [Internet]. Transport for London. Available from: https://tfl.gov.uk/fares/free-and-discounted-travel

[9] Hensher DA. Future bus transport contracts under a mobility as a service (MaaS) regime in the digital age: Are they likely to change? *Transportation Research Part A: Policy and Practice*. 2017;98:86–96.

[10] Free Bus Travel for Under-22s in Scotland Begins. BBC News [Internet]. 31 Jan 2022. Available from: https://www.bbc.co.uk/news/uk-scotland-60087004

[11] Smart Ticketing – School Buses [Internet]. West Lothian Council. [cited 19 Aug 2023]. Available from: https://www.westlothian.gov.uk/smart-ticketing

[12] Rydhero. Rydhero's Vision [Internet]. Medium. 2016 [cited 19 Aug 2023]. Available from: https://medium.com/@rydhero/rydheros-vision-2d264e74f4c9

[13] Chatterjee K, Goodwin P, Schwanen T, *et al.* Young People's Travel – What's Changed and Why? Review and Analysis Preferred Form of Citation [Internet]. 2018. Available from: https://assets.publishing.service.gov.uk/government/uploads/system/uploads/attachment_data/file/673176/young-peoples-travel-whats-changed.pdf

[14] Brand C, Rajé F, Preston J, and Gracieo M. *Transport and Access to Healthcare: The Potential of Information Technology*. University of Oxford/Department for Transport.

[15] Brown V, Barr A, Scheurer J, Magnus A, Zapata-Diomedi B, and Bentley R. Better transport accessibility, better health: A health economic impact assessment study for Melbourne, Australia. *International Journal of Behavioral Nutrition and Physical Activity*. 2019;16(1):89.

[16] Davis A. Claiming the Health Dividend: A Summary and Discussion of Value for Money Estimates from Studies of Investment in Walking and Cycling [Internet]. 2014. Available from: https://assets.publishing.service.gov.uk/government/uploads/system/uploads/attachment_data/file/371096/claiming_the_health_dividend.pdf

[17] Matyas M, and Kamargianni M. The potential of mobility as a service bundles as a mobility management tool. *Transportation*. 2018.

[18] Journey time statistics [Internet]. GOV.UK. Available from: https://www.gov.uk/government/collections/journey-time-statistics

[19] Chinbat T, Fumihiko N, Mihoko M, Shinji T, and Ryo A. Impact Assessment Study of Mobility-as-a-Service (MaaS) on Social Equity Through Nonwork Accessibility in Rural Japan. Asian Transport Studies [Internet]. 1 Jan 2023 [cited 17 Aug 2023];9:100109. Available from: https://www.sciencedirect.com/science/article/pii/S2185556023000147

[20] Khreis H, May AD, and Nieuwenhuijsen MJ. Health impacts of urban transport policy measures: A guidance note for practice. *Journal of Transport & Health*. 2017;6:209–27.

[21] Wolfe MK, and McDonald NC. Innovative health care mobility services in the US. *BMC Public Health*. 2020;20(1):906.

[22] World Health Organization. Coronavirus Disease (COVID-19) Pandemic [Internet]. World Health Organization. 2023. Available from: https://www.who.int/emergencies/diseases/novel-coronavirus-2019

[23] Domestic Transport Usage by Mode [Internet]. GOV.UK. Available from: https://www.gov.uk/government/statistics/transport-use-during-the-coronavirus-covid-19-pandemic/domestic-transport-usage-by-mode

[24] Hu P, Schmitt RR, Robinson R, *et al.* Transportation Statistics Annual Report 2022 [Internet]. United States. Department of Transportation. Bureau of Transportation Statistics, editor. ROSA P. 2022. Available from: https://rosap.ntl.bts.gov/view/dot/65841

[25] Preparing for a Better Future: How Transport Authorities Have Managed the Crisis [Internet]. UITP. [cited 17 Aug 2023]. Available from: https://www.uitp.org/publications/preparing-for-a-better-future-how-transport-authorities-have-managed-the-crisis/

[26] Beck MJ, and Hensher DA. Insights into the impact of COVID-19 on household travel and activities in Australia – The early days under restrictions. *Transport Policy*. 2020;96:76–93.

[27] Uber Medics – Subsidised Rides for NHS and Care Home Staff [Internet]. Uber Blog. 2020. Available from: https://www.uber.com/en-GB/blog/uber-medics/

[28] Gilibert M, and Weymar A. Case study on the use and adaptation of SEAT MÓ motorbike sharing service in Barcelona in COVID-19 pandemic year. *Case Studies on Transport Policy*. 2022.
[29] Hensher DA. What might Covid-19 mean for mobility as a service (MaaS)? *Transport Reviews*. 2020;40(5):551–6.
[30] Soria J, Edward D, and Stathopoulos A. Requiem for transit ridership? An examination of who abandoned, who will return, and who will ride more with mobility as a service. *Transport Policy*. 2023;134:139–54.
[31] Baldassa A, Orsini F, and Ceccato R. Telework and MaaS Adoption in a Post-Pandemic Scenario. Evidence from Municipal Employees of Padua, Italy. Case Studies on Transport Policy [Internet]. 1 Jun 2023 [cited 17 Aug 2023];12:101010. Available from: https://www.sciencedirect.com/science/article/pii/S2213624X23000640
[32] On-Demand Transport: A Tool to Keep People Mobile during the Coronavirus Pandemic [Internet]. [cited 17 Aug 2023]. Available from: https://mobilityinstitute.com/wp-content/uploads/2020/06/On-demand-transport-keeping-people-mobile-during-coronavirus-pandemic_mib_shotl_V1.01.pdf

Chapter 10
The challenge of integration

10.1 Introduction

One of the aims of MaaS has been to replace personal cars. Whether this desire is fuelled by concerns for the environment, congestion, and health (as with many city proponents) or just market share (for instance Uber [1]), there are substantial difficulties in replacing a single multi-purpose vehicle able to carry several passengers (and bulky items) which can make door-to-door trips and choose between multiple potential routes).

Cars have become progressively culturally and economically embedded in societies since the 1950s (with some variation in the region), so much so that our current car dependency is seen as the norm in many societies. The degree of change is only obvious if we take a longer-term, multi-decade, perspective. Looking at trip growth in the United Kingdom (as an example) one can see both the increasing number of trips and the increasing proportion of trips by car (Figure 10.1).

10.2 Factors resulting in car dominance

There are common themes across countries fuelling the growth in car dominance. Covering each in depth would constitute a book on its own, so for this chapter, we will focus on factors of relevance to MaaS.

Population growth is a global trend, as is a general trend of increasing economic prosperity. Social change across the world is more patchy, however, a higher proportion of families have more than one person going out to work (which has correlated with car ownership in many countries). In addition, road building has been a general feature of increasing wealth and development worldwide.

Over this period, cars have become embedded in society both economically (accounting for about 3.5% of the global economy – $3 trillion [3] in a global GDP calculated at around $85 trillion [4]) and culturally – promoted by a multi-billion dollar advertising industry creating social acceptance and cultural momentum around the adoption of personal cars.

The growth in the number of cars on the roads has not come without costs. The impact of cars has been multi-dimensional:

- Environmental – air quality, noise levels, and carbon dioxide emissions
- Public space – both in the form of roads and storage space (parking)

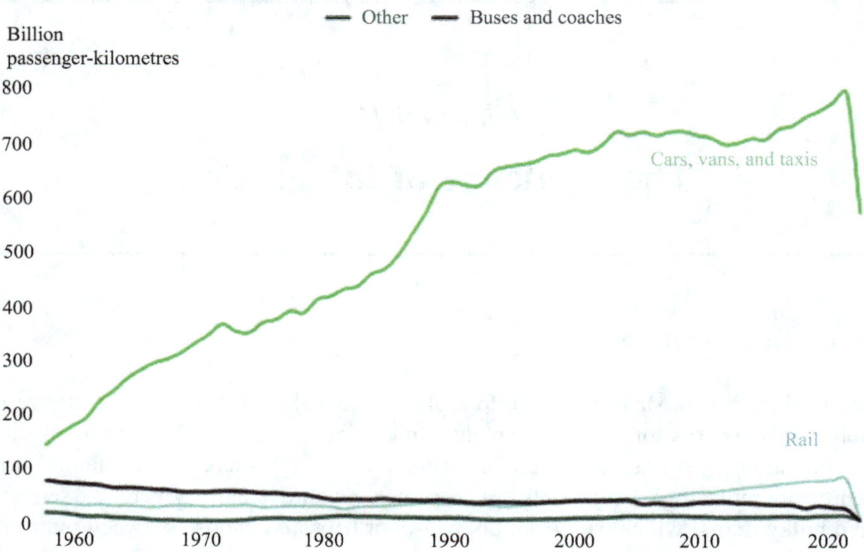

Figure 10.1 Passenger kilometres by mode, Great Britain, 1960 to 2020 [2]

- Public health – directly through poor air quality and noise levels and indirectly as people have become less active

With growing recognition of these impacts, governments have been seeking to reduce traffic, particularly in urban areas where the impacts tend to be concentrated and therefore more apparent. One of the key issues is that, in recent years, the average occupancy of private cars has typically been very low (as low as 1.18 people per vehicle in Italy [5] and around 1.6 people per vehicle in the United Kingdom [6]). The impact is a massive fleet of heavy, resource-intensive, and spatially large vehicles in regular use (the average car is typically between 4 and 5 m long, nearly 2 m wide, and around a tonne in weight).

The alternatives to cars – public transport and active travel – have either contracted (as in the UK case above) or not kept pace with the growth in trips. Public transport cannot – by definition – offer the same door-to-door experience. Active travel cannot cover the same distances as car use. New intermediate modes which can complement public transport and active travel; shared transport modes, have only recently gained traction.

Technology can also have an unintended consequence of resulting in further car dominance. An example is route planning using many popular online navigation systems such as Google Maps, which historically has defaulted to planning routes via car as opposed to by bicycle, on foot, or by public transport. In the case of Google Maps, research has found that Google Maps systemically underestimates the journey time by car, giving users perceptions that car journeys are more reliable and quicker than they actually are [7]. Several changes have been made to map interfaces as a way of encouraging more sustainable travel, notably Google Maps offering lower carbon route options by car [8].

10.3 The challenge for MaaS

The aspirations for MaaS are to enable people to switch from private cars and travel door-to-door by combining alternative modes. The technologies of MaaS enable modes to be combined smoothly and for people to benefit from the positive aspects of mass transit (for instance rapid transport along significant parts of the route).

The biggest challenge to this multi-modal approach is providing transport over the 'last mile' – the distance between the origin of each trip and the rapid transit network and that network and the final destination. Mass public transport operates at scale. The very nature of residential areas, particularly those in the suburbs where the population is less dense, makes this scale difficult to achieve.

Developments in on-demand services, shared transport, and micromobility have enabled different approaches to this thorny issue. These modes have been enabled by technologies, often based on smartphones and IoT capabilities. At the same time, innovation and development of ticketing systems for public transport have moved from paper to other technologies – and the issue of alignment between public transport ticketing technologies and smart transport access has had to be addressed in many MaaS applications.

Although modes and technologies are essential to MaaS, so too is the most basic infrastructure. Pavements and cycleways enable people to leave their homes safely on foot or by cycle to get to their destination or to a mobility hub where they can use public transport/rapid transit for their onward journey. Without people being able to leave their homes safely (and without getting into a vehicle), MaaS is unlikely to work.

There is a degree of agreement among practitioners that MaaS should contribute towards the effective integration of services to provide a meaningful alternative to the private car. In many respects, MaaS has increasingly come to define the experience of providing such an integrated end-to-end service [9]. However, this is underpinned by assumptions as to what the user, commercial, and societal understanding of what constitutes integration is [10], and also an acceptance that collaboration may be required by organisations that benefit from competitive advantage [11].

It can be observed that many of the challenges to the integration of modes that are present in MaaS are not unique to MaaS itself. They have been common grand challenges to the effective delivery of integrated transport services for a number of years, with MaaS simply providing an additional challenge of integrating services into packages to offer to the consumer [12]. Additionally, a challenging assumption of many MaaS offerings is that there is already an integrated service offering with regard to public and shared transport [13].

10.4 Summary of the challenge of integration

The adoption of MaaS has, in most respects, not reduced the challenge associated with the integration of non-car modes. Many of the core elements of this challenge, such as the physical integration of services, are unrelated to the presence of a MaaS. Nor is there any evidence that indicates that the presence of a MaaS offering alone provides a sufficient incentive to use integrated non-car modes of transport.

The emergence of the private car as the dominant mode of transport in many countries has a number of factors behind it. One of the goals of MaaS is that public and active travel is able to compete effectively with the private car as a means of travelling around our towns, cities, and regions. In the following chapters of this book, we will consider the individual modes of transport in greater detail.

References

[1] S-1 [Internet]. Sec.gov. 2019. Available from: https://www.sec.gov/Archives/edgar/data/1543151/000119312519103850/d647752ds1.htm
[2] GOV.UK. Transport Statistics Great Britain: 2021 [Internet]. GOV.UK. 2021. Available from: https://www.gov.uk/government/statistics/transport-statistics-great-britain-2021/transport-statistics-great-britain-2021
[3] Carlier M. Global Automotive Market Size 2030 [Internet]. Statista. 2022. Available from: https://www.statista.com/statistics/574151/global-automotive-industry-revenue/
[4] Statista. Global GDP 2010-2022 | Statista [Internet]. Statista. 2010. Available from: https://www.statista.com/statistics/268750/global-gross-domestic-product-gdp/
[5] Passenger mobility statistics [Internet]. ec.europa.eu. [cited 30 May 2023]. Available from: https://ec.europa.eu/eurostat/statistics-explained/index.php?title=Passenger_mobility_statistics#Passenger_car_occupancy
[6] National Travel Survey England 2014 [Internet]. Department for Transport (UK government); [cited 16 Aug 2023]. Available from: https://assets.publishing.service.gov.uk/government/uploads/system/uploads/attachment_data/file/458414/how-people-travel-car.pdf
[7] Wagner B, Winkler T, and Human S. *Bias in Geographic Information Systems: The Case of Google Maps [Internet]*. Vienna University of Economics and Business. 2021 [cited 5 Feb 2023]. Available from: https://research.wu.ac.at/en/publications/bias-in-geographic-information-systems-the-case-of-google-maps-6
[8] Milmo D. Google Maps to show the lowest carbon route for car journeys [Internet]. *The Guardian*. 2021. Available from: https://www.theguardian.com/technology/2021/oct/06/google-maps-to-show-the-lowest-carbon-route-for-car-journeys

[9] Müller B, and Meyer G. Lecture Notes in Mobility Towards User-Centric Transport in Europe Challenges, Solutions and Collaborations [Internet]. [cited 17 Aug 2023]. Available from: https://ndl.ethernet.edu.et/bitstream/123456789/76072/1/107.pdf

[10] Sochor J, Strömberg H, and Karlsson ICM. Implementing Mobility as a Service. *Transportation Research Record: Journal of the Transportation Research Board [Internet]*. 2015 [cited 29 Apr 2019];2536(1):1–9. Available from: https://research.chalmers.se/en/publication/215088

[11] Merkert R, Bushell J, and Beck MJ. Collaboration as a service (CaaS) to fully integrate public transportation: Lessons from long distance travel to reimagine mobility as a service. *Transportation Research Part A: Policy and Practice*. 2020;131:267–82.

[12] Feasibility Study for "Mobility as a Service" Concept in London [Internet]. 2015. Available from: https://discovery.ucl.ac.uk/id/eprint/1469872/1/FS-MaaS_final%20deliverable.pdf

[13] Utriainen R, and Pöllänen M. Review on mobility as a service in scientific publications. *Research in Transportation Business & Management*. 2018;27:15–23.

Chapter 11
MaaS and mass transit

11.1 Introduction

Most models of MaaS are built around a core public transport offer. These are modes with over a century of history, often growing in line with local and national economies. Public transport is able to move huge numbers of people rapidly and its spatial demands on public space are limited.

From the point of view of cities trying to maximise their land use, mass transit (essentially any form of public transport) enables huge numbers of people to be moved in relatively small spaces. Buses, trains, and trams all outperform any form of car transport, particularly in congested city centres where space is at a premium and huge numbers of people converge. People walking and cycling are also far more efficient than those arriving in cars or taxis – even before the space required for parking and waiting is added into the equation.

Whilst a 3.5 m wide lane can carry 1,500–2,000 people in cars per hour, in buses this rises to a potential 5,000, and with segregated bus lanes able to carry 9,000 people per hour.

With fixed rails this increases several fold: light rail/trams can take 18,000–20,000 people in the same space and time. Trains are able to convey between 40,000–90,000 people per hour.

By comparison, active travel sits (in efficiency) between the two: a 3.5 m lane can take up to 12,000 people on bikes or 15,000 pedestrians per hour (although the distances people travel in this way are much shorter) [1].

Many countries and administrative areas see public transport as an important factor that underpins economic development and provides access to services and opportunities for their populations. Increasingly, in recent years, public transport has become a key element in strategies to reduce carbon emissions and reach net zero.

The utility of public transport for the people using it depends on several factors.

- The proximity of the public transport to the origin and destination point of the journey (e.g., is it accessible from homes, workplaces, shops, universities, and leisure facilities)
- The convenience of public transport (e.g., is there a mode available at the time required, how difficult are any changes between services or modes)

- The speed of the journey (including wait times and changes – these can increase the 'time penalty' for those using public transport)
- The reliability of the journey, which includes the frequency of buses and trains (so if one leg is missed, how recoverable the journey is), their predictability, and running to timetable are all factors in people's reliance on public transport.

Public transport modes cover long and short distances, with different modes better able to serve different journey types and lengths.

11.2 Rail

Rail services vary in speed and distance, from high-speed intercity, international and even intercontinental services that haul passengers hundreds and thousands of kilometres to more local, often 'light rail' services that traverse cities or regions. They are completely separated from the road network. Often the high-speed network forms a national infrastructure run by distinct national companies – as the TGV run by SNCF does in France, RENFE in Spain, or the ICE trains run by Deutsche Bahn in Germany. Whilst these are run as private companies, they are effectively owned by their national governments.

Regional train companies are, in some countries, run in conjunction with administrative areas to further their social or economic aims. They typically connect towns within areas or connect suburban centres within urban areas. In Germany, the Land (a devolved administrative area) is involved in the partnership with regional train companies, and a similar pattern is followed by Départements in France and regional rail. Some countries such as the United Kingdom and Japan have privatised rail networks. Regional rail services are typically distinct from metro and tram services through being surface-based, using heavy rail technology, and serving suburban and even rural hinterlands as well as the major urban areas.

Tram services usually cross cities, providing higher-speed connections to more distant points – they are often moved between sections separated from the road network and sections that run on the road network. Metros are generally underground, meaning that they do not conflict with surface traffic. Most major urban areas of over a million people have some form of tram or metro. Notably Leeds – population 800,000 in an urban area of 1.8 million in the United Kingdom does not, whilst in Europe, cities of smaller size often have tram systems – for instance, Erfurt in Germany has a total population of 231,000 and an extensive tram system. In practice, this enables a pedestrianised city centre with busy commercial streets whilst motor traffic is kept out of this zone.

In some instances, metros and trams can be run in conjunction with one another, and even with rail services. For example, trams and metro trains share some of the same tracks in areas of Brussels, whilst tram-train systems operate in cities such as Karlsruhe (Stadtbahn), The Hague (Randstadt Rail), and Sheffield (Sheffield Supertram).

11.3 Buses

Buses are a more flexible form of transport as they use the road network. Whilst the flexibility of buses is a positive characteristic of bus travel, enabling them to reach suburbs and residential areas without heavy infrastructure, they also have to share roads with other vehicles and suffer from congestion caused by traffic, leading to increased journey times. Train and tram journeys are largely unaffected by road congestion, but bus speeds have been affected as traffic has slowed in city centres.

This has knock-on effects for bus patronage. Analysis by Professor David Begg in 2016 found:

> The net result is a direct correlation between operating speeds and patronage: a 10% decrease in speeds reduces patronage by at least 10%.

In some cities buses are allocated separate priority lanes or even fully guided busways, separate from the carriageway, to alleviate this issue. This goes some way to improving the situation, alongside other measures to reduce car traffic (such as congestion charges and other levies).

The public bus is often the most popular form of public transport in most countries, due to its flexibility (it often only requires the use of the public highway network) and accessibility as a public transport service. For example, in the United Kingdom in the year 2021–2022, 1.363 billion trips were undertaken just in England [2], whereas there were 0.9 billion trips undertaken on the whole UK rail network during the same year [3].

The organisation, ownership, and business models of mass transit vary across countries, from fully privatised mass transit to fully publicly owned. Even in countries where the companies that provide public transport are privately owned, there are usually regulations designed to ensure that they fulfil public functions and contribute to social, environmental, and economic goods. The detail of ownership and organisation of public transport, however, often shapes its relationship with concepts (and realities) of MaaS.

How public transport is organised affects how easily it is incorporated into multi-modal platforms. For public transport companies – rail, tram, metro, or bus which are publicly owned, public policy goals take precedence over profit and competition laws are not applied in the same way as for private enterprise. Privately owned public transport can have different outcomes.

There are, for instance, great difficulties in creating effective MaaS platforms in the United Kingdom where buses have been privatised (outside London) for several decades. Several companies may operate within one urban area, and cannot be required to accept each other's tickets, leading to huge complexity for passengers. Despite 'workarounds' where area tickets can be created, bus companies may not discuss their routes, timetables, and prices with each other as this would be in breach of competition laws in the United Kingdom. Whilst MaaS platforms can also work around the complexity, the prices and trips they can offer may not be the best value or the best route and this makes the platform less attractive (this is further discussed in Chapter 17 on regulation).

The Japanese rail system is privately owned. However, the rules under which it operates mean that incorporation in multi-modal platforms has been possible. Although there are several rail companies, they use a single nationwide network (owned by a single national entity, Japan Railways) and use common ticketing rules. This means that passengers are able to travel across several companies without having to buy separate tickets or change trains.

11.4 The interface with MaaS

City-based multi-modal transport systems already have a long history. In 2003, Transport for London introduced the Oyster card which allowed people to travel on bus, metro, and light rail within a zoned system in London using a single pass. Simple pay-as-you-go zone-based fares were deducted from the balance automatically and capped at day-saver rates for the zones used. These simplified systems were successful in raising revenue and increasing ridership, as set out by Shashi Verma, Chief Technology Officer & Director of Customer Experience, TfL, in response to an inquiry about the potential for MaaS:

> The examples we can cite about the increase in public transport usage through the creation of an integrated ticketing platform in Oyster and contactless are very genuine. On National Rail, the introduction of Oyster pay as you go increased traffic by about 5% or 6%. That is a very substantial amount of increase, especially when it is marginal revenue coming into the National Rail system.

Many of the early MaaS trials originated in city-based multi-modal transport – (discussed in Chapter 6). Alongside the political and practical will in these cities, the move towards MaaS, was enabled by various transport innovations in route planning, ticketing, and payment. Digitisation has enabled mass transit to have a role in multi-modal mobility services and be incorporated into MaaS platforms.

Whilst the vehicles – buses, trams, and trains – have largely stayed the same over the last few decades, the technologies that automated and subsequently digitised timetables and ticketing have played an important role in changing the passenger experience and enabling seamless multi-modal journeys.

The first step of digitisation was the creation of online timetables, a process underway since the turn of the millennium. Ticketing remained largely paper-based for much longer, with RFID cards providing the first smart ticketing. Ticketing on mobile phones and apps followed much later.

Long-distance public transport too – notably national rail operators – have been instrumental in the development of MaaS. The attraction of the MaaS platform for these operators is to create a tool for passengers, enabling them to travel the 'last mile' (or km) between the origin or destination and the mainline stations.

Examples of long-distance rail operators offering MaaS include the Spanish national operator Renfe (see Section 5.3) and Arriva in the Netherlands (which operates 60% of the Dutch rail market) [4]. Glimble, the result of an Arriva partnership with Moovit, is a comprehensive MaaS app that can pay for all public and shared transport countrywide [5].

11.5 Mass transit data as the foundation of MaaS

MaaS is only possible with the digitisation of timetables and ticketing, and the key to enabling this is the provision of transport data – data about services – essentially the routes, times, and prices of services.

To be useful to MaaS apps (and other journey planners), this data depends on sets of standards and rules for a common way of presenting the data, so that it can be used in journey planners and MaaS applications. In Europe the standard is NeTEx. Another widely used format is GTFS, the standard developed for Google to enable transport information to be presented in Google Maps.

NeTEx is intended to provide a European-wide standard for exchanging public transport data for passenger information. It is a general-purpose format capable of exchanging timetables and fares for rail, bus, coach, ferry, air, or any other mode of public transport. It includes full support for rail services and can be used to exchange International Union of Railways data.

NeTEx employs a fully articulated model that represents public transportation concepts as well as well-characterised, layered abstractions; the format is designed for the efficient, updateable exchange of complex transport data between distributed systems. This allows the data to be used in modern web services architectures and to support a wide range of passenger information and operational applications.

The NeTEx schema is free to use and its development is managed by the CEN (European Standard Committee for Standardisation) standards process [6].

For real-time data, protocols that are optimised to exchange data very quickly are required. SIRI (Service Interface for Real Time Information), also managed by CEN, is designed to do this. SIRI is an XML protocol [7] that allows the exchange of real-time information about public transport services and vehicles.

The NeTEx model can understand how the real-time data content of SIRI messages relates to its model.

The General Transit Feed Specification (GTFS), also known as GTFS static or static transit to differentiate it from the GTFS real-time extension, defines a common format for public transportation schedules and associated geographic information. GTFS 'feeds' let public transit agencies publish their transit data and developers write applications that consume that data in an interoperable way. It is a feed specification that enables transport data to appear on Google Maps.

A transit agency can produce a GTFS feed to share their public transit information with developers, who write tools that consume GTFS feeds to incorporate public transit information into their applications. GTFS can be used to power trip planners, timetable publishers, and a variety of applications that use public transit information in some way.

A GTFS feed is composed of a series of text files collected in a ZIP file. Each file models a particular aspect of transit information: stops, routes, trips, and other schedule data. The details of each file are defined in the GTFS reference.

It is useful to note that whilst Google is happy to consume the transport data provided, it does not provide any reciprocal reporting on how it is used by people searching for transport options on Google.

GTFS also has an extension to enable live transit updates – GTFS Realtime. GTFS real-time feed specification supports trip updates (delays, cancellations, and changed routes), service alerts (e.g., unforeseen events affecting services or networks), and vehicle positions (including location and congestion level). The feed is designed to be updated regularly to ensure that it reflects the services underway and provides good information for passengers [8].

An 'on demand' version of GTFS is in development – GTFSflex [9]. However, the use of feeds in the GTFSflex format is not yet supported by Google. The format is still useful for developers integrating on-demand systems with MaaS apps.

11.5.1 Case Study: UK transport data

To give an idea of the complexity and detail of data required by travel planners, this snapshot, whilst not exhaustive, is a guide to the transport data standards and sets used in the United Kingdom. The most comprehensive dataset of scheduled UK public transport is compiled and kept by an organisation, Traveline which makes a national dataset, the Traveline National Dataset (TNDS), available to third parties under the Open Government Licence.

TNDS contains public transport timetables for bus, light rail, tram, and ferry services in Great Britain. It is compiled from local data in the TransXChange format.

TransXChange is the UK nationwide standard for exchanging bus schedules and related data. It is used for:

- the electronic registration of bus routes
- the Traffic Area Network
- the exchange of bus routes with other computer systems such as journey planners and vehicle real-time tracking systems

TransXChange is part of a family of coherent transport-related XML standards that follow GovTalk guidelines.

Foundation datasets, which underlie transport data provision provided by the UK government include:

- UK NAPTAN database: The nationwide system for uniquely identifying all the points of access to public transport in the United Kingdom. Each NaPTAN point is able to contain comprehensive data on street names, localities, stop types, and geospatial coordinates. The dataset is made available in XML, GTFS, or as a CSV.
- National Public Transport Gazetteer (NPTG): this dataset is closely associated with the NaPTAN dataset and contains details of every city, town, village, and suburb in Great Britain. This dataset is based on usage of names, rather than legal definitions and so includes local informal names for places as well as their official names. As a topographic database of towns and settlements, it

provides a common frame of reference for the National Public Transport Access Nodes (NaPTAN).

- National Coach Services Database (NCSD): This dataset contains scheduled timetable data for coach and strategic bus services across Great Britain.
- Transport Accessibility data: Department for Transport data to support accessible journey planning and covers accessible stations/stops and accessible services.
- Registered bus services: Operators of local bus services carried out in accordance with Section 6 of the Transport Act 1985 must register their services with a Traffic Commissioner in the traffic area in which the service operates. The Driver & Vehicle Standards Agency (DVSA) maintains an Operator Licensing Business System (OLBS) to record these registered services. All local bus services in Great Britain in the system are available as open data.

For rail elements, the GB rail dataset is held by Network Rail, and includes a wide range of open data that includes timetables and live information:

- SCHEDULE – daily extracts and updates of train schedules from the Integrated Train Planning System, in CIF and JSON formats
- MOVEMENT – train positioning and movement event data
- TD – train positioning data at signalling berth level
- TSR (Temporary Speed Restrictions) – details of temporary reductions in permissible speed across the rail network
- VSTP (Very Short Term Plan) – train schedules created via the VSTP process which are not available via the SCHEDULE feed
- RTPPM (Real-Time Public Performance Measure) – performance of trains against the timetable, measured as the percentage of trains arriving at their destination on time
- SMART – train describer berth offset data used for train reporting
- Corpus – location reference data
- BPLAN – train planning data, including locations and sectional running times

Bus data is provided on the basis of the Bus Services Act 2018, which made the provision of open bus data law. The government's Bus Open Data Service provides an open data platform for buses. The service is progressively publishing increasingly detailed and comprehensive datasets, moving from service data (timetables) to real-time information and onto open fares data.

Real-time bus information is provided by Traveline using an API. APIs – technically Application Programme Interfaces – are the formats that structure data so that is available at a given online address. Developers can write code to find and retrieve that data from a site (e.g., request timetable information) or post data to that site. APIs are increasingly being provided by transport providers so that developers can incorporate transport information into their applications.

The Traveline Nextbus API provides real-time bus departures by bus stop where they are available, and scheduled where they are not via an API. Nextbus API is based on the SIRI standard, using the SIRI-SM function through a request/response communication mechanism.

In addition, two auxiliary datasets that Traveline maintains enable planning and journey products to give appropriate information in context:

- A national operator code dataset that enables the operator details to be displayed in an understandable way in travel planners. It enables the conversion of local regional operator codes that might exist in exports from the Traveline regions to TNDS, or in real-time feeds to the NextBuses API, to a single universal operator identity and associated code.
- An educational term time dataset is being created as timetables change depending on term times and this will enable the correct schedules to be displayed for the correct dates.

In the United Kingdom, Traveline transforms its datasets into GTFS format, allowing UK transport options to be featured on Google Maps.

In addition to the public transport datasets outlined above, there are various shared transport datasets. These are usually provided as APIs by the operators, often using an open data standard.

Application Programme Interfaces (APIs) are a common way of providing data about transport services. Developers can write code that checks the API address and uses the data it finds there. Transport for London (TfL) launched a unified transport API to simplify its data offering. Historically, TfL shared the data for each mode in different formats and structures – requiring developers to write code for each mode of transport. The unified API is based on the unification of the data for modes of transport into a common format and structure (creating a common canonical data model).

The unified API presents all the data that is semantically similar for each mode of transport in the same format and consistent structures. This enables developers to write once and access all of the same types of data across all the modes of transport quickly, making multi-mode application development easier. The core identifiers for all stations and platforms have been normalised to the national NaPTAN standard. This standard is an identification scheme that is supported nationally across the United Kingdom, allowing the API to integrate data from transport authorities outside of London. The complexity of mapping between multiple identification systems used within TfL has been hidden from consumers of the API. The TfL unified API includes both public transport data (bus, rail, and underground) and the Santander bike share scheme it operates.

Whilst the digitisation of services creates vast amounts of data that describe services, a secondary set of datasets is created by travellers using transport, transport apps, and MaaS data.

Stations and services have traditionally collected operational data about the volume of travellers using their services. Traditionally these are based on ticket sales [10]; however, where significant numbers of season tickets or other types of passes are used, assumptions have to be made. Other sources can refine and augment these counts including mobile phone data.

Whilst the loading of vehicles is an important metric for operators (for business and safety reasons) it has increasingly become of interest to passengers too.

The demand for information about the occupancy of buses and trains increased due to the COVID-19 pandemic. This could be collected from various sources including (for rail) the weight of carriages, electronic ticketing systems (bus), reservation systems, or by analysis of mobile phone data. All these methods have some drawbacks, particularly because the availability of data is patchy. For instance, mobile phone data relies on masts and coverage, giving good data on a station level but not linked to specific services. Whilst buses with connected electronic ticketing machines are good at collecting boarding data (and supplying it in real-time), the tickets sold do not necessarily specify when people will alight (this is particularly the case where passes valid over larger areas are concerned). There are inconsistencies with reservation data as people do not always take up their reservations, so even seat reservations must be regarded as an indicator rather than as hard data.

Real-time occupancy rates have been published by some operators (it is a feature of some bus ticketing products in the United Kingdom – such as Passenger mobile ticketing), however, real-time data does not necessarily give passengers good data about trains or buses they are planning to board in the future.

Different approaches to loading predictions have been tried. FAST – find a space on a train – used a combination of historic loading data, real-time loading data from up-line stations, and contextual information (e.g., weather) and configured a Long Short Term Memory (LSTM) Neural Net designed by Esoterix systems. The team determined that using this combination resulted in better predictions for loading compared to solely relying on historical data [11].

Predicted occupancy rates have also become more frequently used in online journey planners. Deutsche Bahn includes its prediction of how full the train will be when offering online tickets for sale, however, it has attracted criticism because it reserves this data for its in-house platforms without offering it to resellers (e.g., Trainline). This has become the subject of an antitrust action [12].

Besides usage data, one other useful dataset is that of the origin and destination of trips people searched for in MaaS apps and journey planners. This set of data – although not definitive – indicates the trips that people would like to make. In transport planning, there are two types of estimates of demand:

- Revealed Preference Data – This is data on actual trips that have been undertaken. Examples of such datasets include Travel to Work statistics collected through population censuses, public transport ticket sales data (though this typically only provides data between stops or stations), travel surveys, traffic counts, and GPS tracking either through mobile phones or in-vehicle satellite navigation systems.
- Stated Preference Data – This is data of trips which people have indicated an intention to make, or to choose between different travel experiences (e.g., the faster train or the cheaper bus). This has been traditionally collected through stated preference surveys, but newer methods include the analysis of journey planner queries and search engine searches.

Stated Preference Data is the least reliable of the two types of data as it is a statement of intention, as opposed to real-world behaviour. But to a MaaS operator, this may be particularly valuable data where the operator is seeking to provide a tailored mobility experience. For public policymakers seeking to shift travel behaviours, understanding the intentions of people is useful data. Sharing this data would necessitate users giving permission for this highly personalised data to be made available to other parties, as well as the MaaS operator supplying this data on an equitable and fair basis.

A matter that is sometimes of concern is data security. Travel data is highly personalised, and in many circumstances can be personally identifiable (for instance trips from a home to a place of work). But attitudes to sharing such data are nuanced. For instance, in the United Kingdom the proportion of people who would be comfortable sharing information about their ethnicity, gender, or the region they live in, to enable testing of systems for fairness is 69%, 68%, and 61% respectively [13]. But, in 2018 research by the Open Data Institute revealed that across five European countries (Belgium, France, Germany, the Netherlands, and the United Kingdom) between 28% and 36% of people will be comfortable sharing data on their location [14].

A final factor is that of commercial proprietary. For businesses seeking to monetise based on travel habits, ownership of data on travel habits is highly valuable in developing new services. But to develop such services, such data needs to be collected at a significant scale. In many instances, this could be facilitated by sharing data between commercial entities. This is something often stated as a requirement in academic literature [15], and desirable from transport policymakers [16].

The reality of sharing such data is somewhat more complicated. Especially within a content where transport operators operate in commercial competition with each other, such as in the United Kingdom. Accordingly, sharing data with other private entities represents a commercial risk and requires management judgement as to whether this will affect the business operations. This can be overcome through the use of legislation and a requirement to publish anonymised trip data openly as part of the licence or contractual conditions.

Furthermore, the ownership of a significant dataset on journey patterns does not necessarily translate into commercial success. For example, in London, Citymapper (one of the most popular journey-planning apps) has trialled a number of services seeking to monetise its significant user base. Examples of this include creating a new bus service in competition with the extensive public bus network operated by Transport for London. This service was removed after 9 months of operation [17]. Another example is the series of Citymapper passes that provide access and payment for a variety of transport options throughout the city. By 2023, Citymapper had yet to report a profit.

11.6 MaaS and fares

Fares are the second essential element for a comprehensive MaaS platform. However, fare digitisation is not always straightforward. The digital provision of fares (in journey planners and transport apps) depends on the complexity of the fares. Whilst some modes, areas, and systems have fares well codified, this is more difficult to achieve in places where local services have complex rules. It is technically possible to include fares in TransXchange and NeTEx formats 'where they are available'.

The complexity arises because fare digitisation depends on the rules for setting fares being relatively simple and the settlement structures being in place to enable tickets to be sold by third parties.

In the United Kingdom, detailed rail data APIs (including both timetables and ticketing) have been available to route planning and MaaS platforms for some time.

Ticket sales are enabled through the rail industry both, the Rail Delivery Group (RDG), which has a system in place to enable rail ticket reselling. Rail operators can sell each others' tickets (as can a limited number of other resellers). Ticket sales are backed by Lennon (Latest Overview Earnings Networked Nationally Overnight) programme, run by Worldline since 2003. Lennon processes nearly two million rail ticket transactions per night allocating UK rail revenues to Train Operating Companies (TOCs). Without this, UK TOCs would not have the available daily cash flow to meet their obligations.

Worldline works with the RDG as a contract holder and numerous subcontractors to ensure synchronicity between 8,500 ticketing systems and the TOCs which rely on the revenue from those systems. Unsurprisingly, this is a mammoth operation. The system processes every rail ticket sold on the UK rail network (pre-COVID, up to two million transactions per night), producing circa 25 million revenue records. It handles over one million season tickets in a suspense account and provides net settlement values for the €11.1 billion annual rail revenues.

In many respects, it is the reason the UK rail system is 'MaaS ready' and rail ticketing is found in the UK planners and MaaS apps that exist. In contrast, fare digitisation of bus tickets in the United Kingdom has been much less comprehensive. This slow adoption of fare digitisation in the United Kingdom is a reflection of the patchwork nature of bus provision (with multiple private companies) interacting with administrative boundaries.

The complexity of bus fares is one element of this, with, in some areas, theoretically thousands of fares possible, with the full complexity requiring digitisation. Differences in peak and off-peak rules between local authorities mean that fare tables are time and place-dependent as buses cross local authority boundaries. Different fares may be charged to people boarding either side of the boundary (on the same route) and these rules need to be incorporated in digital ticketing. These anomalies make the fare tables excessively complex both to create and to keep up to date with fare changes.

One authority – Transport for the West Midlands – has adopted the approach of negotiating an extremely reduced set of fares with operators within its area. Previously, operators all offered their own fares tailored to routes and times of day, creating a complex set of choices where it was hard to determine the best value. The changes, launched in 2023, bring down the number of bus fares to just four – day, week, 4-week, and monthly passes (at adult or child prices). In addition, daily, three or seven-day fare-capped travel can be paid for using the authority's Swift Go card. The new tickets enable travel on any operator. This simplifies the fare calculations for journey planners and transport apps considerably.

In addition, sales of bus tickets are of another order of magnitude to rail ticket sales. Pre-pandemic there were over 4 billion bus journeys in England alone [18] whilst rail for Great Britain reached 1.8 billion in 2019 [19].

However, the bigger issue is there is no equivalent of the Rail Delivery Group to create and stand behind an open ticket data and reselling system. This means that, at present, MaaS providers have to negotiate individual contracts with operators meaning that there is no parity between apps or areas.

11.7 Digital ticketing

Public transport tickets have become much more diverse over the past two decades. Whilst it is still possible to purchase paper tickets, it is increasingly likely that travellers will buy tickets online, in the app, use account-based ticketing with RFID cards or just their NFC debit or credit cards (when this is the case, ticket inspectors require sight of the card and a 'tap' to as proof that the traveller has a ticket or 'token' giving them permission to travel).

Oyster in London works on a tap-on-tap-off basis with standard fares obtainable with an RFID card or contactless credit or debit card with NFC. UK bus operators are trialling tap-in tap-out Bluetooth beacons and NFC or RFID cards.

Ticketing systems identify the card used each time – the card owner may see a small 'pending' transaction at the time as the system checks the card's validity. The back office system reconciles the number of taps (usually overnight) for the card and charges the card. This set-up enables precise fares to be charged and also enables end-of-day fare capping.

Mobile phones are increasingly used as payment wallets and for purchasing and holding transport tickets. This is either through online/in-app payment for tickets or using their NFC capabilities. The use of the NFC chip for transport ticketing is not fully developed. Whilst iPhones (as other smartphones) have NFC capability, Apple limits the access that third-party developers have to iPhone NFC chips. This practice has been so restrictive that some transport passes and tickets available on Android services have not been available on iPhones (the West Yorkshire Metro card is one such).

In Paris, iPhone users in the Paris Region were able to buy, validate, and top up their RATP transport tickets and 'Navigo' passes from their smartphones a full year after Android users had the functionality. This was reportedly following a protracted negotiation with Apple on both price and developer access to the NFC chip [20]. The European Union launched a challenge to Apple's restrictive practices around access to the NFC chip by third-party developers in 2022 [21].

Tickets on mobile phones that are not bought and stored directly in a wallet (and even some that are), generally depend on embedded QR or Aztec codes that can be read by readers to determine their validity. These are also printable for tickets bought online to be printed out (as per Deutsche Bahn). New digital tickets often require that the person who is travelling is named on the ticket. This means that they will also need to carry valid ID. Digital tickets are generally paid for online using debit or credit cards, a shift away from cash fares.

Ticketing for MaaS usually depends on a mixture of these technologies. Mobile ticketing has to work for both customers and the transport operator. How well an application interacts with ticketing systems (e.g., ticket gates) is a key part of both customer experience (ensuring a smooth passage through stations, gates, and onboarding buses or trams) and also an important factor in operations for transport operators. Efficient boarding can be safety critical – where gate systems need to handle thousands of interactions at peak times to keep stations moving – and also business critical as boarding times affect the performance of services (total bus running time is adversely affected by slow boarding and alighting).

There is an increasing assumption that transport payment will be electronic – current mobile ticketing relies on credit/debit payment standards. Equity considerations mean that public transport operators need to consider how to ensure that people reliant on cash are not excluded from MaaS. The ability to load apps with cash at kiosks, buy MaaS passes for third parties or offer MaaS functionality at public digital information points have been used by operators (e.g., as set out in case studies on Jelbi in Berlin and WYMetro).

11.7.1 Case study: account-based ticketing/MaaS integration

In the Nottingham and Derby area of the United Kingdom, local bus company Trentbarton is integrating its tickets with a regional MaaS app, DerbyGo.

DerbyGo is a MaaS app developed by Mobilleo for KINTO, Toyota's mobility as a service start-up. The app was launched in September 2021, targeted at students returning to the University of Derby and Derby College after a period of remote learning due to COVID-19 lockdowns.

The platform aims to reduce single-occupancy car journeys and emissions and increase active travel and multi-modal trips. In common with other apps, it includes journey planning across modes. It offers tickets for national rail in the Derby City area and digital parking management and electronic payment in its initial phase. Future plans include pre-booked parking spaces for vehicles carrying more than one person.

One of the first full ticketing integrations with a bus company in the United Kingdom is underway. Trentbarton, an independent Nottingham and Derby bus company serves the campus. Its 'Mango' ticketing system requires users to tap a card (either their bank card or an RFID travel card) and tap off the bus on leaving. The ticketing machines (in this instance Ticketer) record these as 'events' which are relayed to the back office system which works out the best price for travel on the day (single, return, day travel, etc.) during an end-of-day reconciliation and charges the traveller's account accordingly. The token (used for the taps) is separate from the account. Other methods of establishing the journey start and finish could be used, such as validation of biometrics or wearables, and linked to the account.

To integrate this system with the DerbyGo MaaS app requires that the app is able to mimic the Mango ticketing process and create both the ticketing token – a unique bar code that the buses' ticketing machines can read – and the 'events' which are then reconciled at the end of the day so that the passengers' accounts can be reconciled and they can be charged for their journeys. The full integration will enable ticket sales within the app, with dynamic ticketing available, journey history visible and account top-ups enabled. Integrations that enable access to the full range of ticketing offered by an operator and guarantee best-value fares have not been easy to implement in the United Kingdom for both technical and commercial reasons.

11.8 Summary of MaaS and mass transit

Mass transit systems such as rail and buses are seen as forming the core transport offer of many MaaS solutions. Each type of mass transit system has its own advantages in terms of providing a public transport offer to residents, and in major cities are considered to be essential elements of the transport mix. In such circumstances, significant effort has been placed into integrating such offerings into MaaS solutions.

Three key elements form the basis of this integration – that of data, fares, and ticketing. Different places have adopted different data standards as well as different levels of technological capability, which essentially results in the MaaS operator providing the integrating role in many cases. This is so the operator can provide the data and interface that is useful for the customer.

When it comes to fares, technology is less of an issue. The issue is the underpinning revenue allocation systems and contracts that are adopted. Different places have adopted different systems of allocation, and in some cases, this requires the consent of operators to deliver. In instances where no such agreements exist, this has resulted in significant challenges for the MaaS operators.

Digital ticketing has been subject to significant development over recent years, with the emergence of technologies like contactless payment and systems such as account-based ticketing. These have been adopted by many MaaS operators, though concerns remain about the assumptions that electronic payments will be the default in the future.

References

[1] Passenger Capacity of Different Transport Modes TUMI [Internet]. TUMI. [cited 19 Apr 2023]. Available from: https://transformative-mobility.org/multimedia/passenger-capacity-of-different-transport-modes/

[2] Bus Statistics Data Tables [Internet]. GOV.UK. Available from: https://www.gov.uk/government/statistical-data-sets/bus-statistics-data-tables#local-bus-passenger-journeys-bus01

[3] Table 1220 – Passenger Journeys | ORR Data Portal [Internet]. dataportal.orr.gov.uk. Available from: https://dataportal.orr.gov.uk/statistics/usage/passenger-rail-usage/table-1220-passenger-journeys/

[4] Arriva – A DB Company [Internet]. www.deutschebahn.com. [cited 16 Aug 2023]. Available from: https://www.deutschebahn.com/en/presse/mp_arriva-6934140

[5] Arriva and Moovit Partnership Brings MaaS to The Netherlands [Internet]. Traffic Technology Today. 2021 [cited 16 Aug 2023]. Available from: https://www.traffictechnologytoday.com/news/mobility-as-a-service/arriva-and-moovit-partnership-brings-maas-to-the-netherlands.html

[6] Overview | NeTEx [Internet]. [cited 16 Aug 2023]. Available from: https://netex-cen.eu/overview/

[7] XML or Extensible Markup Language is a markup language which is used to store, transmit and reconstruct arbitrary data.

[8] GTFS Realtime Overview | Realtime Transit [Internet]. Google Developers. Available from: https://developers.google.com/transit/gtfs-realtime

[9] MobilityData/gtfs-flex [Internet]. GitHub. 2023 [cited 16 Aug 2023]. Available from: https://github.com/MobilityData/gtfs-flex

[10] UK Office of Rail and Road Produces Annual Estimates of Entries and Exits at Each Station in Great Britain: Office of Rail and Road. Estimates of Station Usage | ORR Data Portal [Internet]. dataportal.orr.gov.uk. 2022. Available from: https://dataportal.orr.gov.uk/statistics/usage/estimates-of-station-usage

[11] Transport-Technology Research and Innovation Grants (T-TRIG) 2020 Project Outcomes [Internet]. 2021 [cited 16 Aug 2023]. Available from: https://assets.publishing.service.gov.uk/government/uploads/system/uploads/attachment_data/file/1023585/t-trig-2020-project-outcomes.pdf

[12] Deutsche Bahn Argues that Hogging Data Is Just More Efficient [Internet]. POLITICO. 2022 [cited 16 Aug 2023]. Available from: https://www.politico.eu/article/deutsche-bahn-argue-hogging-data-more-efficient/Miller J, Chazan G. Deutsche Bahn Criticised for Withholding Data during Pandemic. *Financial Times* [Internet]. 20 Aug 2020 [cited 16 Aug 2023]. Available from: https://www.ft.com/content/fa80a1fc-902a-4a90-8ad3-8eed128dd6d0

[13] Public attitudes to data and AI: Tracker Survey (Wave 2) [Internet]. GOV.UK. Available from: https://www.gov.uk/government/publications/public-attitudes-to-data-and-ai-tracker-survey-wave-2/public-attitudes-to-data-and-ai-tracker-survey-wave-2#executive-summary

[14] Attitudes towards data sharing – Europe – 2018 #PUBLIC [Internet]. Google Docs. [cited 16 Aug 2023]. Available from: https://docs.google.com/spreadsheets/d/1_-ntuc7lZEeXxLdvaFsfqrugjNFQUPTI7B377GOXnQo/edit#gid=987804771

[15] Goulding R, and Kamargianni M. The Mobility as a Service Maturity Index: Preparing the Cities for the Mobility as a Service Era [Internet]. 2018 [cited 16 Aug 2023]. Available from: https://discovery.ucl.ac.uk/id/eprint/10063087/1/Contribution_10902_fullpaper.pdf

[16] Polydoropoulou A, Pagoni I, Tsirimpa A, Roumboutsos A, Kamargianni M, and Tsouros I. Prototype business models for Mobility-as-a-Service. *Transportation Research Part A: Policy and Practice*. 2020;131:149–162.

[17] Citymapper Says Its New App-Based Smart Ride Service is Part Taxi, Part Bus [Internet]. Digital Trends. 2018 [cited 16 Aug 2023]. Available from: https://www.digitaltrends.com/cars/citymapper-ditches-london-bus-service/

[18] Statistical Release Local Bus Passenger Journeys Vehicle Miles on Local Bus Services [Internet]. 2020. Available from: https://assets.publishing.service.gov.uk/government/uploads/system/uploads/attachment_data/file/929992/annual-bus-statistics-year-ending-march-2020.pdf

[19] Rail Factsheet [Internet]. Available from: https://assets.publishing.service.gov.uk/government/uploads/system/uploads/attachment_data/file/851082/rail-factsheet-2019.pdf

[20] One Year Later, the Navigo Pass Finally Available on iPhone [Internet]. M2050. 2021 [cited 16 Aug 2023]. Available from: https://lyko.blog/en/ticketing-one-year-later-the-navigo-pass-finally-available-on-iphone/

[21] EU Officially Objects to Apple Limiting Third-Party Access to Apple Pay NFC Capabilities [Internet]. MacRumors. 2022 [cited 16 Aug 2023]. Available from: https://www.macrumors.com/2022/05/02/eu-apple-pay-nfc-objection/

Chapter 12
MaaS and shared transport

12.1 Introduction

The basic concept around which shared transport is defined is the better utilisation of resources and access to shared resources when and where they are needed rather than choosing or being required to have a privately owned car.

Private cars tend to have very low utilisation rates (a 2012 estimate by the RAC Foundation in the United Kingdom found that cars were in use 4% of the time, and only away from their home address a maximum of 20% of the time [1] and in many countries the average number of passengers per car is low). This implies that extensive areas are designated for parking vehicles that are rarely used, detracting from the space that could otherwise be employed for civic or housing needs.

The amenity of having cars available means that people use them for the vast majority of their trips even though other modes might be better suited, choosing the mode in which they have sunk costs and which sits on their doorstep. The availability of shared transport reduces the need for ownership and fosters different patterns of behaviour.

Sharing can boost the proportion of the time a vehicle is in use or the number of people per trip and reduce the total number of vehicles needed. Cars can be shared in a number of ways – as a community resource that can be hired by the hour, by sharing trips, and also by sharing taxi or private hire trips.

Shared micromobility is another way of lowering the barrier to take up alternative modes to the private car. It can include shared bikes, e-bikes, stand-up e-scooters, small sit-down scooters, and even four-wheeled micro-vehicles. The shared micromobility offer is often presented as 'the right size of vehicle for the trip' so that people can scale their mode accordingly.

Bike share in particular offers both a way of improving access to public transport and an increased utilisation of the bikes. It is also a low-cost way for people to try or return to cycling, particularly on e-bikes which present a higher cost-barrier for individuals.

There is general agreement that shared micromobility has the potential to accelerate the transition to low-carbon urban mobility by providing access to transport [2]. However, the design and regulation of the provision are important in ensuring that it fulfils this potential.

12.2 Car sharing

Car sharing (also known as car clubs) provides vehicles that can be rented by the hour or day. Members sign up in advance (as identity and driver's licence details generally need to be verified for credit and insurance purposes) and pay as they use the vehicle (sometimes with a monthly membership fee). They tend to use the car infrequently for trips that cannot be made by other means. Typically members receive a key card to access vehicles or use an app. Smaller community and rural car sharing may use more manual technology such as a key safe (obviously in such schemes the car would need to be returned to its base).

Operationally, there are three broad types of car-sharing schemes:

- Fixed bay – the vehicle is parked in a designated marked parking bay, hired from it, and returned to it after hire.
- Floating – the vehicle can be parked within a designated area. It is usually located by app and can be returned anywhere within the designated zones. How distant that might be depends on the reach and business model of the business. Some versions provide for longer one-way trips.
- Peer-to-peer – people offer privately owned vehicles that can be rented by others via an online platform. The platform takes care of booking, payments, identity checks, and insurance.

To operate smoothly, these models depend on agreements around parking. For fixed bay vehicles, either a private location rental agreement is necessary or for on-street parking, there must be an agreement with the local authority to mark the bay as a car-sharing bay reserved exclusively for those cars (otherwise, if there is no space to return the car because someone has parked there in its absence, the user experience is poor and the management of the service becomes impossible).

In floating zones the authority managing parking authority has to be in agreement, creating a scheme in which there is a flexible parking permit used by the fleet. Finding parking is more stressful for the user in this case, although it does allow them to make one-way trips.

For peer-to-peer, the vehicles may be parked in private driveways or have a permit for on-street parking.

Car clubs and car sharing have grown as community organisations and also as purely commercial enterprises. The case of Norfolk Car Club in the UK charts the move from a community group to a private enterprise.

12.2.1 Case study: Norfolk Car Club UK

The Norfolk Car Club was founded as a Community Interest Company in Norwich, a city of 141,000. It became an exceptional-sized car club for the United Kingdom outside London. For instance, Derby is served by a mere five-car club vehicles. The car club was launched by a private individual and expanded rapidly in 2014 with the collaboration of a national car-sharing organisation, supported by technical agreements with the local authority (controlling on-street parking) and with funding from a national pilot programme. On the retirement of the individual, the car club in the city was taken over by Enterprise Car Club.

The key success factors include:

- Norfolk Car Club was awarded £100,000 by the Developing Car Clubs England Programme administered by CoMoUK in 2014 to cover the cost of new car club infrastructure including the designation and marking up of existing and additional parking bays, the acquisition of telematics equipment and an extensive marketing and promotion campaign to develop the car clubs in Norfolk. Along with funding from Norwich City Council, including contributions from local sources, the total investment in the Norfolk project was forecast to be over £300,000. Funding provided by this not-for-profit entity with a capital injection enabled it to expand at a faster rate than would have been possible without funding. Ongoing costs were met through membership and hire income. The vehicles were asset-locked to ensure confidence. The car club vehicles had an easy online booking and an industry-standard telematics system (supplied by the Co-wheels car sharing organisation) ensuring that they provided an excellent customer experience.

Local authority policy support

- Norwich has central parking restrictions because historically parking demand has been high and has created pressure on the streets for both residents and businesses. However, the car club has been allocated on-street car parking for its vehicles as it has grown, with space allocated whenever it has met certain metrics.
- Levies paid by developers have been used to ensure development supports the car club.

Marketing

- The car club marketing has been very consistent, long-term term, and well-targeted, recognising that people rarely make a snap decision to join the car club. It used both traditional and social media and localised leaflet drops targeting the streets immediately surrounding cars.
- Key messages have been around saving money, easy parking, and 24/7 availability.
- Intensive marketing was successful. Norfolk Car Club grew by 36% over the first 8 months after its Developing Car Clubs investment, with nearly 250 new members joining the car club (2015 membership was in excess of 650 people). The car club also experienced a 25% increase in the number of bookings of vehicles.

There are benefits to achieving a critical mass

- The car club has been proactive in creating a network of cars so that they are perceived as a realistic alternative to car ownership – without the stress of finding parking spaces. As the network has grown, the car club noted a greater enthusiasm for member-to-member marketing. Whilst the cars were perceived as a restricted resource (i.e., just one in a neighbourhood) there was little take-up of member-get-member marketing. Once a network was established that reinforced the perception that if one car was unavailable there would be alternatives (even if slightly further away) member-to-member recruitment increased.

12.2.1.1 Impacts

The 2015 expansion of Norfolk Car Club was examined as a component of the evaluation of the 'Developing Car Clubs in England' programme, which allowed for the assessment of particular effects throughout the program's duration. The number of members increased from 399 in December 2014 to 645 in December 2015, an increase of 246 members.

By comparing 2015/16 Annual Survey of Car Clubs results on the frequency of using private cars before and after joining the car club, it is estimated that these new members in Norfolk have made an estimated 274 fewer private car trips each week. In total, survey respondents reported a net reduction in mileage driven, with each person driving 530 miles less, on average.

The evaluation of the Developing Car Clubs in England programme estimated that for each new car club car, approximately two private cars were removed from the road. About 11% of car club members stated that they had sold or disposed of a car in the 12 months before completing the annual surveys implying that 30 vehicles sold or disposed of. In addition, based on car club members surveyed stated that they would have bought a private car if they had not joined the car club, CoMoUK estimated that 74 car purchases were deferred due to the increased membership levels between 2015 and 2016.

12.2.2 Car-sharing technology

The development of car sharing over the past 20 years has been enabled by automotive technology developments, automated identity verification, and payment systems. In this time informal and community car share initiatives have moved from manual systems often between small groups of friends and neighbours based on trust and shared costs to become adopted as business models in parts of the automotive industry seeking to reinvent itself better adapted to urbanisation and a net zero future. The key elements enabling car sharing are:

- Remote locking technology and digital car keys
- Vehicle telematics using GPS tracking so that the vehicle can be located and also tracked for booking, billing, insurance, and fleet management
- Tracking technology which depends on satellite coverage (2-4G) and mobile connectivity (where mobile-only keys are used, the quality and ubiquity of coverage is important to ensure that cars are not rendered immobile in remote locations)
- Fuel card systems and EV charging platforms

- Identity verification (increasingly using remote platforms) and automation of access to government driving licence data and centrally held insurance records
- Automated booking and billing systems
- Payment platforms

12.2.3 Impacts of car sharing

There is a long-standing body of evidence on the impact of car sharing collected by the UK charity for shared transport, CoMoUK. The organisation has surveyed 'car club' members (also known as car share users in other countries) since 2015 and found a typical pattern of people giving up or deferring car ownership on joining a car club. For instance, in the CoMoUK Annual Survey of Car Clubs for England and Wales 2019 a reduction in car ownership equivalent to 6.1 cars per car club vehicle was observed, and, in addition, per car club vehicle, 12 car purchases were deferred. The same survey found that users, on average, decreased mileage driven by 793 miles (1,269 km) per year.

Vehicle-to-member ratios vary across the United Kingdom but even in areas with relatively modest ratios (5–10 members per car) this means far more people will get to use these vehicles than use each privately owned car – in London, where upwards of 60 members per vehicle have been observed, this effect is greatly magnified.

As car clubs replace and update their fleets more regularly than individual consumers, car club members take advantage of higher-standard vehicles with lower emissions much more quickly than private car owners. The average age of the UK fleet is 8 years old whilst accredited car club fleets replace their vehicles within 3 years (for ICE vehicles – electric vehicles are replaced more slowly).

This is in addition to repeated findings in the Annual Survey of Car Clubs (CoMoUK) which show that car club use 'locks in' other travel behaviours as car club members tend to use public transport, walking, and cycling more than the national average. In the 2021 survey, it was found that 61% of respondents used public transport at least once a week, compared to a national average of only 19% of people who use a local bus at least once a week. In addition, 37% of respondents used a bicycle at least once a week, compared to a national average of 20% [3]. It is interesting to note also that new car club members tend to use the car club most during the initial period of their membership and less so as it goes on and they find other ways to travel.

12.2.4 Car sharing and MaaS

As car sharing is recognised as a good way of ensuring people have access to cars without ownership, conceptually it works well as an element of MaaS that encourages car replacement. Early MaaS models such as Whim and Ubigo bundled forms of car sharing into their offers (covered in Section 5.6).

However, when a MaaS platform integrates car sharing, there are additional levels of security and complexity over and above those required for public transport ticket purchasing. Generally, public transport tickets are bought in advance of use whereas car sharing is billed after use. In addition, driving requires a valid driving licence and insurance and these need to be verified. There is also the matter of loaning a high-value asset to an individual, so companies take steps to reduce the risk that this asset is not returned or returned in a poor state of repair. For this, the platform will generally require a higher level of identity checking, plus driving licence verification and credit checking to access vehicles.

The Jelbi app in Berlin is one MaaS app that includes car share (as well as access to moped and e-scooter share).

There are different payment models for integrating car clubs into MaaS as car share hire is generally post-payment whilst public transport tickets are pre-payment so combining the two requires ways to reconcile different business models. Whim dealt with this by the subscription model generating a single monthly all-inclusive bill (and having agreed contracts with providers purchasing ticketing and shared transport hours across the whole subscriber group).

However, pay-as-you-go MaaS is also popular and achieving an integration in the back office that is able to deal with both pre and post-pay billing is another means to ensure that both business models are handled.

12.3 Ride-sharing or lift-sharing

In contrast to car sharing, car pooling, ride or lift sharing increases utilisation of vehicles undertaking specific trips – in its simplest form it is the modern digitally mediated equivalent of a hitchhiker. Ride or lift-sharing has been increasingly digitised since the turn of the 21st century. The UK company Liftshare was launched in 1998 and French BlaBlaCar followed, launched originally as Covoiturage.fr in 2004.

People register with an online or app-based ride-sharing platform and list the trip they are offering or would like to make – and the platform matches people offering rides with those looking for them. People listing rides are generally able to ask for a contribution to fuel costs (although national regulations tend to affect how this is specified or possibly how much). The platform is responsible for checking identities to provide a basic level of security. It will also mediate any payments.

Whilst online platforms are busy in some countries, adoption is lower in others. The busier the platform the more likely it is that rides and riders will be matched. French startup BlaBlaCar reached 100 million members in 2021 [4] with much growth coming outside of Europe. One of the key tasks of ride-sharing platforms is to foster a sense of trust so that people can share a ride with each other with confidence. BlaBlaCar developed a 'superhero symbol' emblematic of trust within the network. Users have ratings, photos, and a demonstrable network as well as credit card checks. Building the network and brand to normalise trust required a lot of investment – and time.

Besides the 'market place' of rides version of ride-sharing, there is also a trend to create ride-sharing pools for large employers seeking to reduce their parking requirements and carbon footprint. Working with a pre-existing group boosts the elements that make ride-sharing work. People are travelling to a limited number of sites within a reasonably constrained area, there is already a critical mass of trust and, the involvement of the employer communicates the existence and availability of the system.

UK company Liftshare works with over 500 large employers, typically at employment centres grappling with full car parks or even extreme congestion causing widespread traffic issues on local road networks as people arrive at and leave large employment sites by car.

Employers encourage staff to sign up to the Liftshare platform which matches staff trips and also authenticates when staff make shared trips. For example, the company Wolseley in the United Kingdom employs 360 people at their Warwick Technology Park site. There, 61% of employees are signed up to the platform with 33% making regular authenticated shared journeys. Whilst this helps the company save on parking costs – reducing demand by 80 spaces has reduced parking costs by an estimated £100,000 per year – it also reduces the amount of traffic entering the site which improves congestion on local roads.

Ride-sharing associated with employers makes it easier to establish trust. In an interview in 2019, Ali Clabburn the founder of Liftshare noted that:

> Within a work environment 'the fear of being let down' is a lot lower – there's a perception of people you work with being less likely to let you down. Risk is not just of danger but also of arriving late or social awkwardness.

Whilst 'closed loop' ride-sharing offers benefits to employers, there are also ways to extend these benefits to wider transport networks. Ali Clabburn reports that:

> When we set up a Liftshare scheme in a business (unless there's a compelling security or business reason not to) there's usually the option that people can make their journey public or just offer it within the organisation. This helps boost the pool of people to draw from. 70% of users are willing to share more widely – although most end up sharing within their business network.
>
> Boosting use of business schemes – we see schemes that work perfectly well in situations where there isn't any issue over parking availability or reserved parking. However, if there is pressure on parking (e.g., through reserved parking) the percentage of people liftsharing increases.
>
> We find that people start sharing usually because of an employer scheme – but they continue sharing because they've discovered savings, made friends.

12.3.1 The technology of ride-sharing

To be successful, a great deal of the technology of ride-sharing depends on mobile phone functionality, from camera phones to GPS connectivity and satellite coverage (plus other location data derived from masts, WiFi, and local networks) to Bluetooth.

For instance, Faxi (now Kinto Join) designed a ride-share validation solution using Bluetooth verification to confirm that two (or more) phones were connected and were travelling together. In one employer-based trial, cars with multiple occupants would be allocated reserved car parking spaces. Car park barriers were raised automatically if the car had multiple occupants (as validated by the app).

A partnership between Liftshare and digital kerb and parking management provider AppyWay enables companies to manage their car parks to prioritise people who rideshare and provide them with reserved parking. AppyWay sensors in each parking space connect with the driver's mobile phone – parking permission is then managed through the AppyWay app. The integration means that verified Liftshare users get free parking or permission to use carpool reserved bays when their phones pair with the sensors in each bay.

In addition, camera phones are now involved in checking identity documents (in conjunction with access to relevant credit-checking services). Plus social network mapping and other soft data can be used to support the authentication of people's identity and reduce the risks of getting into a stranger's car.

12.3.2 Case study: Faxi pilot project

Faxi is a corporate car-pooling service that was used by Santander, Vodafone, and Northamptonshire County Council in 2017.

Parking at these organisations was increasingly contentious and 'one person one car' commutes had become untenable in many areas. Faxi provided a real-time solution using mobile phone technology to link to site car park barrier systems to reward car sharers with free parking in guaranteed 'sharer' spaces, whilst charging – or barring – those driving in alone.

The app also tackled some of the barriers to car sharing. Faxi found that the main barrier was finding people with similar work/travel patterns. The app provides real-time location services so users can see where potential sharers are. They could also chat with co-workers to arrange a ride or alternatively log when they will leave and ask if others would like to be picked up. Others could then simply accept or reject the invitation. The Faxi app estimated a contribution towards costs, which could be paid directly through the app.

In Northampton, the scheme was extended to the park and ride enabling employees to buy a ticket from their Faxi app. The scarcity of parking has driven ride-sharing adoption, freeing up parking spaces and reducing congestion [5].

Faxi entered a cooperation with Toyota Financial Services and, after 4 years of collaboration, became Kinto Join at the end of 2020 [6].

12.3.3 Ride-sharing and MaaS

Ride-sharing complements MaaS as it improves choice for travellers and also increases the number of passengers per car. The provision of rides is dynamic – depending on rides offered at the discretion of drivers – so MaaS platforms have to be able to incorporate dynamic elements into route planning.

In France, OUI.sncf, the booking portal for the national railway, SNCF, will search for journeys as with any other rail journey planner. However, the site will also return alternative travel options, including those from the ride-sharing platform BlaBlaCar if someone is making a similar trip (or your journey covers part of their trip). It can show multi-modal travel options combining rail, bus, or rideshare.

Behind this interesting development is a business cooperation between SNCF and BlaBlaCar in which SNCF sold its loss-making long-distance bus service Ouibus to BlaBlaCar in exchange for an equity stake in BlaBlaCar [7] in 2018.

It is also possible to anticipate MaaS integrations where ride-sharing and parking are combined in employer-centred MaaS applications.

12.4 Bike sharing

Bike sharing started more as a social movement than a transport mode (with the infamous '50 white bicycles' painted and left for people to use freely in central Amsterdam in 1965) and rapidly moved to funding itself as rolling advertising hoardings. Early municipal and campus-based bike share schemes were trialled in the 1990s in this way but they found a firmer footing with public sector involvement in the early 21st century.

Large-scale municipal bike share schemes were launched in cities across Europe, from Citybike Vienna, launched in 2003, to Vélib in Paris (2007) and Santander Cycles in London in 2010. These schemes were large-scale systems, in which bikes were rented from and returned to a network of bike docking stations. Several schemes were operated by advertising agency J C Decaux which sold on-bike advertising and sponsorship. Although the larger systems quickly became embedded in metropolitan transport networks, the path was not as smooth in smaller cities. The Welsh capital city, Cardiff, had a bike share scheme that preceded the London one and operated from 2008 to 2010, only relaunching 8 years later after a long evaluation and learning process.

Commercial models of bike share have been in operation for some time. German operator nextbike, launched in 2004, put small fleets of rental bikes (for instance 200 in the city of Hamburg) that did not require docking stations on city streets in Germany [8]. A much larger scale version of this rose in China in 2015. Operators flooded Chinese cities with thousands of brightly coloured, specially designed connected bikes that could be rented and left anywhere within the city. These 'dockless' bike share schemes grew rapidly, funded by huge inflows of venture capital [9] and expanded to European cities. The bikes were supposed to create revenue streams from ride fees and advertising. However, this business model failed to make money and was undermined further by the costs of vandalism and higher than anticipated fleet management and maintenance costs. By 2018, the sector was undergoing a rapid contraction and many operators had exited the market by 2019 [10].

Bike share has increasingly been sought and welcomed by local authorities as it provides sustainable last-mile travel and has demonstrated that it attracts new cyclists and widens the demographic of those cycling [11]. In addition, besides enabling multi-modal journeys, there is some evidence of a modal shift from car use to bike use enabled by bike share.

12.4.1 Operational model of bike share

As outlined, bike share operations have different bike share models; they may be 'docked' schemes with on-street station infrastructure from which bikes are hired and to which bikes are returned, or 'dockless' where bikes are not anchored to bike stations. Dockless bike share schemes use apps and mobile technology to manage bike hires and have varying rules about where bikes may be used and returned. There are also hybrids where bikes can be hired from both stations and also hired from or left in designated areas without on-street station furniture. Some schemes include electric bikes (where batteries must be recharged either by a team swapping batteries or by charging integrated with docking systems).

Schemes are operated by a number of bike share operators usually in conjunction with the local authority either through a tender process or other agreement. However, commercial operators are sometimes set up in cities where they judge the market as sufficiently large and unsatisfied by public systems.

The commercial models are funded using investor capital (or loans) and derive income from charging users or monetising the system in other ways (including commercial sponsorship deals and advertising). Such schemes are generally dockless as the investment in infrastructure and partnership required to build docking stations requires a different approach.

Other schemes are funded through a blend of partnerships, grants, subsidies, advertising, sponsorship, and user charging depending on the context, infrastructure requirements and agreements with local authorities.

For instance, Berlin is home to a number of bike-share operators – and has seen a number of high-profile failures. Public bike share was launched in 2011, and operated by Call a Bike (part of Deutsche Bahn). It remains on the streets although the Berlin public bike share tendering process awarded the contract (worth €7.5 million over 5 years) to nextbike in 2016 [12]. In 2017, the city became a key target market for startup bike-share companies such as Mobike, Jump, and ofo. Problems including billing errors, vandalism, and the ill-suitedness of some bikes for taller riders led to most of these schemes closing down by the end of 2020. The original Call a bike scheme continued to operate, as did nextbike, Lime, and Copenhagen-based Donkey Republic.

Competition has created market failures in some areas. The proliferation of dockless schemes in Oxford including start-up ofo was cited as the reason for the withdrawal of Oxonbike. Oxonbike was a docked scheme run by Hourbike in Oxford – which included electric bikes and was supported by the local authority. It enabled travel between some key NHS and authority sites which were generating traffic and parking problems [13]. After Hourbike was withdrawn, however, ofo reduced the area served by its dockless scheme, creating concerns that the remaining operators may not provide a stable scheme covering a sufficient area to meet transport goals or provide travellers with sufficient options. These concerns were vindicated when the dockless bike share schemes collapsed in 2019 [14].

One of the more recent bike share forms to appear on the market is that of long-term leasing. Bikes (and particularly e-bikes which have a higher cost barrier to ownership) are rented for a monthly fee. This enables people to try bikes over a period before committing, or rent one for a few months (for instance whilst living in a place for a few months). Often the monthly rentals include maintenance and insurance, meaning that people are able to ensure they have help with their bike if needed. Leasing such as this has yet to be integrated into any MaaS apps.

12.4.2 Impact on car travel

Whatever the framework for the bike-share scheme, there is evidence for the reduction in car use and its role in multi-modal transport. The CoMoUK Annual Survey of Bike Share, in 2021 reported that 53% of respondents said that they would have made their last trip by car or taxi if bike share had not been available [15].

The survey suggests that bike share is often used as part of multi-modal trips, with 70% of respondents saying that they combine their use of bike share with other modes. This is generally combined with public transport to start or complete the trip –39% of users stated they used a bus and 28% stated that they used a train as part of their overall journey with bike share (these figures are similar to those from surveys pre-dating the COVID-19 pandemic when, over the 4 years of previous surveys (2016–19), on average, 23% of users combined their trip with bus and 35% with train travel).

Another interesting feature of bike share users is their overall interest in other shared mobility services – in the 2019 CoMoUK survey, 28% of respondents were likely to use car clubs, 23% rideshare services, 28% shared taxis, and 44% a MaaS account [16]. The use of bike share following the integration of e-bike share into the Uber app (Section 5.3) provides real-world validation of this finding.

More recently, bike share has included e-cargobikes and electric bike share. The latter opens up bike share to a broader group of people, ensuring more equitable access and higher utilisation as demonstrated in the case study below.

Analysis of the United Kingdom's first fully electric bike share scheme in Exeter found a more equal proportion of women and men used it compared to the general cycling population – 62% men compared to 75% men in general. Combined studies of multiple e-bike share schemes indicate an even more equal gender balance among users, of 55% male, 45% female [17].

Surveys of users have found that shared e-bikes are particularly appealing to new and occasional cyclists. In one UK-wide survey, one in three respondents rarely or never cycled before they started using shared e-bikes. They also report cycling further than users of conventional bike-share schemes – an average of 5 miles compared to 3 [18].

12.4.3 Vandalism

Vandalism is a key issue for bike-share schemes, with some cities seemingly more prone than others. Besides adding to management costs, vandalism can make the scheme unusable if large numbers of bikes are out of action. The Cologne system in

Germany fell victim to a TikTok challenge craze in 2022, in which the challenge was to wrestle a KVB bike out of its docking station [19]. In 2022, over half the fleet of 3,000 bikes was out of action at one time, and bikes were being vandalised at a rate of 300 (or 10% of the fleet) every 2 days.

A range of actions is necessary to address vandalism, including behavioural change measures, docking and locking design, as well as proactive monitoring, maintenance, and fleet rebalancing.

12.4.4 Cargobikes

There is a strong argument for integrating e-cargobikes into bike share fleets. The weekly shop is often cited as a reason for needing to own a car and as an argument against the usefulness of cycling. Cargobikes, trikes – and their e-versions, in particular, make this assertion less compelling. A trip to the supermarket or the school run is not a problem for utility bikes with a bit of additional torque. Whilst 78,000 cargobikes were sold in Germany in 2020, the German government has also invested in e-cargobike sharing and several German cities have launched e-cargobike share schemes including Stuttgart and Berlin. Shared e-cargobikes are particularly important where housing cannot necessarily accommodate parking an e-cargobike securely (for people living in apartments for instance).

The pattern of e-cargobike distribution is somewhat different to bike-share schemes which tend to focus on city centres. E-cargobikes are generally required for trips to and from homes (for shopping or the school run) so need to be sited in residential areas. Some providers – e.g., Sigo in Germany [20] have focused on working with housing developers to provide shared e-cargobikes in new developments. Other providers such as Cargaroo have taken a similar approach but 'on street', partnering with cities to park e-cargobikes in marked parking bays in residential areas. The bikes are hired using the app and returned to their original bay. Cargaroo launched in the Netherlands in 2017 and rapidly added bikes in Belgium, the United Kingdom, and Germany.

12.4.5 The technologies of bike share

The technology that powers bike share involves connecting the bikes to management platforms so that they can be found (using GPS units and the mobile phone network) and IOT technology to lock and unlock them, besides mobile phone technology to manage the customer interface.

Early mobile-based bike unlocking systems worked via combination codes which were communicated to the user by SMS (requiring only basic or 2G mobile connections). The combination codes were randomly generated which required a simple cryptographic system.

Some systems have used RFID cards to lock and unlock bikes. This meant that they could be integrated with public transport travel cards (the pre-cursors to MaaS in some cities). For instance in Cologne, Germany, the bike share system is branded with the local transport provider, KVB, as KVB Rad. Once registered, people with a KVB 'chipkarte', an RFID travel card, can use it to lock, unlock, and pay for bike share.

The second-generation bike unlocking systems transitioned to mobile-based customer interactions for managing their bikes. Typically the bikes would be labelled with a unique QR code, which would be scanned by the phone to identify the bike, open the app or site, and alert the management platform of a hire request (whilst confirming proximity). After validating the request (i.e., a legitimate registered user with credit), the management platform then communicates with the bike lock and opens it. In some systems, this is via Bluetooth operation.

Heavily used bike share schemes in major cities such as Paris and London use secure docking systems anchored to the ground and designed to be extremely resistant to vandalism. The locking mechanisms are built into the docks rather than on the bikes. This means that the docks are connected and can use wired infrastructure rather than relying on mobile networks.

On less heavily used systems, or for flexible bike share, rear wheel locks mounted on mechanical bikes and triggered remotely are a popular way of securing the bikes. In 2021, the Belgian Rail SNCB bike-share system 'Blue-bike' moved from being managed by a key vending system (which unlocked the bike dock) to being managed through connected rear-wheel locks manufactured by Sentinel. These on-bike smart locks can be unlocked using Blue-bike issued RFID cards or be opened by the Blue-bike app [21].

E-bikes in bike share systems are usually locked at the motor level (with or without an additional docking station, cable, or rear lock). E-bikes that are moved when their motors are locked will also deploy additional sounds and activate remote alerts and tracking on the operation platform so that the operations team on the ground can recover the bike rapidly.

E-bikes can also be 'geofenced' so that they are locked if they travel beyond the permitted zone. Geofencing can also be used to slow cycles down if they enter a zone in which a lower speed is required. This requires more sophisticated geolocalisation with trackers that are able to transmit location data in sufficient quantity and detail so that the platform can locate the bike more precisely with respect to its operating boundaries and manage its movements. The most sophisticated systems use 4G and narrow band IOT to locate and connect the bike. Additional Bluetooth anti-theft trackers may also be installed as a backup to make theft unattractive.

In addition, bike share bikes often have sensors that connect to maintenance software to alert operator managers of maintenance issues arising on the hardware – from brakes to tyre pressure. This extreme connectivity is helpful for optimising preventative maintenance.

Bike share operators increasingly use algorithms and machine learning to predict and optimise rebalancing. This ensures that bikes are at docking stations when and where they are most likely to be used – and that there is space at docking stations for the return of bikes when and where this is likely to be required.

Bike share operations are often expensive for heavily used schemes so this is an important component of system management.

12.4.6 MaaS and bike share

Bike share has long been recognised as an essential element of MaaS. In 2018 people working in the transport sector in the UK ranked bike share the third most essential mode of a Mobility-as-a-Service system (after bus and rail) in the Annual Survey of Mobility as a Service [22].

The integration of bike share into journey planning was strongly aided by the development of the GBFS – global bike share feed specification. GBFS was launched in 2015 by the North American Bike-share Association (NABSA) to provide real-time data feeds in a uniform format publicly available online, with an emphasis on findability. GBFS is intended as a specification for real-time, read-only data – any data being written back into individual shared mobility systems are excluded from the specification. It is designed to provide the status of the system at the given moment (not to provide historical data). GBFS is an open-source project developed under a consensus-based governance model. Contributors come from across the shared mobility industry, public sector, civic technology, and elsewhere. In 2022, responsibility for the oversight of the GBFS was taken by MobilityData, a non-profit organisation aiming to ease travel through shared data and information [23].

Since 2015, it has been used by most of the major operators and bike share systems around the world, including nextbike, Motivate, Social Bicycles (and therefore Jump and latterly Lime), Vélib, Citi Bike, etc. Some e-scooter operators also adopted it (Spin, Lime, and Jump) [24,25].

The availability and standardisation of bike share APIs have made them easy to incorporate in journey planners and other mappings. Citymapper was an early adopter of the bike-share API for London. It also incorporated the growing number of floating bike-share systems in 2018 [26]. Google Maps began showing real-time bike-share information [27] and serving multi-modal options for trips in 2019 [28].

GBFS is specifically designed to show where bikes can be found – and whilst it now includes a deep link field so that users can be connected with bike booking interfaces, there is still some variation between the degrees of integration between bike share and MaaS apps (for instance, users are often moved from the MaaS app platform in order to book and pay). Seamless booking and payment from within the MaaS platform depend more on business models than technical difficulty.

One of the interesting things about larger providers is that they will simultaneously maintain their own branded app whilst integrating with other journey planners and MaaS providers. For instance, people who sign up for (most) bike share schemes managed by nextbike can also use their accounts to book and pay for nextbike run schemes in other cities. This presents the conundrum of whether it is more important for people to sign up for their local MaaS app or their local bike share scheme. Operators can use 'technology layers' (software as a service) providers that network shared transport offers. The effect of such SaaS layers is that existing users can use their familiar apps to access similar services in territories not served by their operator. A case in point is nextbike, which is also directly integrated with super-apps via the Splyt interface [29]. This integration supports the entire end-to-end experience, meaning that several apps users can search, book, unlock, and pay for bike sharing and scooters, directly from within apps they already know and use. Splyt handles user verification and other features required for bike share (like parking zones) which reduces complexity for its super-app partners.

Whether there is a tension between MaaS take-up and the growth of operator user bases is still to be explored. What is certain is that a number of bike share integrations already exist in MaaS apps, from the city Velo share in Antwerp's Whim to the offer of Cargaroo e-cargobikes in the Dutch MaaS provider Gaiyo, so their bikes can also now be booked through Gaiyo's app [30].

12.4.7 Case study: Cardiff bike share

Cardiff bike share is a docked bike share scheme launched in May 2018 operated by nextbike as the result of an extended tendering process run by Cardiff City Council. The scheme covers a 22 km^2 central area of the city (which has a total population of 361,000 people). It has a diverse funding base including partnership, sponsorship and income from use, and an agreed initial period of operation.

The scheme was an almost instant success with more than double the UK average number of trips per bike per day. Bikes in the Cardiff scheme were used 4.27 times per day – just over twice as often as the UK average of 2.10 in the first 3 months after the launch.

Key success factors

The high usage of the nextbike Cardiff bike share is unlikely to be due to one element alone. The team and partners in the project were fortunate in being able to establish a virtuous circle between the people, place, and partners which built a collaboratively designed network. The partners worked hard to create a broad sense of ownership and a fantastic narrative around the bike share scheme which facilitated excellent use of the scheme.

The success factors can be distilled into:

- Cardiff City Council has a strategy of keeping cars out of the central area as much as possible. The bike share scheme and other cycling infrastructure measures are part of this and are prioritised by officers.
- A strong partnership was built between the Council, the principal sponsor (and one of the key users) Cardiff University, and the operator, nextbike. Other civic partners were involved from before the launch including cycling groups and the police.
    - The maintenance and redistribution team and the police have supported the sense that this is a scheme that is looked after and cared for.
    - Partnership between nextbike, Cardiff City Council, and Cardiff University has attracted a loyal membership with excellent usage of the bike share scheme.
    - Funding was obtained from the Welsh Assembly which purchased the assets (bikes) and means that there was minimal capital requirement from the operator or city.
    - Ongoing costs are largely met by sponsorship (the University in effect pays for membership for all students and staff through an annual

contribution) and bike hire income. The operator is able to concentrate on maintaining the quality of the scheme not just on revenue.

- nextbike has provided a comprehensive bike share system with booking and back office infrastructure which enables the reporting, customer service, and health of the scheme to be monitored and maintained. The scheme is a docked scheme, and whilst this is neither a guarantee of success nor an absolute prerequisite for it, it does allow a certain amount of communication (i.e., signage) around the bikes to help people understand the scheme, and it also helps people identify out of place bikes (and for the scheme operator to retrieve them) helping ensure the feeling of community ownership around them.
- The network works – docking stations are sited in high profile and visible locations which combine with the city cycling culture and infrastructure (which has been developed over the past 10 years by the Council) to enable a broad cross-section of the population, including visitors to Cardiff, to use the scheme.
- Communication has been particularly strong and consistent over a long period. From the initial announcement of the tendering of the scheme through to launch (over a year). All partners were concerned about countering scepticism after the previous scheme was withdrawn in 2010 and have worked hard to remain on message.

Whilst Cardiff is a smaller city, Paris is a global capital. In recent years the transport network in the city core has been transformed by the mayor, Ann Hidalgo, to increase active travel by building hundreds of kilometres of cycle lanes and reducing access by cars. The bike share system is among the initiatives supported to increase cycling and connectivity in the city. The introduction of e-bikes to the scheme has had an outsize impact on utilisation and provides insights for other cities considering schemes.

12.4.8 Case study: Paris, France

The Vélib bike share scheme in Paris is the biggest in the world outside China. It has around 24,000 bikes (including e-bikes), with 1,400 docking stations for multiple bikes, which provide over 46,000 e-bike charging points.

The first scheme was launched in 2007, operated by J C Decaux. At the end of the contract, the scheme was launched with a new operator and new bikes in January 2018, with a majority of mechanical bikes. At launch, the fleet consisted of 30% e-bikes and 70% mechanical bikes. Across the different subscription packages, e-bike rides are charged at a higher rate compared to mechanical bikes. For instance, e-bikes are charged per minute from the start of hire whereas mechanical bikes are free for the first 30 min of the hire for the V-Libre and V-Plus packages.

The Paris Vélib scheme's monthly usage is typically 600,000 unique users and over 5.5 million journeys per month (2021).

Data published by Fifteen, one of the operating partners of the Vélib scheme indicated that, in 2021, the mechanical bikes were used for four trips per bike per day, whereas the e-bikes made 10 trips per bike per day. In addition, each trip was, on average, much longer for the e-bikes – 25 min average journey duration per trip – compared to 17 min average journey duration per trip for mechanical bikes.

Effectively, 30% of the fleet is used to make more than half the total trips per day. Even more revealing, 30% of the fleet is used for 61% of the total time the fleet bikes are in use.

By analysing search preference data, the operator can also conclude that two out of three people arriving at a docking station are searching for an e-bike. Not all of these people will be able to find their first preference.

The Vélib case study highlights the potential for e-bikes in bike share schemes and within MaaS.

12.5 Micromobility

Following in the footsteps of bike share, e-scooter hire companies began introducing fleets of small, light electric stand-on scooters in urban centres in 2017. These are very much 'digital native' vehicles, introduced as fully connected, app-operated as well as electric.

The electric motors have advertised top speeds of around 20 miles per hour. The handlebars steady the device, steer it, and provide brake mounts. Scooters are intended for short journeys – although their typical range per charge is 15-20 miles. Their introduction started in the US cities from Los Angeles and Salt Lake City but spread to Europe over the subsequent year. Designs for sit-on scooters are also emerging, from super light folding devices to much sturdier trikes and motor-cycle style vehicles at the larger end of the scale. These latter are more likely to fall into different vehicle categories – for instance, requiring driving licences and helmets (limiting their appeal and the size of their potential user group).

Small personal vehicles that provide short-hop travel between public transport nodes and origins and destinations potentially have a role assisting people to opt for public transport. They are attractive to cities trying to encourage fewer cars from entering central zones. They have the potential to save space and provide last-mile solutions without requiring expensive infrastructure. However, the experiences of cities where fleets have appeared overnight have not been completely straightforward.

Complaints have been generally over street litter and the behaviour of users. Large numbers of poorly parked scooters can make streets difficult to navigate for pedestrians or those in wheelchairs and are particularly dangerous for blind and partially sighted people.

The speed of scooters – up to 20 miles per hour – can also lead to accidents and, significantly, to pedestrians feeling less safe walking. If widespread use of e-scooters discourages people from walking because they feel vulnerable and unsafe, their adoption could disadvantage already vulnerable groups and reduce their access to active travel.

In Madrid, three companies, Lime, Wind, and Voi launched in 2018 but by the end of the year, authorities had revoked their licenses in order to regulate their operation more closely. Scooters were banned from pavements early on, after an accident in August in which one person died. Since then the city has brought forward new regulations to manage scooters more closely.

The use of electric scooters there is now restricted to bike lanes and minor roads (those with speed limits of 30 kilometers per hour or below). The operators' licences were suspended to enable the city to review whether they were able to comply with the new rules.

Brussels adopted a more regulated approach from the outset. Scooter operators must register with the city authorities, run their scooters on green energy, and enforce rules so that the scooters are not parked blocking entry or exit from public transport.

In 2022, Paris, having reduced the number of operators to three and restricted their activities, has considered banning them entirely after a number of high-profile accidents [31]. In 2023, Paris residents voted overwhelmingly to ban shared scooter fleets [32].

12.5.1 LADOT and the MDS

Alongside concerns about e-scooters is a consideration that city transport authorities provide the infrastructure along which they operate – roads and paths – and therefore should have at least some insight into those operations and potentially the ability to shape them positively.

Los Angeles has required that, in order to operate in the city, all scooter companies provide their data in real-time using their Mobility Data Specification. This is an open-source specification with full documentation available on Github. It is described as:

> MDS—"Mobility Data Specification"—is a digital tool that helps cities to better manage transportation in the public right of way. MDS standardizes communication and data-sharing between cities and private mobility providers, such as e-scooter and bike share companies. This allows cities to share and validate policy digitally, enabling vehicle management and better outcomes for residents. Plus, it provides mobility service providers with a framework they can re-use in new markets, allowing for seamless collaboration that saves time and money.

The specification is a way to implement real-time data sharing, measurement and regulation for municipalities, and mobility as a service provider. It is meant to ensure that governments have the ability to enforce, evaluate, and manage providers.

MDS [33] is currently comprised of two distinct components:

- The provider API is to be implemented by mobility as a service provider, for data exchange and operational information that a municipality will query. This is how the provider presents the historical view of operations.
- The agency API is to be implemented by municipalities and other regulatory agencies, for providers to query and integrate with during operations. In this way, the agency provides tools to inform and permit future operations.

Cities and regulators can choose how best to implement Agency and Provider either separately, concurrently, or by endpoint.

As cities have pushed back at a 'free for all' attitude from some operators, other providers are working on ways to manage scooter use. Both light infrastructure – which has a tendency to provide information and use instructions that augments apps – and also technical control through geofencing have the potential to manage the behaviour of scooter users. Life-cycle analysis of early fleets also found that e-scooters were in use for a matter of months before being beyond repair either through wear or vandalism and fleet operators have increasingly tried to make their vehicles more durable. However, there are still huge concerns about the short lifespan of the scooters – with many ending up in local rivers and canals.

Whilst early geofencing – such as that used by Mobike – would only manage behaviour after docking (for instance by fining people who parked out of the operational zone), newer, more sophisticated, geofencing can manage user behaviour whilst the scooters are in use. Their speed can be regulated according to their location so that they can go faster in some areas and are limited in others – and indeed they can be brought to a halt if they stray beyond the operational area.

The United Kingdom remains one of few countries in which it is not possible to ride an e-scooter legally on the road or pavement. In 2020 the Department for Transport announced trials of rental e-scooter fleets in a number of areas and with specific conditions. The UK city trials of rental e-scooter fleets are only possible under strict conditions – including the rental fleet operators agreeing to extensive data sharing. This has enabled the introduction to be closely monitored. The Solent area hosted a fleet that reached over 1,800 e-scooters across three areas by the end of November 2021 and surveys suggest that at least 40% of the trips made by the e-scooters would otherwise have been made by car or taxi. The trial in Nottingham includes take-home e-scooters on monthly rental in addition to the on-street rental and reports positive feedback on cost and journey time savings from key workers who have taken up the exclusive hire offer.

12.5.2 Technology of e-scooters

E-scooter share has much technology in common with bike share. E-scooter hire generally launched with smartphone-reliant systems in which users are required to register by app to set up their accounts and connect payment methods. The apps generally used the phone's systems – camera and or Bluetooth to recognise scooters and trigger the locks. In addition, the phone's camera may be required to photograph the scooter at the end of the hire and upload the image to ensure that the scooter is properly parked.

E-scooters generally have GPS and mobile-connected trackers to enable location and management. Geolocation has become an important component of fleet management as e-scooter fleets are required to comply with city regulations on parking and speeds in pedestrian areas. E-scooter motors can be limited by engine management systems in certain areas or users who park outside permitted areas can be identified and charged additional fees.

Battery management is another important component of e-scooter technology, and battery levels must be communicated to the operations platform to ensure that battery swap or charging operations are optimised.

12.5.3 Micromobility and the environment

The International Transport Forum published an analysis of the equity and sustainability of micromobility in 2021. This compared the carbon footprints of micromobility with mass transport and with traditional and electric automobiles (Figure 12.1).

Whilst the report noted that city authorities were in general agreement that micromobility enriches urban mobility and has the potential to change behaviour, it also highlighted that the carbon footprints of some micromobility – principally e-scooters – are not quite as small as those of mass transit. This is particularly the case in terms of the embodied carbon in e-scooter vehicles. It also highlights that the green house gas emissions of taxi-type on-demand modes (which also figure in the modes of MaaS) are relatively high.

In addition, these figures do not take into account the spatial requirements of impacts of modes. Whilst car travel requires space to be allocated to parking and has encouraged road building, other modes can move people into more compact spaces. The more compact cities are, the more walkable they are – and the fewer vehicles in total are required to move people within them [34].

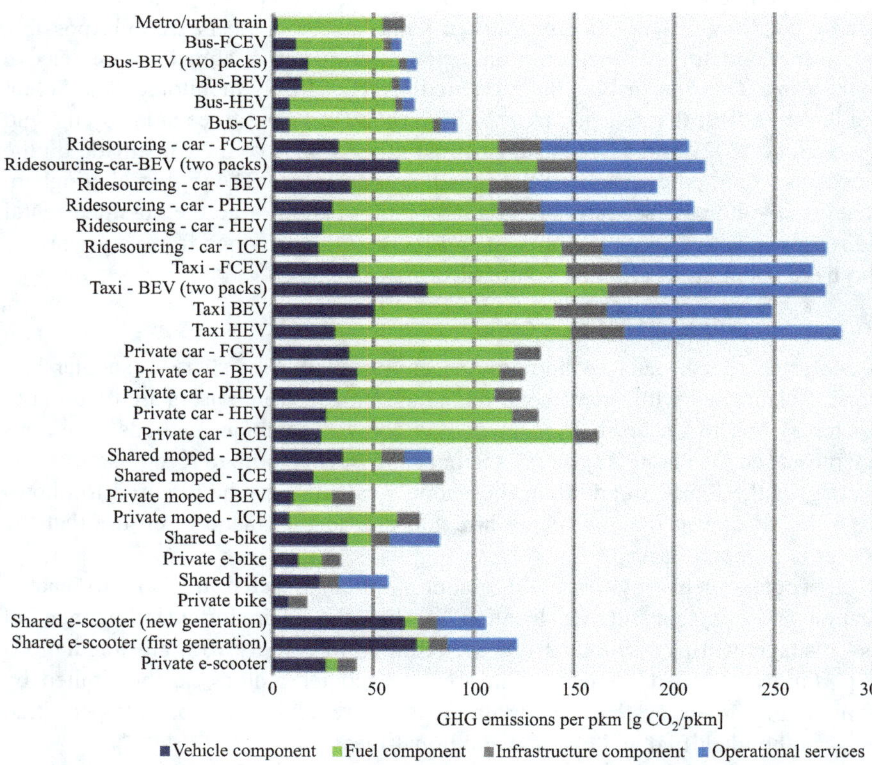

Figure 12.1 Central estimates of life-cycle greenhouse gas emissions of urban transport modes per passenger kilometre [35]

12.5.4 E-scooters and MaaS

The more recent addition of e-scooters to the shared mobility mix means that they were launched as digitally managed systems from the start, and as such were generally 'MaaS ready' on their arrival on city streets. The use of the popular GBFS schema for scooter location data by several operators made their appearance in journey planners and on google maps straight-forward. E-scooters appeared in Google Maps in 2019. They have also been incorporated in MaaS apps from both commercial (e.g., Whim) and city (e.g., Berlin's Jelbi and Vienna's WienMobil) platforms.

12.5.4.1 Case study: Knot City, Strasbourg

Knot City provides a docked scooter service in Strasbourg. The scheme operates 30 scooters parked at eight stations located at the exits of sites run by city parking operator PARCUS. They are slightly larger, sturdier, and more robust than scooters used more typically in other schemes. These scooters have bigger wheels intended to cope with a variety of surfaces.

The Knot City collaboration with PARCUS works well. PARCUS has access to city centre land and the scooters are provided as a service to people parking on the sites – they can use scooters for their last mile journey using their parking ticket. Of the total usage, 15% is from PARCUS car park users and 85% general public.

The scheme initially had an uneasy relationship with city authorities. Co-founder Polina Mikhaylova said:

> City authorities at first were very reticent and not willing to communicate on any kind of scooter sharing installations – partly because of the relatively high usage of bicycles (for France) and existing conflicts between bikes and pedestrians. After months of discussions, the situation is changing – although it is inevitable that there will be quotas limiting the number of scooters in coming months.

The station-based nature of the scheme reduces operational costs (Knot reckons that their costs are around 0.5 cents per charge compared with an estimated 8€ per dockless scooter) and secures the scooters. The Strasbourg scheme launched as an open network in September 2018 and in its first 6 months of operation scooters are ridden 4–6 times per day with no scooters broken or destroyed due to vandalism.

Knot City has continued to develop its offer and created a 'universal' docking system that will work both for its scooters and some models of e-bike.

12.6 Taxis, private hire, and ride-hailing

Whilst Uber can genuinely lay claim to launching the process of digitising taxis/private hire vehicles and turning them into ride-hailing (and is extensively covered in Sections 5.3 and 5.10), it is not the only platform to do this – nor is it the only business model.

Ride-hailing platforms sprung up worldwide – including Lyft in the United States, Didi Chuxing in China, and Yandex in Russia. Whilst they were app native, they also prompted more traditional taxi businesses to digitise, enabling smaller companies, territories, and independents to network together. Much of this digitisation was achieved through the automation of despatch software, which has increasingly moved to a software-as-a-service business model, digitising all parts of the private hire business from fleet management to booking, trip routing, and payment.

Key to 'discoverability' however, is the provision of taxi APIS through these platforms, what was once only within the overview of the taxi despatcher can now be published to the world, enabling real-time availability of taxis to be used in journey planners and maps. For those services, links mean that journey plans can be communicated to taxi and private hire providers' platforms, and bookings made with only a few clicks.

As of the end of 2022, the list of 'ride service' companies discoverable through Google Maps has grown (although the actual choice of ride-service providers that individual users will see depends on their country or region). The companies Google can send data to include:

- 99
- Beat
- Bitaksi
- BlaBlaCar
- Blablacar Daily
- Bolt
- Cabify
- Careem
- Chauffeur Prive
- DiDi
- Freenow
- Gett
- Go-jek
- Grab
- HK Taxi
- JapanTaxi
- Juno
- Lyft
- Meru Cabs
- MyTaxi
- Ola Cabs
- Sixt
- Texi.eu
- Uber
- Waymo
- Waze Carpool

There is an interesting side-note to this list. In January 2017, Uber was integrated very deeply with Google Maps – enabling users to search for a trip, and book and pay for an Uber to make it directly from Google Maps. Although many services were supported by Google Maps, Uber was the only one for which users could pay for the ride and complete the trip within Google Maps However, this functionality was removed from Google Maps in 2018 [36]. It is not clear whether it was a result of concerns on Uber's side (users did not have to download the Uber app) or Google's choice. It does show that Google Maps has the potential to be a powerful journey facilitator with many of the features of MaaS, should it choose to do so.

As of 2022, people searching for journeys are able to find options for ride-hailing (taxi), dockless vehicles (bikes and scooters), docked bike share, and public transit. For instance, the process for finding such solutions on the popular route-finding app Google Maps is, as of August 2023:

1. On your mobile phone or device, open the Google Maps app. Make sure you are signed in to your Google Account.
2. Search for your destination or tap it on the map.
   - For a rideshare, make sure you select both your starting point and destination.
3. At the bottom left, tap Directions.
4. At the top, tap Rides A list of services, prices, and times appears.
   - **To check the price for a ride:** Select a ride-service provider.
   - To search for dockless vehicles near you: Tap Transit or Cycling.
   - To search for docked bikeshares near you: Tap Cycling.
5. Choose the service you want to use.
   - If you have the service's app on your device: the app will open.
   - **If you do not have the service's app:** the App Store will open to download it.
6. To finish your ride request, follow the instructions in the provider's app

12.6.1 Ride-hailing technology

Whilst ride-hailing technology is largely based on smartphone capabilities – both for rider and driver – much of the functionality depends on the underlying maps. Smartphones are critical for managing the locations of both rider and driver (using GPS and possible local area networks). They are used in identity validation (for instance through cameras connecting individuals to validation platforms or by connection to banking data) and payment (either through wallets or connection to payment processors). In addition, despatch platforms assess the availability of vehicles, provide both time and cost predictions and routing, and assign rides to riders.

Taxi and private hire despatch platforms, in particular, have enabled aggregated availability and price to be available to journey planners and MaaS apps via APIs. Each despatch SaaS platform (the likes of iCabbi and Autocab) can publish all the companies that use it to a single API, creating an aggregate fleet with a much broader reach. This then can be incorporated into route planners.

However, private companies have not been the only means to this. The French government has sought to create a single taxi API to improve the discovery of taxis across the nation. This, in their view, is an extension of the real-world visibility – the illuminated taxi signs on each vehicle – that is demanded as part of taxi registration. The concept is to require taxis to be as visible digitally as they are in the street. It is also a recognition of the role that taxis can play as an adjunct to public transport and an enabler over the last mile. In France, municipalities limit and release taxi licences in response to local demands [37].

12.6.2 Mapping and GIS

The other key element of ride-hailing technology is mapping. Underlying maps – and their quality – are key to providing routes, and journey time predictions and integral to optimisation.

Digital mapping and Geographic Information Systems (GIS) have been evolving since the 1960s, first commercialised in 1969 with the founding of geographical information systems software provider, Esri [38]. Over this time, maps have become increasingly sophisticated, moving from being digital representations of paper maps to incorporating increasingly detailed sets of information which has made them into extremely powerful tools and a huge resource that enables both MaaS and many of the modes offered within it.

One of the first essential developments was adding the rules of the road (one-way streets, motorways, and pedestrianised areas for instance) and even parking restrictions. This enables more reliable routing – and generally, these instructions (in satellite navigation devices and latterly via smartphones) were the first popular use of GIS systems. Details of public transport services (and subsequently shared mobility) have augmented their utility as journey planners. Images and satellite layers showing landscapes from above complement the offer – allowing people online access to view the world from many angles before they set foot outside their homes.

The public's primary experience with digital mapping is through smartphones, with the majority using Google Maps and a minority using Apple Maps. Other maps are available; however, Strategy Analytics, in its Location Benchmarking Report: 2022 [39], only recognises a small number of companies as having the capabilities and resources to make and maintain maps at a global level. These include Google, HERE, TomTom, and Mapbox. Whilst Google fulfils a very broad and high-profile consumer-oriented role, HERE, TomTom, and MapBox are focused on providing mapping layers for third parties – whether specialist logistics operators or journey planning or logging services directed at particular groups. Between them, they also power automotive industry maps (for instance built into car navigation systems).

Almost certainly the most comprehensive maps accessible to the general public are those supplied by Google – and its technology lies at the heart of many applications.

Google has invested in an immense amount of mapping technology and also collecting map data itself. Google launched Street View in 2007 having commissioned a fleet of cars to drive along the road network capturing 360° views of the environment at regular intervals [40]. It also ingests satellite and aerial photographs to create the 'Google Earth' views of places in three dimensions. Creating these vistas requires enormous computing power to normalise data from different sources – essentially stitching together footage, tying it to geographic locations so that it provides seamless maps and views, calculates distances between places along roads and paths and predicts journey times.

Whilst Google is comprehensive, it has drawn criticism for locking people into its ecosystem. For instance, it crowd-sourced information about disability access which was then only made available to users within an Android/Google Maps environment (with no option to access via iOS or online) [41].

There are also still gaps in Google's mapping when it comes to walking and cycling infrastructure. Whilst the option to select journey plans for walking and cycling routes exists, the results are often served with a disclaimer that the pedestrian and cycling routes may not, in fact, be passable.

There are some proprietary alternatives. Apple's own maps have not yet proved to have the same level of detail as Google thus far. Launched in 2012, they started to collect their own map data in 2018 (however, 4 years later richer maps are still limited to the United States) [42].

Other suppliers – primarily HERE, Mapbox, and TomTom – provide specialist mapping to various industries including automotive navigation services. These suppliers power well-known systems like sports app and route planner Garmin (HERE) [43] or athletes' app Strava (Mapbox) [44]. They are well-used and well thought of.

One of the key components of the alternative to Google (or indeed Apple)'s proprietary maps is an enormous crowdsourcing project: OpenStreetMap. OpenStreetMap is built by a worldwide community of mappers that contribute and maintain data about roads, trails, and other features (including cafes). This emphasis on local information has created a rich set of information about local areas and local transport combined with openly licenced data from national mapping agencies and other sources.

OpenStreetMap is widely used by applications that depend on mapping. It is a source of mapping data ingested by both the TomTom Maps [45] and Mapbox [46] platforms. It is also the map for journey planner Moovit [47] and for outdoor trip planner Komoot [48]. OpenStreetMap powers the A Better Route Planner website and app, which uses extremely granular data about electric vehicles (specific to the model and even using data from vehicles directly) to plan journeys including charging stops [49].

12.6.3 Maps and real-time information

Maps interacting with real-time data are powerful components of journey planning and journey time prediction – for instance, taxi routing depends on traffic information, which is a feature of many maps.

Traffic insight is generally sourced from a variety of sources – from sensors and traffic management systems to 'crowdsourcing' – using data generated by mobile phones. The aggregate positions and speeds of mobile phones can be provided by mobile network operators to assess how free-flowing traffic is. This data is cleansed and used to indicate how congested roads are and to calculate journey times. One of the key providers of 'clean' traffic data is INRIX, founded in 2004 [50]. Google Maps began indicating traffic and congestion in 2007.

Rather than just using data, the founders of Waze adopted a social approach. Waze was launched as a company in 2008 [51] – which set out to create a community of drivers tipping the best routes and letting others know about specific problems, and, by installing the Waze app on their phones, contributing their own real-time data to the general overview of traffic data as they drive around [52]. With over 100 million users in 2020, Waze's 'community-powered navigation' could collect and analyse huge amounts of crowdsourced data and provide drivers with tailored route planning and accurate predictions – in general an extremely powerful resource [53].

Google bought Waze in 2013 [54] and began to incorporate features from Waze into its mapping (whilst retaining Waze as a stand-alone app with more 'social' features than Google generally supports).

Within transport apps (e.g., by taxi despatch platforms and Uber), traffic information is further treated to indicate when rides are likely to be available and how long journey times are likely to be.

In an acknowledgement of the importance of mapping, Uber noted, in its IPO filing in 2019 that Google was a key supplier of its mapping technology.

> We rely upon certain third parties to provide software for our products and offerings, including Google Maps for the mapping function that is critical to the functionality of our platform

Google Maps technology appears when a person enters a destination in the Uber app; the technology shows the current place, the destination, and a route, along with estimates of arrival times.

This relationship is longstanding – although not without bumps in the road and possibly concerns about over-reliance on it. Uber said that it entered into an agreement with Google around map services in October 2015 noting that:

> From January 1, 2016 through December 31, 2018, we have paid Google an aggregate of approximately $58 million pursuant to this agreement,

Uber also recruited Manik Gupta, then the director of product management for Google Maps around the same time. Maps are, however, so crucial that Uber also acquired mapping start-up deCarta in 2015, bringing deCarta's founder, J. Kim Fennell, onto its staff; he is the person who signed two contracts with Google for mapping technology on behalf of Uber.

It is not clear that Uber is entirely comfortable with this dependency, highlighting that Google has very much captured the market:

> We do not believe that an alternative mapping solution exists that can provide the global functionality that we require to offer our platform in all of the markets in which we operate. We do not control all mapping

functions employed by our platform or Drivers using our platform, and it is possible that such mapping functions may not be reliable.

Uber also said between the beginning of 2016 and the end of 2018 it paid Alphabet's affiliates around $631 million for marketing and advertising services [55]. This highlights the advertising power that Google leverages with its maps and searches.

No discussion of mapping and traffic data is complete without noting that it can be gamed. In 2020, Simon Weckert, an artist in Berlin, tricked Google Maps into displaying congested routes by loading 99 phones onto a pull-along cart and wheeling them around streets in the city [56].

12.6.4 Ride-hailing and MaaS

There is a close association between ride-hailing and MaaS. It is clear that MaaS requires the incorporation of ride-hailing to be able to cover last-mile trips that can not be completed in other ways. This is reflected in the fact that MaaS applications have incorporated ride-hailing from early on. For instance, Whim in the West Midlands integrated with Gett, a platform for local taxi companies and fleets that combines them into a single app. Gett has partnered with over 2,000 of what it terms 'ground transport providers' – boasting over a million drivers – in 10 countries [57]. Gett is particularly focused on managing travel for business users, however, its fleet in the United Kingdom is also open to individuals via its app (hence the option to integrate with Whim).

Other despatch software has similar functionality. Autocab and iCabbi enable smaller private hire companies to join a more comprehensive API. Autocab was acquired by Uber in 2020, enabling Uber to increase its coverage in the United Kingdom. The metamorphosis of Uber from a ride-hailing app to a MaaS app is dealt with in Chapter 5.

12.6.5 Case study: iCabbi

Dublin-based iCabbi is a cloud-based despatch system that enables local taxis to offer an app-based customer booking and payment experience. It was launched in 2009 and claimed to power over 100,000 taxis worldwide and to have facilitated 1 billion bookings by 2022. The business model is Software-as-a-Service – with licence subscriptions sold per vehicle in use. The end-to-end platform manages all aspects of running a taxi operation. It includes the core despatch system and both passenger and driver app plus various business services and payment management necessary to running a private hire business. Key to integrating in journey planners, however, is its creation of an open API, and a marketplace of partners.

iCabbi is now part of the Mobilize ecosystem, which gathers all the mobility and energy-related services of the Renault Group. It operates in Ireland, the United Kingdom, the United States, Canada, Brazil, Australia, New Zealand, and Finland.

12.6.6 Integrations of ride-hailing

The more taxi companies that can be incorporated into a single MaaS platform, the better the coverage will be and the more likely that users will find a taxi when they need one. However (apart from the French government's ambitions to create a unified API), there are few truly comprehensive offers. Aggregating multiple APIs is one solution – this happens in some places on Google Maps and with some taxi APIs that draw from more than one source. In addition, integrator Splyt allows the integration of ride-hailing (and other on-demand) services to be integrated into mobility platforms, super-apps, and other service platforms and is discussed in Section 5.7).

Over the preceding 5 years, there have been various integrations of ride-hailing into journey planning and mobility platforms in Europe. For instance, Citymapper launched running its own taxis (and buses) in 2018 as a way of monetising its free journey planner, however, this proved short-lived. Whilst it seemed smart to use the data generated by journey plan requests to design the services, the regulatory environment was difficult and operations proved unprofitable. They were withdrawn in 2019 and instead, taxis and private hire bookings were facilitated from links in the planner [58].

The examples below highlight a trend of mobility start-ups being wrapped into established businesses, particularly in the automotive sector.

Such developments included French ride-hailing start-up Kapten being financed by Daimler AG and BMW Group in 2019 – leading to it being wrapped into the FREE NOW offer. This typifies a trend of automotive industry investment in new mobility as a service and on-demand models. Also in 2019, FREE NOW also moved to offer its ride-hailing in Italy through journey planner Moovit, integrating the availability of 4,500 FREE NOW taxis across Milan, Turin, Naples, Catania, and Cagilari into the Moovit app [59].

Spanish railway operator Renfe similarly examined how to link its services over the last mile and, to this end, set up a collaboration with ride-hailing operator Karhoo. This integrated Karhoo's offer of 30,000 taxis and private hire vehicles with Renfe's 'Renfe as a Service' app. This RaaS app – piloted in 2020 – proved to be a precursor for a more comprehensive MaaS platform launched in 2022 – dōcō [60].

At the same time, the payment methods for ride-hailing have diversified. For instance, Polish taxi-hailing marketplace, ITaxi, introduced direct carrier pay (which allows payments to be charged through the mobile phone bill or prepaid SIM card) [61]. In 2021, transport authorities in Sydney, Australia, extended the ability of the digital version of the travel card – the 'Opal Card' – to be able to pay for both public transport and ride-sharing in a trial [62].

German car rental company SIXT has also moved to broaden its services and offers by developing what it terms a MaaS platform – this offer incorporates a broad range of automotive services including rental and SIXT rides. This latter is SIXT's range of taxi, transport and transfer services, offered via the app and in cooperation with more than 1,500 partners worldwide, providing access to more than 1 million drivers worldwide.

SIXT has a focus on business bookings and has an integrated corporate payment function to enable it to target its services at corporate travel. It has grown its network to extend its services internationally, on-demand (less than 10 min) rides in more than 250 major cities in Europe and the United States, and advance bookings in 500 cities in over 60 countries worldwide.

These partnerships allow it to claim one of the world's largest taxi services and ride-hailing networks – all booked and paid for via the app. In the United Kingdom, SIXT partners with Addison Lee, whose services in London can be accessed via the SIXT app. Further car rental, car sharing, and travel brokerage services are provided via SIXT's ONE platform – which combines reliable partners, to provide all its customers with a comprehensive range of services. This platform is presented as mutually beneficial – providing SIXT's partners with access to its 20 million customers and has enabled SIXT to expand internationally and scale its product portfolio worldwide [63].

It is important to note that whilst European MaaS can generally be typified as starting with transport and aiming to integrate the transport offer, in many countries in Asia, transport choices are often presented within super-apps (like AliPay and WeChat) that cover multiple aspects of life, from finances to communication. Some – like Grab (offering financial services, delivery, and transport) and Go-Jek – started life as ride-hailing apps. Ride-hailing has been one of the most obvious integrations, however, these apps are more aimed at commercial success through convenience than designed to reduce car ownership or fulfil social or environmental goals.

12.6.7 Taxisharing

One of the adaptations of recent times – found in both the global south and northern regions – is the concept of taxi sharing. This has the potential to reduce costs per passenger and better utilise resources. In the north, it is mainly possible through the development of ride-hailing technologies or the adaptation of on-demand bus technologies. And as such, it is often 'MaaS ready'. Taxi sharing converges with informal transport in the global south and is treated in more detail in the next chapter.

One of the early examples was the UberPool service in which Uber passengers could opt to offer to share their ride and would pay a reduced fare. As with many shared services, use was affected by the COVID-19 pandemic and it was withdrawn in 2020. Uber announced it would return to certain cities in 2022, and rebranded as UberX Share [64].

Some of the applications of shared taxis have been more specific and targeted. In the United Kingdom, a start-up, Ride Tandem, launched after working closely with communities where there were few fixed-line buses or they did not operate in line with working patterns (for instance shift workers at industrial sites commuting to and from their homes in nearby towns). It aimed to provide reliable services at consistent fixed fares for people getting to and from work.

Tandem partners with local taxi and private hire operators creates groups of passengers and provides routes for drivers. This enables it to provide a mix of flexible routes with key pick up points, core fixed journeys, and multiple drop offs. Its key market is private sector partners helping to underwrite budget-friendly commutes for their staff.

The UK regulation around taxis and private hire is very different from that for buses, vehicles used in DRT schemes that can carry up to eight passengers can be licensed as Private Hire Vehicles (Slide Bristol, MyFirstMile, Tandem), and those carrying nine or more passengers require a Bus Operators Licence (ArrivaClick,

HertsLynx). This is not an insignificant administrative detail. It has implications for the design, operation, and scaling of services and for the cost base. For instance, employing bus drivers and managing a fleet is quite different from subcontracting to private hire drivers (who are generally self-employed). Local private hire licenses will differ from county to county meaning scaling from area to area can be complex. And those services with fewer than nine seats attract 20% VAT on fares whilst bus operations are exempt. The different costs and regulatory frameworks also make scaling up or down across the vehicle sizes between vans and buses within an area much more complex.

For this reason, Tandem is currently not part of the public transport network, however, Alex Shapland-Howes, Tandem CEO, also sees that their technology has a role in a broader transport system working with bus operators and local authorities.

> We're beginning to work with operators. Our team is able to model the actual demand, so we have a case where a large port has a highly subsidised bus service with inconsistent usage. We can tailor the service to provide one which requires lower subsidies.
>
> In another instance, we're working with a bus operator to create a 'blended' model of a large bus (with booking) at peak and a flexible taxi fleet to ensure trips can be made outside core hours.

Shared private hire platforms effectively are able to flex 'their' fleet because they do not directly own vehicles or employ drivers. They can commission services from private suppliers rather than bearing the overheads of fleet and drivers which enables them to match their vehicles to demand.

Whilst Tandem aspires to become part of the public network, there will need to be a revolution in regulation, fares, and ticketing to enable bus tickets and private hire fares to be integrated with the wider network – it is unlikely to happen without a Mobility-as-a-Service approach to public transport.

12.7 Informal transport services

In many countries across the world, there is often little in the way of a formalised public transport service. This is especially common in cities in the Global South. Such services are characterised by there being no formal timetable, no official stops, and no fixed departure times. Such services are also provided solely by private operators, although in some areas there may be a legal recognition of such services although they operate under informal conditions. In many countries in the Global South, informal transport services are considered to be the norm, as opposed to the exception that they are in the Global North.

Whilst often not considered part of the MaaS literature, such services have a significant social value attached to them. Historically they have been difficult to rationalise from the public policy perspective precisely because of their informal operations. Such services are driven by market need and have been described as 'consummate gap fillers' in the absence of any formalised public transport offering [65]. Another description of them is the purest form of laissez-faire transport provision, where deals are struck between the person with the willingness to pay and the operator with the willingness to provide [66].

For the purposes of this book, a strict definition of vehicle types for informal transport is not useful. Because of the organic nature of informal transport services, the type of vehicles used varies between countries, and even between cities in countries. From motorcycles to minivans, the range of types of vehicle are endless. Regardless, UN-Habitat has identified a number of common characteristics in relation to services that distinguish informal transport services from other types:

- Entrepreneurialism: Informal transport is without exception the domain of the private sector – owned and operated by private freelancers. Drivers sometimes own vehicles, though in many instances vehicles are leased by absentee owners for a set fee or a share of daily proceeds. Also, services are designed and priced according to what the market will bear.
- Use of small, aging vehicles: Universally, small vehicles dominate the informal sector – small, at least in comparison to conventional buses and other modes of the formal sector. Usually, vehicles seat anywhere from 1 to 16 passengers. Small vehicles enjoy several advantages over bigger ones: they take less time to load and unload, they arrive more often, and they stop less frequently. They are also more manoeuvrable in busy traffic and can accelerate and decelerate faster.
- Low-performance services: A disadvantage of small vehicles is their low power-to-weight ratios, and thus slow average speeds. Slow speeds pose problems mainly because they are at odds with what is rapidly becoming the mobility standard of the world – automobile travel.
- Competitive, niche markets: Operators compete in fiercely competitive marketplaces, relying on hard work and low ratios of capital-to-labor to eke out a living. Ease of market entry often leads to a surfeit of service providers and over-competition. Dependency on vehicle leasing and restricted access to credit to purchase capital limit the ability of most operators to save income and get ahead.

Such services often fill a gap in transport provision where a more formal public transport service is not provided, and even where such a service is provided they effectively provide competition to it. An example is the Boda Boda, which are bicycle and motorcycle taxi services that are ubiquitous across Eastern Africa, although in Kenya they are more often called piki pikis. Within Kampala, Uganda, it is estimated that there are 300,000 Boda Bodas in operation, although there are no official statistics on the number of vehicles and operators.

An interesting development over recent years has been the increasing digitisation of such services, correlating with the increasing adoption of smartphones across the Global South. Even new entrants into such markets from the Global North, such as Bolt and Uber, have attempted to adopt informal transport services alongside what is considered their more traditional ride-sharing offer. Both operators though have faced significant criticism over poor service delivery in the markets in which they have entered.

This should be seen in the context of the ability to adopt such technologies that are possible within the context of many cities in the Global South. In a study of informal workers in the transport sector in Ahmedabad (India), Durban (South Africa), and Lima (Peru) [67] revealed that a diverse range of tools are adopted by informal workers in the transport and other sectors. But what is interesting is how they are adopted by workers for specific tasks. For example, in terms of intensity of use of technology in work, on a scale of 1–10 (1 being not used and 10 being used extensively), mobile phones ranked 5.5 (Ahmedabad), 8.5 (Durban), and 7.8 (Lima). And, in terms of intensity of use in organising, mobile phones ranked 6.2 (Ahmedabad), 3.9 (Durban), and 6.7 (Lima). It should be noted that some technologies like mobile phones and radios are almost ubiquitous, whilst most others are not.

Significant effort has also been placed into mapping such services. What such activities have revealed is quite how extensive such services are in practice. For instance in Mexico City and Nairobi, whilst there are a few formal public transport routes, informal transport services cover a more significant proportion of the city. In the former, 74% of all public transport trips are by 'colectivo's' (a range of vehicles from vans to pick-up trucks) whilst for the latter 70% of all commuters that use public transport are by 'matatus' (privately owned minibuses). Another notable example is Whereismytransport, which collects location data and route information from both formal and informal transport sectors to provide route planning information for passengers.

A notable, and unexpected, change in the experience of using such services brought about by the use of technology has been the differences in experiences of the drivers compared to passengers. Whereas in much of the Global North, where the adoption of such technologies has led and improved service offerings for passengers and lower incomes for drivers, there is some evidence in the informal transport sector that this is not the case.

For example, in Jakarta, Indonesia, ojek's are informal transport services provided by motorcycles. The introduction of smartphone technology and the ability to book rides via apps has resulted in increased incomes for drivers, driven in part by longer distance trips and increasing coverage areas. Passengers using the apps, meanwhile, report higher instances of feeling safe and a higher level of satisfaction compared to those who use ojek's traditionally [68]. Within this context, Go-Jek emerged as a smartphone-based service and began to offer an on-demand service for motorcycle taxi rides in Jakarta, targeting new users such as professionals and expatriates in the Central Business District who were frustrated by the terrible traffic and yet were unaccustomed to the motorcycle transport of the masses. The company was valued at $4 billion in 2018 [69].

The establishment of technologies within the sector varies from much of the Global North in one key aspect. In the Global North, the rollout of intelligent transport systems technology has been based on technology investment in the physical infrastructure first before the adoption of people and vehicle-based technology later. Meanwhile, in the informal transport sector largely in the Global South, this process has proceeded straight to the adoption of people and vehicle-based technology [70]. This can achieve the same outcomes in terms of the experience of the user, and monetisation by the operator for a much cheaper price compared to the investment in physical infrastructure to transport data purely for the purposes of transport infrastructure operators.

In comparison to the Global North, the adoption of MaaS within contexts where there is a significant proportion of services provided by informal transport is relatively under-studied. In this regard, this depends on the definition of MaaS that is used, explored earlier in this book. Regardless of how MaaS is defined, there are some signs that at least elements of MaaS are being adopted in the informal transport sector.

A key assumption underpinning many of the definitions of MaaS is a strong role for the public sector, whether it be as a coordinator, procurer of services, or any other role. Within the context of the informal transport sector, this will be more challenging due to the nature of the sector typically being outside the public sector's direct control, and sometimes interest. In attempting to formalise such sectors, the public sector can have the unintended consequence of destabilising the sector, where traditionally such transitions take an extended period of time [71].

One of the barriers to the adoption of MaaS-style solutions within a context of high informal transport use is the fact that the informality of the transport operations is one of its strengths as a service offering. When ride-sharing applications have moved into these markets seeking to adopt a ride-sharing approach used in the Global North, they have often met with resistance from local operators. The policy responses though, for example, the banning of the use of such apps, have also had the unintended consequences of stifling the development of apps by local firms [72].

One of the requirements for common visions of MaaS is the ability to provide an integrated service offering. Within the context of many places where there is a high use of informal transport services, this is a notable challenge, with many city development plans containing insufficient measures to integrate such services within a wider mobility service offering [73]. Furthermore, current technologies that have been developed lack consideration of other forms of public transport (with the exception of mapping).

Some work has been undertaken to conceptualise an integrated transport system in a high informal transport use context. For example, in Kampala, Uganda, a principles-based framework for the integration of informal transport services is proposed called iDRT [74]:

- Seamless multi-modal integration. At the core of the iDRT is the need to integrate the three main modes of public transport in Kampala (minibus taxis, boda-bodas, and buses) so that they can seamlessly work together and provide an efficient service to the commuters. A fundamental reorganisation of vehicles is required. Buses would be scheduled to operate on trunk routes at high frequency during peak hours and moderate frequency during other times.

- Demand Responsiveness. A Demand Responsive Transit service provides transport on-demand from passengers using fleets of vehicles scheduled to pick up and drop off passengers per their needs. The DRT on-demand service consisting of minibus taxis and motorcycle taxis operating on sub-arterial and access roads could work in such a way that motorcycle taxis feed into the minibus taxis which in turn feed into the high-frequency buses on the trunk route.
- ICT as an enabler. In the conceptual ICT-enabled iDRT, a robust central ICT system could be used to service passenger requests (trip booking, ticketing, and payment), driver requests (vehicle scheduling and routing, salary and commissions), operators needs (fleet monitoring and management), and the needs of the city authorities (monitoring the performance of operators in different zones, general demand and supply). Cognisant of the fact that the urban poor may not have access to smartphones and the Internet, the platforms could be digitally sufficient to provide terminals for trip booking and several payment methods.

A potential area for the development of MaaS-style services is through existing informal partnerships and networks. For instance currently, in India coordination between individual intermediate public transport owners (i.e., taxis, auto-rickshaws) is through their corresponding local and regional unions. Furthermore, the existing systems are very simple, informal, and use low-technology systems [75].

There has been some work to transition many operators into the realms of smart ticketing and some MaaS-lite style services, primarily on the promise for local operators that they will get a better understanding of the performance of their fleets. In 30 Indian cities, Chalo is a mobile app that tracks the real-time position of buses and other vehicles on behalf of operators and offers real-time positional and departure information for passengers. It also offers integration with ticketing systems such as smart cards.

12.8 Summary of MaaS and shared transport

Shared transport services constitute a wide variety of different types of transport services. Among the most popular, especially in developed nations, are car-sharing services and bike-sharing services. The variety of models that have been deployed in both markets have extensive experience in operation. Furthermore, there is evidence to indicate that both solutions have been able to achieve wider policy aims, such as reducing the use and ownership of private cars. Much like other transport services, they have been enabled by a variety of technologies that have become more integrated as the solutions have developed and matured.

A relatively new player is that of micromobility. Enabled by significant investment by the private sector, this has undergone significant growth and has presented a further challenge in terms of integrating an additional mode – one primarily operated by private companies – into a wider mobility offering, and MaaS platforms. Their emergence has also spurred work in an area critical to enabling MaaS – that of common data standards.

The impact of ridesharing platforms such as Uber has received much discussion. But Uber is simply one means of providing a ride in a vehicle that can be hailed. The emergence of such platforms has acted as a spur for many traditional companies like taxi operators to invest in their technological capability and provide an enhanced service to customers.

Finally, informal transport is a matter not often discussed by professionals, but is critical to many people in the world, especially in developing countries, in how they get around where they live. Whilst the formal adoption of MaaS style systems may not be widespread, this sector has been subject to a significant amount of technological innovation with the view to improving the experience of the customer.

References

[1] Spaced Out parking report [Internet]. Racfoundation.org. 2012. Available from: https://www.racfoundation.org/research/mobility/spaced-out-perspectives-on-parking

[2] ITF. *Micromobility, Equity and Sustainability: Summary and Conclusions*, ITF. Roundtable Reports, No. 185, OECD Publishing, Paris, 2021.

[3] Car Club Annual Report United Kingdom 2021 [Internet]. Available from: https://uploads-ssl.webflow.com/6102564995f71c83fba14d54/632885c07c790d2577d1445f_CoMoUK%20Car%20Club%20Annual%20Report%20UK%202021.pdf

[4] BlaBlaCar reaches 100 million members for its 15th anniversary [Internet]. BlaBlaCar. Available from: https://blog.blablacar.com/newsroom/news-list/blablacar-reaches-100-million-members-for-its-15th-anniversary

[5] Lynch, Faxi T. Interview for the Annual Survey of Mobility as a Service 2017. Landor Links. 2017. NCC case study – KINTO Join helps councils resolve parking issues [Internet]. KINTO Join. [cited 14 Aug 2023]. Available from: https://joinuk.wpengine.com/ncc/

[6] About KINTO Join [Internet]. KINTO Join. [cited 14 Aug 2023]. Available from: https://joinuk.wpengine.com/about/

[7] SNCF to sell Ouibus to BlaBlaCar in multimodal partnership [Internet]. *Railway Gazette International*. [cited 14 Aug 2023]. Available from: https://www.railwaygazette.com/business/sncf-to-sell-ouibus-to-blablacar-in-multimodal-partnership/47545.article

[8] Hamann TK, Güldenberg S, and Renzl B. Overshare and collapse: How sustainable are profit-oriented company-to-peer bike-sharing systems? *Die Unternehmung* [Internet]. 2019 [cited 14 Aug 2023];73(4):345–73. Available from: https://www.jstor.org/stable/26995450

[9] Dowling S. Here's how much bike-sharing startup ofo raised before it crashed [Internet]. *Crunchbase News*. 2019. Available from: https://news.crunchbase.com/venture/heres-how-much-bike-sharing-startup-ofo-raised-before-it-crashed/

[10] Yang Y. Chinese bike-sharing start-up Ofo on verge of bankruptcy. *Financial Times* [Internet]. 19 Dec 2018 [cited 16 Aug 2023]. Available from: https://on.ft.com/3qqU5dNEverett H. Mobike pulls out of Newcastle & Gateshead [Internet]. Cycling Industry News. 2019 [cited 14 Aug 2023]. Available from: https://cyclingindustry.news/mobike-pulls-out-of-newcastle-and-gateshead/

[11] CoMoUK Annual Bike Share Report UK [Internet]. [cited 14 Aug 2023]. Available from: https://uploads-ssl.webflow.com/6102564995f71c83fba14d54/644679f0f1a12cebdfca2390_CoMoUK%20Annual%20bike%20share%20report%20UK%202022.pdf

[12] Berlin und Nextbike unterzeichnen Vertrag für neues öffentliches Fahrradverleihsystem [Internet]. orion.senstadt.verwalt-berlin.de. [cited 14 Aug 2023]. Available from: https://www.stadtentwicklung.berlin.de/aktuell/pressebox/archiv_volltext.shtml?arch_1607/nachricht6107.html

[13] Oxford bike share scheme shut down "with immediate effect" [Internet]. *Oxford Mail*. 2018 [cited 14 Aug 2023]. Available from: https://www.oxfordmail.co.uk/news/16438945.oxonbikes-shut-with-immediate-effect/

[14] Lucas L, Bradshaw T, Yang Y, Liu N, and Spero J. Ofo's demise highlights risks of Chinese tech model. *Financial Times* [Internet]. 27 Jan 2019 [cited 16 Aug 2023]. Available from: https://on.ft.com/2Uukd3T

[15] Document > Bike Share Annual Report GB 2021 [Internet]. www.como.org.uk. Available from: https://www.como.org.uk/documents/bike-share-annual-report-gb-2021

[16] Bike Share Users Survey 2019 [Internet]. CoMoUk; 2019 [cited 16 Aug 2023]. Available from: https://uploads-ssl.webflow.com/6102564995f71c83fba14d54/6138af3ba54d1741ee01f033_CoMoUK%20Bike%20Share%20Survey%202019.pdf

[17] Shared Electric Bike Programme Report 2016 Findings and Recommendations from Eleven Shared Electric Bike Projects [Internet]. [cited 14 Aug 2023]. Available from: https://uploads-ssl.webflow.com/6102564995f71c83fba14d54/62dabeef4b3f6d4db2ce7042_CoMoUK%20Shared%20Electric%20Bike%20Programme%20Report%202016.pdf

[18] Shared Electric Bike Programme Report 2016 Findings and Recommendations from Eleven Shared Electric Bike Projects [Internet]. [cited 14 Aug 2023]. Available from: https://uploads-ssl.webflow.com/6102564995f71c83fba14d54/62dabeef4b3f6d4db2ce7042_CoMoUK%20Shared%20Electric%20Bike%20Programme%20Report%202016.pdf

[19] https://www.rundschau-online.de. TikTok-Challenge: Die Zerstörung der KVB-Räder in Köln geht weiter [Internet]. *Rundschau Online*. 2022 [cited 14 Aug 2023]. Available from: https://www.rundschau-online.de/koeln/tik-tok-challenge-die-zerstoerung-der-kvb-raeder-in-koeln-geht-weiter-302764?cb=1670505551289

[20] Sigo E-Lastenrad Sharing [Internet]. sigo | E-Lastenrad Sharing. [cited 14 Aug 2023]. Available from: https://sigo.green/

[21] Slim naar Antwerpen [Internet]. www.slimnaarantwerpen.be. [cited 14 Aug 2023]. Available from: https://www.slimnaarantwerpen.be/en/succes-stories/together-we-can-support-the-progress-of-shared-bikes-in-flanders

[22] Kubitz, B *et al. Mobility as a Service: The Second Annual Survey of MaaS in the UK*, 2018, Landor LINKS, ISBN: 978-1-899650-85-9

[23] MobilityData takes over micromobility open-data standard [Internet]. Smart Cities Dive. [cited 14 Aug 2023]. Available from: https://www.smartcitiesdive.com/news/mobilitydata–micromobility-open-data-standard-GBFS-bike-share-scootershare/633473/

[24] General Bikeshare Feed Specification [Internet]. GitHub. 2023. Available from: https://github.com/MobilityData/gbfs

[25] The evolution of GBFS [Internet]. Shared Micromobility. 2020 [cited 14 Aug 2023]. Available from: https://shared-micromobility.com/the-evolution-of-gbfs/

[26] Citymapper adds dockless bikes and scooters to its urban transport app [Internet]. VentureBeat. 2018 [cited 14 Aug 2023]. Available from: https://venturebeat.com/transportation/citymapper-adds-dockless-bikes-and-scooters-to-its-urban-transport-app/

[27] Real-time bikeshare information in Google Maps rolls out to 24 cities [Internet]. Google. 2019. Available from: https://blog.google/products/maps/real-time-bikeshare-information-google-map/

[28] Travel your first and last mile with Google Maps [Internet]. Google. 2019 [cited 14 Aug 2023]. Available from: https://blog.google/products/maps/travel-your-first-and-last-mile-google-maps/

[29] News EIN, Amin V. Splyt integrates nextbike, Europe's Leading Bikeshare Provider, into Its Micromobility Network [Internet]. EIN News. 2022 [cited 14 Aug 2023]. Available from: https://www.einnews.com/pr_news/580878604/splyt-integrates-nextbike-europe-s-leading-bikeshare-provider-into-its-micromobility-network

[30] Cargo bike sharing: A fun way for families to move – movmi [Internet]. movmi.net. 2022. Available from: https://movmi.net/blog/cargo-bike-sharing-a-fun-way-for-families-to-move/

[31] Clercq GD. Paris considers electric scooter ban over safety concerns. Reuters [Internet]. 15 Nov 2022 [cited 14 Aug 2023]. Available from: https://www.reuters.com/world/europe/paris-considers-electric-scooter-ban-over-safety-concerns-2022-11-15/

[32] Paris residents vote overwhelmingly to ban rental e-scooters. Le Mondefr [Internet]. 2 Apr 2023 [cited 14 Aug 2023]. Available from: https://www.lemonde.fr/en/transport/article/2023/04/02/paris-residents-vote-overwhelmingly-to-ban-rental-e-scooters_6021483_216.html

[33] Mobility Data Specification [Internet]. City of Los Angeles GitHub. 2023 [cited 14 Aug 2023]. Available from: https://github.com/CityOfLosAngeles/mobility-data-specification

[34] ITF. *Micromobility, Equity and Sustainability: Summary and Conclusions*, ITF. Roundtable Reports, No. 185, OECD Publishing, Paris, 2021

[35] ITF. *Micromobility, Equity and Sustainability: Summary and Conclusions* ITF. Roundtable Reports, No. 185, OECD Publishing, Paris, 2021.

[36] Amadeo R. Google Maps removes Uber integration [Internet]. Ars Technica. 2018 [cited 14 Aug 2023]. Available from: https://arstechnica.com/gadgets/2018/06/google-maps-removes-uber-integration/

[37] Le registre des taxis: comment une API permet d'adapter la mission de service public des taxis au numérique [Internet]. blog.beta.gouv.fr. 2021 [cited 14 Aug 2023]. Available from: https://blog.beta.gouv.fr/general/2021/05/05/letaxi-service-public/

[38] Helft M. The Godfather of Digital Maps [Internet]. *Forbes*. [cited a4 Aug 2023]. Available from: https://www.forbes.com/sites/miguelhelft/2016/02/10/the-godfather-of-digital-maps/

[39] Location Benchmarking Report: 2022 Report Snapshot [Internet]. 2022 [cited 14 Aug 2023]. Available from: https://www.here.com/sites/g/files/odxslz256/files/2022-06/final_2022_strategy_analytics_location_benchmarking_report_2022-2.pdf#page6

[40] Celebrate 15 years of exploring your world on Street View [Internet]. Google Maps Street View. Available from: https://www.google.com/streetview/anniversary/

[41] Google Maps crowdsources info for wheelchair-accessible places [Internet]. Engadget. [cited 14 Aug 2023]. Available from: https://www.engadget.com/2017-07-06-google-maps-crowdsource-wheelchair-accessible.html

[42] Peters J. Apple Maps turns 10 – and it's finally worth using [Internet]. *The Verge*. 2022 [cited 14 Aug 2023]. Available from: https://www.theverge.com/23323550/apple-maps-10-year-anniversary-iphone-google

[43] Popa B. Say goodbye to Google Maps as Garmin launches new GPS units with large displays [Internet]. autoevolution. 2021 [cited 14 Aug 2023]. Available from: https://www.autoevolution.com/news/say-goodbye-to-google-maps-as-garmin-launches-new-gps-units-with-large-displays-171656.html

[44] Design Custom Maps with Mapbox Studio [Internet]. www.mapbox.com. [cited 14 Aug 2023]. Available from: https://www.mapbox.com/mapbox-studio/

[45] TomTom Maps Platform [Internet]. TomTom. [cited 14 Aug 2023]. Available from: https://www.tomtom.com/tomtom-maps-platform/

[46] Add OpenStreetMap data to your Mapbox project | Help [Internet]. Mapbox. [cited 14 Aug 2023]. Available from: https://docs.mapbox.com/help/tutorials/overpass-turbo/

[47] Major OpenStreetMap Consumers – OpenStreetMap Wiki [Internet]. wiki.openstreetmap.org. Available from: https://wiki.openstreetmap.org/wiki/Major_OpenStreetMap_Consumers

[48] Improving the komoot map using OpenStreetMap [Internet]. Komoot.com. 2023. Available from: https://support.komoot.com/hc/en-us/articles/360022830972-Improving-the-komoot-map-using-OpenStreetMap

[49] ABRP [Internet]. abetterrouteplanner.com. [cited 14 Aug 2023]. Available from: https://abetterrouteplanner.com/

[50] INRIX. INRIX AI Traffic [Internet]. Inrix. 2019. Available from: https://inrix.com/products/ai-traffic/

[51] Levy-Weiss G. The insider story of Waze [Internet]. NFX. 1 Jun 2021. Available from: https://www.nfx.com/post/the-insider-story-of-waze

[52] How does Waze work? – Waze help [Internet]. Google.com. 2020. Available from: https://support.google.com/waze/answer/6078702?hl=en

[53] What is Waze and how does it work? [Internet]. Lifewire. 2019. Available from: https://www.lifewire.com/what-is-waze-4153570

[54] Google bought Waze for $1.1B, giving a social data boost to its mapping business [Internet]. *TechCrunch*. Available from: https://techcrunch.com/2013/06/11/its-official-google-buys-waze-giving-a-social-data-boost-to-its-location-and-mapping-business/

[55] Novet J. Uber paid Google $58 million over three years for map services [Internet]. CNBC. 2019. Available from: https://www.cnbc.com/2019/04/11/uber-paid-google-58-million-over-three-years-for-map-services.html

[56] Artist manipulates Google Maps Traffic View with cart full of smartphones [Internet]. *PCMag*. [cited 14 Aug 2023]. Available from: https://www.pcmag.com/news/artist-manipulates-google-maps-traffic-view-with-cart-full-of-smartphones

[57] Gett Cities [Internet]. Gett UK. [cited 14 Aug 2023]. Available from: https://gett.com/cities/

[58] O'Kane S. Citymapper has been granted a license to operate taxis in London [Internet]. *The Verge*. 2018 [cited 114 Aug 2023]. Available from: https://www.theverge.com/2018/2/20/17033412/citymapper-bus-route-uber-taxi-london; Vincent J. Transport app Citymapper gives up on reinventing the bus [Internet]. *The Verge*. 2019 [cited 14 Aug 2023]. Available from: https://www.theverge.com/2019/6/12/18662577/citymapper-london-transport-app-shut-down-smartride-ride-to-focus-on-pass

[59] Moovit and FREE NOW partner to offer integrated mobility services in Italy [Internet]. Intelligent Transport [cited 14 Aug 2023]. Available from: https://www.intelligenttransport.com/transport-news/84585/moovit-and-free-now-partner-to-offer-integrated-mobility-services-in-italy/

[60] Silicon Canals. London-based Karhoo collaborates with Spanish rail giant Renfe for first and last-mile travel | Silicon Canals [Internet]. 2019 [cited 14 Aug 2023]. Available from: https://siliconcanals.com/news/london-based-karhoo-collaborates-with-spanish-rail-giant-renfe-for-first-and-last-mile-travel/

[61] Skeldon P. iTaxi launches direct carrier billing with Centili and Play | Billing & Payments [Internet]. Telemedia Online. 2020 [cited 14 Aug 2023]. Available from: https://www.telemediaonline.co.uk/itaxi-launches-direct-carrier-billing-with-centili-and-play/

[62] Rabe T. Opal card to expand to Uber, share bikes in world-first trial [Internet]. *The Sydney Morning Herald*. 2021. Available from: https://www.smh.com.au/national/nsw/opal-card-to-expand-to-uber-share-bikes-in-world-first-trial-20210329-p57eyd.html

[63] Middleton N. Addison Lee services now available through Sixt mobility app [Internet]. International Fleet World. 2019 [cited 14 Aug 2023]. Available from: https://internationalfleetworld.com/addison-lee-services-now-available-through-sixt-mobility-app/SIXT launches world's first comprehensive MaaS app [Internet]. Fleet Europe. 2019 [cited 14 Aug 2023]. Available from: https://www.fleeteurope.com/en/maas-fleet-strategy-smart-mobility/global-europe/news/sixt-launches-worlds-first-comprehensive-maas

[64] Uber brings back shared rides as UberX Share [Internet]. www.masstransitmag.com. [cited 14 Aug 2023]. Available from: https://www.masstransitmag.com/alt-mobility/shared-mobility/article/21272370/uber-brings-back-shared-rides-as-uberx-share.

[65] Cervero R, and Golub A. Informal transport: A global perspective. *Transport Policy*. 2007;14(6):445–57.

[66] Informal Transport in the Developing World [Internet]. Available from: https://unhabitat.org/sites/default/files/download-manager-files/Informal%20Transport%20in%20the%20Developing%20World.pdf

[67] Chen MA. Technology, informal workers and cities: Insights from Ahmedabad (India), Durban (South Africa) and Lima (Peru). *Environment and Urbanization*. 2016;28(2):405–22.

[68] Medeiros RM, Duarte F, Achmad F, and Jalali A. Merging ICT and informal transport in Jakarta's ojek system. *Transportation Planning and Technology*. 2018;41(3):336–52.

[69] Temasek investing in Indonesia's Go-Jek as ride-hailing rivalry deepens: Sources. Reuters [Internet]. 18 Jan 2018 [cited 14 Aug 2023]. Available from: https://www.reuters.com/article/us–fundraising/google-temasek-investing-in-indonesias–as-ride-hailing-rivalry-deepens-sources-idUSKBN1F70GS

[70] Booysen M, Andersen S, and Zeeman A. Informal public transport in Sub-Saharan Africa as a vessel for Novel Intelligent Transport Systems [Internet]. [cited 14 Aug 2023]. Available from: https://scholar.sun.ac.za/server/api/core/bitstreams/91d41567-f560-463e-8dfb-f316bf36478f/content

[71] Sunio V, Argamosa P, Caswang J, and Vinoya C. The state in the governance of sustainable mobility transitions in the informal transport sector. *Research in Transportation Business & Management*. 2020;100522.

[72] Sopranzetti C. Shifting informalities: Motorcycle taxis, ride-hailing apps, and urban mobility in Bangkok. *Geoforum*. May 2021.

[73] Thinus M, Booysen, Ndibatya I, and Booysen M. Transforming paratransit in Africa's congested cities: An ICT-enabled Integrated Demand Responsive Transport (iDRT) approach [Internet]. 2020 [cited 14 Aug 2023]. Available from: https://api-ucudir.ucu.ac.ug/api/core/bitstreams/4d199538-324d-4439-9f53-6298b2e81953/content

[74] Thinus M, Booysen, Ndibatya I, and Booysen M. Transforming paratransit in Africa's congested cities: An ICT-enabled Integrated Demand Responsive Transport (iDRT) approach [Internet]. 2020 [cited 14 Aug 2023]. Available

from: https://api-ucudir.ucu.ac.ug/api/core/bitstreams/4d199538-324d-4439-9f53-6298b2e81953/content

[75] Wright S, Emele CD, Fukumoto M, Velaga NR, and Nelson JD. The design, management and operation of flexible transport systems: Comparison of experience between UK, Japan and India. Research in Transportation Economics [Internet]. 2014 [cited 14 Aug 2023]. Available from: https://edisciplinas.usp.br/pluginfile.php/3411406/mod_folder/content/0/WRIGHT%20Bus%20on-demand%20Jap-UK-India%20ResTranspEcon%202014.pdf

Chapter 13

MaaS and demand responsive transport

13.1 Introduction

Dynamic demand responsive transport (known as DDRT or DRT) has begun to occupy a niche between ride-hailing and fixed-route buses. DRT is generally a part of public transport, usually using small buses to create a tailored response to transport provision in low-density, peri-urban, and rural areas.

The key features of DRT are that the bus services run on demand (rather than to timetable), and they are booked by app, online, or via a call centre. DRT services generally charge fares on a fixed basis – per person per set distance – because they are part of public transport, rather than varying fares (or fares split between passengers) as a taxi would.

The aim of DRT is to maximise the number of passengers per vehicle, and DRT platforms manage shared journeys that may have overlapping passenger pick-ups and drop offs. Services generally have virtual infrastructure (no physical bus stops or paper timetables) – although they still need real-world marketing and signage to help raise awareness. Passengers can have accounts that manage fare payment and concessionary fare validation (although some platforms enable on-bus cash or card payments).

Optimal journey patterns, routes, and timetables are calculated by algorithms to meet passenger demand. Aggregate service demand over time is managed via algorithms that predict vehicle and driver requirements to ensure efficient planning to meet demand patterns. This optimisation is important as it is aimed at maximising the number of people per trip to minimise the cost per passenger.

The sophisticated platforms of DRT were pre-dated by 'dial-a-ride' transport, often targeted at specific groups (social or community transport or people with mobility issues) and booked by telephone, and managed largely by analogue systems. In addition, other services like the German 'Sammelbus' or 'Rufbus' (a shared taxi type arrangement which ensured that people arriving by the last train would be able to get home) or the rural fixed route buses which could be pre-booked (by phone) to visit certain stops were ripe for automation.

The United Kingdom hosted some early trials of DRT platforms including the RATP service Slide Bristol (powered by the Padam Mobility platform), which was launched along a corridor in the city in 2016.

13.1.1 Case study: Slide Bristol

Slide Bristol was targeted as a commuter bus service; with journeys booked in advance through the app up to ten minutes before travel and users picked up close to their starting point, taking out the stress of driving and parking. At the time, its offer of free WiFi on board was revolutionary. Seats, pick-up, and drop-off points are guaranteed, as was a fixed fare depending on distance (which is between £4 and £7). The cost was designed to be cheaper than a taxi and a just little more than a bus.

The vehicles were permitted to use bus lanes with routes optimised to transport passengers travelling on a similar journey. Journey times were predicted and passengers were assured that they would not be delayed more than 15 min by variations to pick-up fellow passengers. Slide Bristol was featured on the TravelWest platform, which provided travel information and journey planning for people living in the Bristol area.

Early DRT had to prove it could achieve multiple paradigm shifts at once. From proving to passengers that digital booking (and sometimes payment) would work and buses would turn up on demand, to managing flexible routes and training drivers with new skills. In addition, (largely) new vehicles were commissioned to give a better experience with WiFi and nice seats. Passengers with conflicting requirements (e.g., needing a pram or wheelchair-accessible space) were accommodated through the booking process.

13.1.2 Case study: ArrivaClick Sittingbourne

As DRT developed, it became important that it could integrate with public transport systems. Another early example of DRT in the United Kingdom was designed to run over an area beset by roadworks.

The ArrivaClick service was a flexible minibus service operating over a 4 × 6-mile area, booked and paid for by app (using the Via platform) launched in 2017. The service area was designed to enable the buses to serve the Kent Science Park on the outskirts of Sittingbourne where demand had outstripped parking space on the site and to enable buses to follow their own routes. They could bypass a year-long road remodelling scheme that slowed down linear bus trips along the main corridor. The ArrivaClick buses were routed according to traffic conditions as well as passenger pick-up and set-down requirements. Although the app was extremely popular (over 5,000 downloads representing 10% of the local population) trips could also be booked via a call centre to enable universal access to booking.

DRT schemes had to prove that they would tap into groups that were previously not using buses and expand the overall market rather than cannibalising other services' existing ridership. Whilst this was clearly the case to some extent, an ArrivaClick survey found that 52% of customers switched from private motor transport, including taxis (22%), driving their own car (18%), or being a passenger in a car (12%) [1]. Some users migrated from other bus routes and the new customers do not seem to have done so in sufficient numbers to create pure 'for profit' services. Whilst the numbers would probably be attractive in other business models, the deregulated bus environment in the United Kingdom mitigated the scheme.

In areas where access to buses is the priority (with profitability as less of a driver), DRT has been used to ensure universal access to public transport. In France, municipalities are required by law to ensure that citizens have access to public transport [2]. This has given rise to a number of rural and peri-urban DRT services aimed at ensuring that the legal requirement is met.

The integration with public transport services in France (which are all run by the Department) means that the network is designed not to compete with itself. Specific rules are put in place to ensure that the DRT service serves fixed-line routes rather than displacing passengers from them. Rural areas in France have been among the first to trial new DRT services in order to meet their obligations.

13.1.3 Case study: Le Cotentin, France

The port of Cherbourg sits at the northwest tip of France within the commune (metropolitan area) of Cherbourg-En-Contentin. It is part of the department of Manche, within Normandy. Until recently, public transport has been limited to the town and its immediate surroundings. As the municipal area which the Cherbourg Transport Authority is responsible for has been expanded to include its rural surroundings, it must ensure that the population has equitable access to transport. Creating regular 'fixed line' buses was deemed too high cost so it decided to trial a flexible, on-demand bus service.

The area is rural, with agriculture, shipping, naval shipbuilding, and two nuclear power stations as its primary economic activities. The other settlements in the area are small towns and villages. The total population is 79,200 (2017) over an area of 68.5 km^2. Whilst the area-wide population density is in the order of 1,200 people per km^2, this is concentrated in the port area with low-density rural areas surrounding it.

The initial proof of concept DRT trial was planned to cover three 'communes' (villages), Digosville, Bretteville, and Le Mesnil-au-Val. The trial area covered a total of 4,000 habitants and 30 km^2.

A total of 18 stops were designated as pick-up/drop off points. Most of them were in places that were not served by public transport, but three (along the coast) were on lines where services are infrequent outside of peak hours. The on-demand service was specified to connect the on-demand stops to three connection points with the Cherbourg public transport network, Northeim, La Place, and Collège Diderot (which is also a high school). The design ensured that the majority of people would walk less than a kilometre to the stop and could connect to onward travel.

The trial service consisted of a single 8–9 seat minibus bookable by app running between 8 a.m. and 7.30 p.m. 6 days per week (Monday to Saturday). Trips cost 1 euro 30 (around £1). Travel times between the most distant points of the area and the interchange stops were 15 min. On average passengers waited up to 30 min (and at most 45 min) from booking the service. The 6-month trial ran between July and December 2019.

The single bus made 1,000 trips per month. It was very popular and highly rated by people using the service. The largest group using it were young people and college students for travel outside of school bus times – for instance to attend activities or to be able to travel home if their school day had free periods etc. It was greatly appreciated by parents who found that they did not have to operate as 'parent taxis' for their children, thus reducing car journeys. The fact that the trial was bookable by app appears to have increased take-up by young people.

The service was highly subsidised, with costs per trip calculated in the region of 20 euros. But it was recognised that it was the only way – and the most efficient way – to provide public transport to the population. The authority used this trial to plan services covering all 135 villages in the region served by the transport authority.

Whilst the transport authority expects the costs of DRT to fall as usage rises, it also takes the view that DRT is specifically for these areas where fixed routes are not economic. Where usage grows and repeat journeys occur to the point that economies of scale are possible, the authority plans to develop fixed route services.

Perhaps the largest DRT is the system around Paris which connects areas in the Île-de-France to the rapid transit network. Paris is increasingly well-known for its change to road layouts to reduce traffic and increase active travel through the provision of cycle lanes and pedestrian routes and areas. However, beyond the core administrative zone which is under the administration of Paris mayor, Anne Hidalgo, people living in the Île-de-France area feel disconnected, and are often angry that it is now less easy to drive into central Paris.

One of the interventions of the authorities has been to upgrade on-demand buses in the area, making them easier to book online and via an app, moving them onto a single platform, and integrating them. This creates a unified passenger experience (even though the small buses may be operated by different companies) and connects people better to rapid transport.

A similar example, which has upgraded people's access to transport whilst reducing inefficient services is in Orleans. The city transport authority has been piloting DRT since 2018, adopting a trial-and-adopt model, which has involved a gradual increase in size and scale in each iteration.

The initial pilot in 2018 converted a legacy 'dial-a-ride' style system to a modern DRT platform. The previous scheme had been complicated to book, with strict advance booking deadlines and little real-time information about the bus arrival times. The April 2018 pilot covered an area of northeast Orléans, which trialled on-demand rides that were booked either using an app or by calling a call centre. Real-time bus information enabled users to book transport at short notice, and uptake of the app increased, enabling call centres to give a better service to passengers booking by phone.

As the technology was proven, the areas served were expanded. First, in September 2019 it was expanded to nine areas which almost encircled the city, and then again in 2021, the areas were adjusted to cover a population of over 95,000 people and an area of 161 km^2. The service comprised 19 vehicles that offered connections within the areas and to rapid transit nodes, allowing people to travel into central Orleans from 6.00 a.m. to 9.00 p.m.

A further expansion brought the area covered up to 300 km^2 and the population to 175,000. The number of vehicles was doubled to 40.

Whilst the service area and population were not quite doubled, the number of people travelling per week almost tripled from 2,900 to 8,500. In addition, the providers claim that 33% of their passengers have switched from private cars, whilst 19% previously had no access to transport.

The city authorities, having committed to ensuring that people have access to public transport even in the outskirts of the city, were seeing the costs per person travelling falling satisfactorily. More people were travelling without requiring additional subsidies.

Similar results have been found across Europe, with DRT used in the areas surrounding Barcelona to connect outlying areas with transit services. For instance, in Sant Cugat, to the northwest of Barcelona, an on-demand route was established to help connect commuters in the Can Barata neighbourhood with the train station in 2017. The second service started in March 2019 in the outlying hilly area of Les Planes. And just a few months later, in December 2019, Shotl deployed a new route in Can Trabal that marked the third DRT service within the city boundaries. Demand for the Can Barata service saw a five-fold increase compared with the previous traditional service. In Les Planes, ridership tripled in just 10 months, whilst in Can Trabal it increased by 12% in just 6 weeks.

Ridership for these services was a healthy mix of students, young people, and domestic cleaners commuting to nearby cities for work. One interesting piece of feedback was that the DRT service meant that parents were able to take their children to school and arrive at work to schedule. The service has also meant that older and younger generations are able to travel independently.

13.2 The technology of DRT

DRT combines a user interface (an app, a call centre, a website) for booking the trip, a platform which, at its core, matches trip requests with available vehicles and manages communication with the passenger (e.g., showing the position of vehicles, arrival times, delays) and with drivers (to route them). Passenger apps can include payment or not, depending on the preferences of the operator. Drivers also have a navigation system to route them to pick-ups and drop offs, they also receive notifications to ensure that the correct people board and alight, and where fares are payable on board, the correct fare for the trip.

The passenger and driver apps are based on smartphone functionalities in use in journey planners (and MaaS apps). The platform that manages the trips uses algorithms to plot the trips it can offer to passengers and to manage multiple passengers' trips and specific needs (e.g., wheelchair or pushchair space). Sophisticated algorithms can be configured to offer more personalised services depending on the passenger's needs. For instance, it may offer door-to-door for passengers with mobility needs and stop-to-stop for those without whilst ensuring that the pick-up times and journeys offered will work in the real world. It combines real-time data from vehicles and other sources with mapping, journey rules, transport, traffic, and congestion data.

DRT becomes more complex depending on the rules the operator or public transport authority creates. For instance, some technology providers can design 'non-compete' rules so that if there is a fixed-line service, passengers will not be offered a DRT bus and may be directed onto the fixed route instead. These rules can be modified further – so for instance, if the overall journey includes excessive wait times or would take much longer using fixed transport, the governing algorithm will still offer a DRT bus. To ensure these rules are observed, the DRT platform also needs to ingest transport data from timetables or real-time if they exist. DRT providers usually work with local data standards (e.g., GTFS, NeTeX) and real-time data extensions to check for existing transport before offering passengers journeys that they can book.

The degree of complexity is down to the objectives the operator is trying to achieve in terms of passenger experience and operational efficiency. For instance, tweaking the algorithm to increase the acceptable deviation from a route to pick up additional passengers can help increase the loading of the bus, however, ensuring that this does not increase the journey times too much to be acceptable to passengers is also important.

13.2.1 Case study: connecting with rapid transport – Pays de la Loire

Pays de la Loire is one of the 18 regions of France, in the west of the mainland. It comprises five departments: Loire-Atlantique, Maine-et-Loire, Mayenne, Sarthe, and Vendée. The total population in the region of 3.5 million people in an area of 32,082 km^2. Whilst the population density of the capital, Nantes, is over 1,200 people per km^2, the metropolitan area is much less dense and the departments comprise large rural areas scattered with small towns and villages.

The public transport operator for the Pays de la Loire, Aléop, runs a network of trains, trams, fixed bus routes, and boats. These have been augmented by a limited transport on-demand service in areas with low demand where it is not economical to run a fixed-line bus service. They are generally designed to help ensure people can get to key services and the users are generally older. Journeys are booked by telephone.

These services are gradually being switched onto a DRT platform run by Padam Mobility which enables the bookings to be managed within the overall network context. This means that the platform ensures connections with fixed-line services can be made (ensuring that passengers will make their mainline train service or get home in the evening).

The operator, Aléop, uses the GTFS feed (Google's transport data standard) as its data standard and data from Aléop fixed services is parsed by the platform to enable it to make bookings that ensure these connections with other services. This function also ensures that the DRT service does not replace trips that could be made on fixed lines (it will not offer the option for an on-demand bus if there is a fixed-line service running that would meet the caller's needs).

The platform selects appropriate vehicles for trips, combines routes and riders, and optimises them. It has integrated electric vehicles into the fleet, enabling the first steps to decarbonise rural transport. The aim for the authority is that journeys are optimised so that they meet both the passenger's needs and enable the optimal utilisation of the network.

Pre-booking DRT is one element that can be positive for both passengers – providing them with certainty that they will get to work or their appointments on time – and ensuring that the DRT operator has some advanced idea of services required on the day. Where there is a good level of pre-booking, the DRT platform can communicate with passengers to help increase its grouping rates, encouraging them to travel in similar time slots along similar routes. Passenger parameters such as 'arrive by' time bookings enable their needs to be met whilst the DRT platform calculates combining with other trips.

13.3 DRT and MaaS

One of the key characteristics of DRT is that it is generally configured to fill in transport gaps beyond city centres, ensuring that the population beyond the urban core is connected to rapid transport networks and can make local journeys.

The integration of DRT into MaaS is critical to ensuring that MaaS works in these areas, enabling people living in or travelling to suburbs, peri-urban, and rural areas to use public transport.

A Finnish bus consortium, Matkahuolto, commissioned a multi-service journey planner and booking app. This integrates the consortium's services – which are primarily intercity bus services – with local transport. The journey planner is based on the Kyyti MaaS Platform, incorporating its standard elements for ticketing (for the Matkahuolto consortium routes), payments and GTFS feed configuration. Over time the app has added GTFS feeds for all regional and local public transport routes in Finland, ticket integrations for several cities, additional long-distance routes, and a demand-responsive transport service in one city, Porvoo [3].

MaaS developers are increasingly integrating DRT services with fixed line and shared transport. Integrating MaaS and DRT is an interesting process as MaaS apps use primarily fixed data from public transport timetables augmented with real-time updates to suggest journeys. When there is no on-demand transport, the journey planner will match the theoretical services with real-time updates and return journeys based on that information.

DRT data is much more dynamic with many more potential stops and many more potential routes between them, creating huge sets of potential combinations and variations of routes. This means that once DRT is included in MaaS there are many more theoretical journeys available – either for the whole journey or as a component of the journey. This latter is particularly important for multi-modal connections between transport-rich and less densely served areas (as in the peri-urban examples above).

The precise order in which the MaaS app handles the combination of fixed line and flexible data can be different, depending on the way in which it is set up. For instance, it could be configured to maximise the use of fixed line or DRT services or to minimise interchanges, journey times, or costs.

The order in which it works through the options available will affect the potential transport services presented to the traveller and will depend on the rules within the MaaS application and of the DRT service. This is dependent on both practical and commercial concerns. For instance, some areas regard the DRT services as services of last resort and algorithms are written to prioritise travel on fixed routes, whilst others are aimed at keeping passenger trips per vehicle per hour maximised on the on-demand services.

In general, the MaaS algorithm checks all the parameters around the journey (e.g., is it within the rules for the journey to be made by DRT, or partially by DRT), where the criteria for 'made by DRT' are met, it creates a set of theoretically possible journeys which it then checks against the DRT service's server for availability. Only once the possible combinations have been worked out are trip suggestions returned to the customer in the journey planner.

Malta is an island with a resident population of half a million people which increases by 5 million tourists annually. This creates variable flows of passengers with different needs – some of which have been met by creating a variety of DRT services. These services have become popular with both inhabitants and tourists.

The island's MaaS application, built by Spanish MaaS provider Meep has upgraded the public transport app from fixed route public transport only to include DRT, ferries, and bikes in the journey planner. In this application, any trips that fall within the DRT area will check the availability of DRT services first and foremost. The DRT services are prioritised by the trip planner whether there are fixed-line services or not. Whilst there is an element of convenience for passengers, there is also a willingness by operators to sacrifice fixed routes to on-demand services because the DRT fare box is higher [4].

In other cases, the parameters can be set so that only the fixed-line services are served when a journey is requested. These parameters are often set by the public transport operator wishing to ensure that DRT does not cannibalise existing services.

At present, Google Maps does not serve DRT options, limiting its potential to become a MaaS solution. Without flexible options that ensure that people are not stranded at the margins of cities or the end (or beginning) of the day, public transport cannot claim to have a full solution to people's mobility needs.

13.3.1 Case study: on-demand fills the gaps in Brussels

Brussels Mobility licenses taxi services in the central Brussels BCR region. There are strict rules governing taxi licensing, and it is used as a complement to public transport. Licensing is conditional on operators owning the vehicles, paying taxes related to the operation of the service as well as salaries and social security contributions, and being able to cover maintenance costs for their fleet. All licensed vehicles are required to be available to the public, with a driver, for the time equivalent of a full-time job per year. Both operators and drivers are subject to conditions with respect to police records. Drivers must obtain professional certification and attend training every 4 years.

Collective taxis are licensed to operate between the hours of 23:00 and 06:00 to substitute for public transport services when these are out of operation at night. Passengers book a ride at least 30 min before departure via a switchboard and the vehicle is permitted to carry up to three passengers. Collective taxis must pick up clients in designated areas though they may be dropped off at any address given at the time of booking. BCR sets a fixed price for collective taxi passengers depending on the pick-up and drop off location. If the amount shown on the taximeter is higher than the total contributions paid by passengers at the end of the ride, the Region pays the difference to the operator.

Collective taxi vehicles must provide real-time data to their switchboard including geographic coordinates, vehicle ID, time of embarkation of the first customer and time of disembarkation of the last customer, distance travelled, price on the taximeter, and the number of passengers. A monthly statement of each operators' rides is shared with Brussels Mobility to facilitate reimbursement of the difference between the cost of rides and the price paid by passengers [5].

Transit is one of the original journey-planning apps (launched in 2012) which has morphed into a MaaS platform. Whilst it offers public transport information for many corners of the globe, it also works with transport authorities to provide specialised MaaS for their areas. This is often in the form of real-time journey planning and integrated ticketing for fixed-line services. However, as of 2023, 15 US transit agencies use the GTFSflex specification to integrate their on-demand services with other transport in the Transit MaaS app. The pilot projects are in rural counties in Minnesota where fewer transport options are available and on-demand buses provide essential services. Integrating on-demand services required alterations to the Transit app's architecture to include services that do not have a fixed route component. The resulting service means that riders can discover demand-response services, plan multi-modal trips, quickly call to book a ride, and pay their fares. The development means that Transit has become a very active contributor to the GTFSflex specification process [6].

DRT generally covers the areas that are hardest and most expensive to serve – first and last-mile trips in low-density and rural areas. Whilst the provision of these services is necessary for MaaS (and indeed the public transport network) to function, operators and authorities are under constant pressure to review and reduce costs. One of the proposed solutions is the use of autonomous vehicles in these situations, and several pilot programmes have been launched to assess the potential of autonomous vehicles within last-mile situations. These proof of concept trials are effectively testing several elements at once, including the autonomous driving technology, the journey booking and management, how people view and behave with respect to self-driving vehicles as well as the potential for cost reductions. Recent trials in the EU are discussed in Section 5.9.

13.4 Summary of MaaS and demand responsive transport

Much like other areas of transport, demand-responsive transport has been much in the way of experimentation with new means of operating, enabled by significant developments in technology. Notably in terms of real-time positioning, booking systems, and internet connectivity. Whilst integration with MaaS is seen as critical there are some unique challenges with integrating DRT services into MaaS offerings.

The most notable challenge is the nature of their operation. DRT services often fill gaps between a fully flexible service (e.g., taxis) and standard fixed-route public transport services. When MaaS operators require access to consistent data, such as that public transport timetable, this makes integration with the technical and commercial aspects of MaaS operators tricky. Regardless, there is evidence of successful integration with MaaS offerings.

References

[1] Jackson P. ArrivaClick ends Sittingbourne Pilot [Internet]. CBW. 2019. Available from: https://cbwmagazine.com/arrivaclick-ends-sittingbourne-pilot/

[2] La Loi LOM: la Loi d'Orientation des Mobilités 2019, Gouv.fr. 2021. Available from: https://www.legifrance.gouv.fr/dossierlegislatif/JORFDOLE000037646678/

[3] Nationwide MaaS Application for Matkahuolto [Internet]. Kyyti. 2021 [cited 14 Aug 2023]. Available from: https://www.kyyti.com/nationwide-maas-application-for-matkahuolto/

[4] Campomar, Meep G. 2023.

[5] Corporate Partnership Board CPB Developing Innovative Mobility Solutions in the Brussels-Capital Region Case-Specific Policy Analysis [Internet]. Available from: https://www.itf-oecd.org/sites/default/files/docs/innovative-mobility-brussels-captial-region.pdf

[6] Bringing Rural Riders into the Fold [Internet]. Transit. 2023 [cited 20Aug 2023]. Available from: https://blog.transitapp.com/case-study/mndot-gtfs-flex-bringing-rural-riders-into-the-fold/

Chapter 14
Active travel

14.1 Introduction

Journeys inevitably begin and end with a walk – or, for wheelchair users, a wheel. The distance that people are able to walk, wheel, or cycle to attain their destination or a mode of transport that will get them to their destination is not just bounded by their physical capability, but also by their physical safety.

The increase in motor traffic over the preceding century has meant that roads have become less hospitable to those on foot, cycling, or using wheelchairs. The increase has been both in volume and speed of traffic. Whilst pavements (or sidewalks in the United States) have a long history, they have become increasingly important for the protection of pedestrians. Whilst popular in cities, roads crossing rural areas and linking towns are much less likely to have infrastructure designed for active travel alongside them.

14.2 Delivery of active travel infrastructure

There have been periodic attempts at building active travel infrastructure in several countries. UK journalist Carlton Reid found 80 wide, protected, cycleways along the sides of major roads built in the United Kingdom between 1934 and 1940 [1]. In the United States, the bikeway construction boomed in the 1970s [2]. However, in general, road building for motor traffic has far outstripped these (fairly isolated) periods of cycle enthusiasm – until more recently. Countries across Europe have varying approaches, with embedded cycle lanes having a longstanding impact on cycling rates in the Netherlands and northern Belgium. Germany has a national cycleway building programme underway connecting areas with long-distance routes. German cities are also developing their own schemes. The French government announced a €2 billion programme in 2023 to be spent over 4 years on doubling the cycle network to 100,000 km plus schemes encouraging cycling.

Consumer research by bike component manufacturer Shimano, in its 2021 'State of the Nation' report examining attitudes towards e-bike usage in 12 European countries, indicates that consumer attitudes to e-bikes and e-cargo bikes are dependent on the perception of safety. The research finds that, in the United Kingdom, cost is the biggest reason given not to buy or use an e-bike (49%), but 'not feeling safe' is not far behind (41%).

However, in countries with more developed infrastructure, safety is much less of a concern. Concerns about safety are low as 12% in Denmark, 17% in Germany, and 21% in the Netherlands whilst cost is a much bigger deterrent (42% in Denmark, 54% in Germany, and 64% in the Netherlands). Across the 12 European countries researched for the report, on average 24% of people cite safety as a concern – only just over half the UK figure (41%), demonstrating quite how far the United Kingdom has to go.

This is not surprising. In the United Kingdom, there has been an ongoing battle to establish a useful network of cycle lanes. During the pandemic, the (extremely sparse) cycle infrastructure provision was augmented with pop-up cycle lanes but many were removed over the course of the following year.

In contrast to cycling, making space for walking has not experienced the same degree of controversy, although that does not mean that investment in walking has been significant. Similar to cycling, investing in the walking environment has significant benefits in terms of health, the environment, and even the economy [3].

The challenge facing walking specifically is that the lack of it is seen as primarily a health problem, but the solution rests within the transport sector [4]. Aside from some potential benefits from the modal shift for very short trips over which walking is a realistic alternative in terms of travel time, the primary benefits of more walking are in terms of physical health, something that has only recently been quantified in transport decision-making through cost-benefit analysis.

Over the last 15 years, there has been particular interest in the concept of placemaking in towns and cities. This is where through improving the place value of streets as opposed to primarily their movement value, they can become more attractive for walking and cycling. This also gives a significant amenity value on these streets which is beneficial to those who walk and cycle [5].

14.3 Integration with MaaS

MaaS depends on a safe active travel infrastructure. For people to walk or wheel to the nearest bus stop, railway station, or mobility hub they need safe pavements. Equally for crossing towns and cities between modes.

During lockdowns due to the COVID-19 pandemic, temporary pavement widening, pedestrianised areas, and outdoor hospitality areas were introduced to mitigate risks. In many places, these proved popular. There has been a shift in emphasis to creating spaces for walking and wheeling and for cycling in many European cities during this time, accelerating a trend that had already started prior to 2020.

Active travel infrastructure supports MaaS – directly in the case of some modes such as bike share but indirectly by enabling people to feel safe whilst on foot or wheels for parts of their journeys.

14.3.1 Case study: Paris

Central Paris has been transformed in the last decade. Green bike and scooter lanes line major streets, filled with a steady stream of bikes, e-bikes, and scooter traffic mixing people's own bikes with a plethora of types of rental scooters and bike share vehicles. Meanwhile, the traffic lights for pedestrians and cyclists across side roads are generally set to green, assuming an active travel priority. There are fewer cars and they have less opportunity to dominate the streets.

In concrete numbers, the mode share of private cars has declined by half since 1990, dropping to 12% of journeys. Meanwhile, cycling has doubled, albeit from a much lower base than many European cities, and is beginning to take its place as an everyday mode of transport.

The shift away from cars in Paris is inextricably linked with its mayors, who are elected by residents from inside the Périphérique, rather than the city's outer suburbs, where car ownership is much higher. The Mayor of Paris has a substantial degree of control over traffic management and road layouts within the city, which has been used to gradually curb car dominance and carve out space for cycling, walking, and public transport [6].

The city's rise in cycling arguably began in 2007, when the Vélib system of hire bikes was introduced. Now the largest city cycle hire scheme outside China, Vélib has also gone on to become one of the most successful in Europe, racking up 209,000 rides in a single day. At the same time, new cycling infrastructure and temporary road closure initiatives, such as 'urban beaches' on the banks of the Seine, began to become a feature of life in the city.

Sustainable transport received another boost in 2014 when Anne Hidalgo was elected mayor on a platform that included tackling the city's notorious air pollution problems. Diesel engines were identified as a major contributor – in 2015 74% of the city's cars were diesel models, thanks in part to a lower tax rate on diesel fuel.

The Paris Respire scheme began in May 2016, banning cars from certain areas on the first Sunday of every month. In 2017, the first permanent low emission zone was introduced, the Zone de Circulation Restreinte (ZCR). Enforcement of this was made more straightforward by the introduction of the previous year of Crit'air, a scheme where every vehicle in an air pollution control area is required to display a tax disc-style sticker confirming what emissions standard it adheres to.

Following the success of the ZCR, in 2019 the first metropolitan Low Emission Zone (LEZ) was set up, banning diesel cars made before 2006 from the city centre. This was enforced by automated monitoring, penalties, and Crit'air stickers. Meanwhile, a 46-day transit workers' strike helped to double the number of cyclists between 2018 and 2019, instead of pushing residents back into cars. The current Low Emission Zone permits only vehicles with the best Crit'Air stickers into the area from Monday to Friday, between 8 a.m. and 8 p.m.

Also in parallel, the first 'Plan Vélo' (2015–2020) created cycle 'expressways' along the north-to-south and east-west axes, the banks of the Seine and to access the parks of the Bois de Boulogne and Vincennes. In addition to the expressways, it includes a secondary network to ensure coverage of cyclable routes at the local level. Importantly it also provided over 10,000 cycle parking spaces. This represented a large investment in cycling, totalling 150 million €, and including 63 million € in cycle lanes, 30 million € to finance 'Paris at 30 km/h' (including two-way cycle lanes along the grand axis routes that remained at 50 km/h and

7 million € in cycle parking. In addition, subsidies of 10 million € were provided to assist in the purchase of e-bikes and cargo bikes.

In her 2020 campaign for re-election, Mayor Hidalgo unveiled her now-famous plans for a '15-minute city', removing 72% of on-street parking and making all streets cycle-friendly. Later that year in response to the pandemic, the city created an additional 60 km of temporary cycle lanes dubbed 'coronapistes', many of which will be made permanent as part of 'Act Two' of Hidalgo's Plan Velo.

Air pollution in Paris is still a problem, with reductions from the LEZ offset by a rebound in traffic levels following the lifting of travel restrictions. However, levels of NOx in the centre of Paris are now half as much as on the outer ring road. Cycling mode share has grown and is now estimated at 15%.

Anne Hidalgo was re-elected despite vocal opposition – demonstrating a sharp disconnect between the media narrative and decisions made through the ballot box. She has announced further bold measures to restrict traffic, including a new Low Traffic Zone, active from 2022, which will prevent non-residents' cars and vans from crossing the city centre at all, and a high-profile plan to remodel the Champs-Elysées as a pedestrian and cyclist-friendly linear park. In a March 2020 interview with *The Guardian*, Hidalgo made her position clear:

> We've given vehicle owners plenty of notice so they cannot complain they were not warned. The place for the car in our city will be reduced even more with more alternative mobility available like bicycles, more buses, and car sharing.

14.3.2 Case study: Waltham Forest, London

Waltham Forest 'Mini Holland' is the longest-standing of the schemes to create Low Traffic Neighbourhoods (LTNs), and, because early measures are now well bedded in, the best studied. It is also an example of how a high-profile vocal opposition was not necessarily typical of the concerns of residents and how measures to reduce traffic, even in the face of a number of loud critics, were worth pursuing.

Originally awarded funding in 2013, Waltham Forest 'Mini-Holland' was one of the first traffic schemes in a London borough to use modal filters extensively, in a way that typifies LTNs. With a budget of £27 million from Transport for London, the scheme initially faced vocal opposition, including protests and a 6,000-signature petition opposing the measures.

The use of consultation tools, including Commonplace, and engaging with visitors to the area as well as residents, allowed project officers to counteract arguments from the business community that removing parking would be commercially disastrous.

This also enabled the scheme to proceed with evidence of widespread public support for its outcomes, such as better pavements, slower traffic, and more pedestrian crossings.

The scheme broke ground in January 2015 and was completed in 2021. It comprises over 50 modal filters, 26 km of segregated cycle track, pocket parks, alterations to crossings, and public realm improvements including tree planting.

Parts of the Waltham Forest scheme have been in place long enough to be independently evaluated by researchers from the University of Westminster [7], who found that residents living in areas with high numbers of LTN measures were statistically more likely to choose active travel and less likely to be car owners. A year on, people living in the borough were on average walking and cycling for 41 min a week more than those living in comparable areas.

The report did emphasise that LTNs take time to create behaviour change, with benefits becoming clear around a year after implementation. The same team also analysed STATS19 collision data and concluded that walking, cycling, and driving becomes 3–4 times safer per trip within an LTN.

A separate report found no impact on emergency service response times.

With regard to walking specifically, there is a challenge to MaaS operators in that the walking route may provide the most value to customers – being the quickest or shortest route for example – yet it is something that cannot be directly monetised like other modes. The concept of Walking as a Service (WaaS) takes this further and proposes that in the case of walking, rather than customers providing direct revenue for the purchasing of a service, through their use of navigational aids (e.g., the MaaS or a navigational app of their smartphone) users provide value through the data on their travel patterns and details of the geographies they frequently visit or use [8]. Whilst walking is less easy to monetise in MaaS models, it has been shown to create economic value where it is facilitated [9].

14.4 Summary of MaaS and active travel

The interface between MaaS and active forms of travel like walking and cycling are often considered in terms of the interface between bike share systems, in other words, a technological interface. If MaaS is to achieve its ambitions in terms of encouraging a modal shift away from the private car, MaaS needs to consider factors outside of purely technological. Many of the key enablers of active travel, notably the development of infrastructure, are outside of the direct control of MaaS operators. It is these factors that will influence the adoption of MaaS as a means of encouraging active travel.

References

[1] Reid C. How 80 forgotten 1930s cycleways could transform UK cycling. *The Guardian* [Internet]. 9 May 2017 [cited 14 Aug 2023]. Available from: https://www.theguardian.com/environment/bike-blog/2017/may/09/how-80-forgotten-1930s-cycleways-could-transform-uk-cycling

[2] Reid C. Pedal-ins and car burials: what happened to America's forgotten 1970s cycle boom? *The Guardian* [Internet]. 16 Jun 2017. Available from:

https://www.theguardian.com/cities/2017/jun/16/pedal-ins-patchouli-bikeology-americas-forgotten-1970s-cycle-boom

[3] Sinnett D, Williams K, Chatterjee K, and Cavill N. Making the case for investment in the walking environment: a review of the evidence. tridtrborg [Internet]. 1 Jun 2011. Available from: https://trid.trb.org/view/1113034

[4] Milton K, and Grix J. Public health policy and walking in England—analysis of the 2008 "policy window." *BMC Public Health*. 2015;15(1).

[5] Tsai P. Australasian Transport Research Forum [Internet]. 2019 [cited 19 Aug 2023]. Available from: https://australasiantransportresearchforum.org.au/wp-content/uploads/2022/03/ATRF2019_resubmission_38.pdf

[6] Paris: The 15-minute city makes timely progress [Internet]. www.smarttransport.org.uk. Available from: https://www.smarttransport.org.uk/case-studies/europe/paris-the-15-minute-city-makes-timely-progress

[7] Aldred R, Croft J, and Goodman A. Impacts of an active travel intervention with a cycling focus in a suburban context: one-year findings from an evaluation of London's in-progress mini-Hollands programme. *Transportation Research Part A: Policy and Practice*. 2019;123:147–69.

[8] Lyons G. Walking as a service – does it have legs? *Transportation Research Part A: Policy and Practice*. 2020;137:271–84.

[9] The business case for better streets and places: THE PEDESTRIAN POUND [Internet]. Available from: https://www.livingstreets.org.uk/media/3890/pedestrian-pound-2018.pdf

Chapter 15
Mobility hubs and MaaS

15.1 Introduction

Mobility hubs have come to be thought of as the physical embodiment of MaaS, as evidenced by the dovetailing of MaaS with these physical groupings of transport modes and information in some cities (the example of Berlin's Jelbi app and hubs is detailed in Section 5.5). Mobility hubs reinforce users' understanding, expectations, and use of MaaS by providing infrastructure that groups together and signposts modes found in MaaS platforms.

The emerging European Mobility Hub model brings together multiple transport modes to encourage people to easily interchange modes on multi-modal journeys. They may bring together buses, trains, and taxis, and include elements that enable active travel (bike parking and bike share, seats and shelter), shared transport (car clubs and bike share), or even services that enable people to avoid journeys (clicks and collect lockers) or WiFi. In effect, almost any mode of MaaS (or that could be connected to the MaaS platform) can be found at the mobility hub.

15.2 Potential elements

There have been many attempts to define the key elements of what constitutes a Mobility Hub. Historically, they have been considered in terms of their primary function – namely a transport interchange. Consequently, by looking purely at this functional element, there are many potential elements of a mobility hub:

- Car club bays
- Bike share docking stations
- Demand Responsive Transport stops
- Taxi ranks
- Public transport stops and waiting facilities
- WiFi
- Delivery lockers
- EV charging and parking bays
- EV bike charging and parking area
- Bike parking
- Wayfinding and digital portals

Mobility hubs have developed in parallel to MaaS, and although this development was independent of it there are elements that resonate with MaaS development, very much as the modes they contain have developed in parallel to both MaaS and the hubs. In addition, both have been integrated into plans and policies to help reduce car dependency and associated problems such as parking and congestion.

The earliest development of a mobility hub was recorded in Bremen, Germany, in 2003. The city decided to address issues caused by on-street parking, which was clogging streets causing parking problems, traffic congestion and access difficulties for emergency vehicles, and other services (such as street cleaning, delivery, and waste disposal). The idea of the mobility hub was developed to provide a place for shared mobility where people could book and find shared cars. They also provided bike parking and bike share was available at many of the locations. At the time, the urgent requirement was to reduce the number of cars in the city by encouraging people to use shared resources. The mobility hubs in Bremen were designed as attractive public spaces, with a branded totem to indicate an official shared mobility area.

By 2020 there were 10 of the large 'mobil.punkte' and 33 of the smaller 'mobil.pünktchen'. These latter are largely located in residential areas so that they are within easy reach of homes.

The growth of the mobility hub network in Bremen has been extensively studied, and its impacts evaluated. A study for the EU North Sea Region calculated that each car club vehicle present on a mobility hub has replaced 16 private cars (with users reducing their car ownership by the equivalent of seven vehicles and because they have avoided acquiring vehicles due to the availability of the car club cars). The net reduction in parking requirements as a result of the mobility hub network is estimated at 7,500 vehicle spaces by 2023. In addition, people who are members of car clubs are more likely to use public transport, walk, and cycle than people who own privately owned vehicles [1].

Whilst Bremen's network grew from car-sharing, in Budapest, mobility hubs have grown from public bike sharing. The MOL Bubi public bike-sharing scheme was launched in 2014 by the BKK Centre for Budapest Transport. By 2021, the network had grown to include micromobility hubs at 100–150 m intervals across the city centre, 'neighbourhood hubs' known as mobility points including shared cars at wider intervals, and 'mobility stations' located near mass-transport stations which included more services such as bicycle repair stations, electric chargers, and parcel delivery pickup services.

More ad hoc approaches to bringing car club vehicles into developments have enabled less city space to be given over to private cars. Policies and practices to encourage and compel car-free development have integrated car clubs and shared transport.

15.2.1 Case study: car club and car-free housing development case study – Millar Crescent, Morningside, Edinburgh

A small 'car-free' development of 14 apartments was brought forward in an area well served by public transport and in which council policy discouraged on-street parking. Planning permission was granted on the basis that residents were legally restricted from applying for on-street parking permits, and that the developer contributed to the establishment of two car club vehicles parked immediately beside the development. These cars, and those in nearby locations, have been well used.

Edinburgh is the capital city of Scotland, home to around half a million people. It is a university city, with a dense urban core and a strong public transport network serving residential areas and connecting it within the region.

The City of Edinburgh Council has a long history of political support for car clubs. The establishment of the first on-street car club in the United Kingdom was approved by the council in 1998, with cars deployed the following year by City Car Club (which has since become Enterprise Car Club) which worked in partnership with the council from 2002. By the end of 2015, it could boast 158 car club vehicles and over 5,000 members.

Planners have allowed developers more flexibility on parking ratios if car club on-street parking is increased locally through the development of supplementary planning guidance on car clubs. The result is that over the past decade, the application of developer agreements through S75 of the Town and Country Planning (Scotland) Act 1997 has contributed significantly to the creation of on-street car club parking infrastructure.

The Millar Crescent development is situated in Morningside and lies about two and a half miles southwest of central Edinburgh. It consists of 14 newly built apartments in an established residential area with pedestrian access to a range of local services. It is close to several bus routes and has benefitted from council investment in improved dedicated cycling routes.

The developer describes the apartments as offering a combination of space, style, quality, and high specifications. The living space in each is predominantly open-plan with large windows and lots of natural light. Some apartments were designed with separate dining areas, and were sold as flexible – they could also be used as a home office or an extra bedroom. Practical features common to all homes included generous fitted storage, the latest security features, and cutting-edge heat recovery units.

Planning permission was granted in 2008 and work began on the development in 2010. It was completed in 2011 and all apartments were sold within 14 weeks of release to the market.

The planning agreement set a precedent. There is no parking provided for the development and a covenant prevents residents from applying for parking permits, making it a 'car-free' development.

Instead, the planning agreement provides support for the expansion of the car club through two on-street bays to be implemented by the council and for the first year leasing costs of two cars, which add to the network for all local residents. This required the developer to make a modest contribution of £12,500 to kick-start the additional on-street network locations in the immediate vicinity of the development.

The developer's sales office provided all those considering purchasing an apartment with details of the car club and incentives in the form of free membership for the first year and driving credits.

The two car club cars in Millar Crescent, parked together in bays beside the development, were also augmented by three other cars within close proximity (between 5 and 15 min walking time). All cars were well utilised.

Although the development was largely successful, parking management has not worked as well as hoped because the development was outside the area zoned as a Controlled Parking Zone (CPZ). Although initial residents and any succeeding owners are precluded from holding a residents' parking permit through a covenant on their lease, there have been examples of this restriction being breached. Residents have found sponsors living elsewhere in the area to secure a parking permit in the zone that operates close to the development. Others have used parking in on-street space that is outside the zone to overcome the restrictions. These breaches can only be tackled by the council expanding parking controlled areas (CPZs).

The Millar Crescent development worked well for all stakeholders and demonstrated the existence of a market for car-free housing in well-connected locations with accessible local services where a car club is provided. The acceptance of this type of situation is likely to be helped because awareness of the car club, which at the time had over 150 cars, was higher in Edinburgh than in any comparable UK city.

The attractiveness of the car club to new members in this location is in the choice of vehicles in the four adjacent locations and the availability of vans in strategic places in the city.

Although there have been some residents breaching the car-free restrictions on the development, their opportunities for doing so will be restricted by the expansion of controlled parking zones.

15.2.2 Mobility hubs and eHUBS

What is common across much of this practice is how the Mobility Hub concept has moved away from purely that of a transport interchange function into something that considers its role as part of the wider area in which it sits. Consequently, they may include in addition to their transport elements:

- Placemaking and community building to ensure a welcoming and useful experience through public art, landscaping, retail outlets, parklets, interactive design, and connections to the neighbourhood
- Drinking fountains and/or bathrooms to enhance comfort and desirability of space
- Storage and/or urban freight facilities such as package lockers to relieve pain points of travelling with or sending/receiving medium- or larger-sized goods

Although not specifically developed as a physical embodiment of MaaS initially, the modes gathered together on mobility hubs very often coincide with modes integrated within MaaS platforms. They are also generally accessed by apps or online – with some MaaS apps already being used to access those specific modes in some cities (for instance in Berlin, in particular, physical mobility hubs contain the shared transport offered in the Jelbi MaaS platform). In Flanders, Belgium, MaaS is being developed from the mobility hub network, clearly demonstrating the link between mobility hubs and MaaS. The network developed by a partnership between car club Autodelen.net, Mpact, and Infopunt Publieke Ruimte since 2019 has been taken and further reinforced by the Flemish government. These mobility hubs and 1,000 forthcoming hubs will be branded under a new 'Hoppin-punten' brand, along with the mobility as a service app, a call centre, and pricing policy for on-demand services [2].

The development of mobility hubs with wider offers has not been uniform across Europe.

Hubs were suggested as part of MaaS development in the United Kingdom in the mid-2010s. By 2017, a pilot in Essex town of Southend combined electric car club, bike, and e-bike hire from single 'hub' locations, accessible with a single membership, in a 'Motionhub'. The service was aimed at local users in the Southend area, commuters and visitors to the area looking for the most cost-effective and sustainable routes to complete their journey.

The aim was to create a viable platform for integrating transport services to provide convenient and sustainable transport choices for passengers and enhance end-to-end journey options. The project partners include an electric car club (E-Car Club), bike share (via Hour Bike), plus EValu8 Transport Innovations, Transport Systems Catapult, and Southend-on-Sea Borough Council [3].

Together they sought to provide online access to:

- Advanced journey planning systems encompassing all public transport, bike hire, car club, and car hire vehicle options
- Booking systems allow all parts of the journey to be purchased from a single site
- 'Mobility as a Service' piloting, permitting some/all of the transport modes to act under the MaaS umbrella.
- Behaviour change functionality to support people to make the shift from conventional transport modes to more sustainable choices.

Although the demonstrator site was launched with shared mobility options, it proved difficult to add extra functionality and better integrate public transport modes because of a reluctance by privately owned public transport operators in the United Kingdom to participate in integration at the time. However, it demonstrates an early desire to replicate the MaaS platform concept – including first/last mile transport options available from a digital platform which also offers train, bus, and other local transport options – on the ground. The demonstrator ensured that visitors to the hub had both digital and physical access to the means to make their journeys from that point.

Although this pilot did not turn out to be as comprehensive as hoped, subsequent hubs have developed along the same lines, to promote e-mobility. An EU Interreg project launched in 2019 was established to pilot eHUBS sites in six partner cities (Amsterdam, Leuven, Arnhem, Kempten, Manchester, and Dreux) to develop knowledge, best practices, and a blueprint for other cities and regions. The eHUBS are described as 'on-street locations that bring together e-bikes, e-cargo bikes, and/or e-cars, offering users a wide range of options to experiment and use in various situations. The idea is to give a high-quality and diverse offer of shared electric mobility services to dissuade citizens from owning private cars, resulting in cleaner, more liveable, and pleasant cities'. The project also supports different sizes and location types – smaller and positioned in residential areas (e.g., just one or two parking spaces) or bigger and positioned close to stations and major public transport interchanges. The key quality, however, is that should always be 'where supply and demand meet' [4].

The eHUBS project is partly about the transition to electric vehicles, and partly about reducing the number of cars parked and driven in urban areas. Nijmegen, Arnhem, one of the pilot cities specifically noted that:

> Electric mobility is important, but it is also needed to decrease the amount of parked and driving cars, because we don't want to be stuck in a sustainable traffic jam. Therefore, the aim is to stimulate active mobility (cycling and walking) and public transport as the most obvious choice for mobility. It is more space-efficient and keeps people healthier. When people are willing to give up their second or first car, we gain parking space and have the opportunity to improve the quality of the public space and create a more attractive living environment [5].

The eHUBS in Nijmegen includes e-bikes and e-mopeds as well as e-cars.

In the United Kingdom, mobility hubs with 25 e-cargo bikes and 10 electric vehicles, are being implemented in three areas in Greater Manchester: Bury, South Manchester, and Ancoats. Whilst two areas – Bury and South Manchester were simpler pilots colocating electric car club cars and e-cargo bikes for a year, the third in Ancoats was planned on a longer timescale due to its complexity and size. After the year of pilot funding ended, the e-cargo bike element of the Bury and South Manchester was withdrawn, but the e-car club vehicles remained in place.

The third area, Ancoats, is a much more ambitious project. The Ancoats hub is integrated with a new development of 1,500 new homes at the same time as new cycle routes have been established in the areas. It is an ambitious building that required a full planning process which was granted for:

- 150 secure bike parking spaces and changing facilities
- 102 electronic vehicle charging points, with the potential for more
- Up to 30 spaces for car club/car share schemes
- 408 car parking spaces for residents and visitors
- Parcel delivery hub
- Over 400 rooftop solar panels
- Space for a new café
- A new commercial unit
- Over 400 sqm of green wall and 950 sqm of planted façade
- 1,370 sqm of the new public realm, connected to the adjoining Ancoats Green area.

The hub is intended to take vehicles off local roads at the same time as more housing is built in the area [6].

In Barcelona, mobility hubs have grown out of parking transformation, also focusing on electrification. BSM, the municipal agency of Barcelona, is converting 42 off-street parking structures into 'multi-service platforms'. These combine public electric vehicle charging points, locker-based 'click and collect' points, micromobility parks for bikes and scooters, battery swapping pilots, sharing services, and micro-distribution centres as well as traditional car parking spaces [7].

15.3 Mobility Hubs and MaaS

To date, the role of Mobility Hubs in MaaS has primarily been indirect in terms of improving transport options in a manner that could be complementary to MaaS. As other cities and areas have developed mobility hubs as tools to improve transport options, reduce car parking requirements, and encourage more sustainable transport, different versions of the mobility hub have emerged in different contexts and adapted to different areas. In general, they bring together different forms of public transport and shared mobility on a scale relevant to their situation. Whilst they commonly facilitate multi-modal journeys by increasing the visibility of shared modes and making them easier to find and book (mostly via sign-posting although occasionally with kiosks or smart street furniture), there are different scales and layouts required to serve different scenarios from the urban core to suburbs and even much smaller settlements.

The following five types are defined by the UK shared transport charity CoMoUK:

Large interchange or city centre hub
This is a mobility hub at the nexus of different transport types where high foot traffic and multi-modal travel are anticipated. It is usually sited at an interchange of bus, rail (or light rail or tram) with new mobility such as bike share, taxis, etc. accommodated to help people make onward journeys. Good signage with real-time information is key, along with public realm improvements including seating, shelter, and green space.
Transport corridor hub
Points along corridors provide opportunities for wayfinding and the provision of shared transport along bus/tram routes. They can also provide seating (which improves the experience for people walking around towns or between points of interest) and improve the public realm.
Business park or new housing development hub
Mobility hubs at destinations or designed into new housing developments help people find sustainable transport options. They bring together bike share, car clubs, electric vehicle

(Continues)

(*Continued*)

Business park or new housing development hub
charging, and bike parking into an easily identifiable, safe, and attractive space, usually colocated with public transport. These hubs are intended to encourage people to travel by more sustainable means, and when designed well provide attractive greened space instead of a standard tarmac parking grid.
Suburbs or mini hub
Siting mobility hubs in residential areas tend to focus mainly on the provision of shared transport options – bike parking, e-cargo bike share, and car clubs – where the housing types and road layouts mean that people might not have access to these facilities otherwise. They may effectively be quite small hubs, more distributed through residential areas so that people only have a short walk. They reflect the fact that as travel patterns change, with MaaS platforms and shared transport use increasing, people need to be able to find these modes close to their homes – the origins of their journey.
Small market town or village/tourist hub
Small towns with a mixture of homes and businesses could also benefit from mobility hubs. The rise of the 'parklet', adding small areas of green wherever possible to town centre streets, increasing facilities for pedestrians and encouraging people to linger has been one of the key motifs of the last decade. Small town mobility hubs, useful for both tourists and visitors and residents are another element that brings a transport dimension to these public spaces. Mobility hubs in these areas integrate shared transport and public realm improvements to improve economic and social indicators.

Whilst these five types of mobility hubs broadly cover where and how mobility hubs are being developed, it is important to note that they play important roles in the rise of the shared mobility element that is key to MaaS.

First, they are not solely focused on urban centres – there is specific recognition of the role of mobility hubs in residential areas where they provide and highlight access to shared mobility close to people's homes where the 'last mile' element of transport must necessarily be situated if it is to play any role connecting people from their homes to their destinations. This mirrors the increasing realisation that MaaS too cannot just focus on urban areas – the transport it offers must be available to people living in suburbs, peri-urban, and even rural areas. In addition, designing and planning new homes has started to move away from requiring or allocating a set number of car parking spaces per home to integrating shared transport elements or hubs into new developments. There are also notable car-free developments emerging (see case study).

Mobility hubs also fulfil useful roles within the urban core. As micromobility 'clutter' has met with push-back, mobility hubs have become an attractive idea for gathering, organising, and in some cases, re-charging, e-scooters, and bike-share bikes so that they do not interfere with pedestrian thoroughfares. This is very important in maintaining attractive routes for people walking and crucial for those with reduced mobility and in wheelchairs. It is also essential to keep pedestrian areas tidy for people who cannot see well. Not doing so increases inequality and means that public space and the least expensive way of getting around (on foot or in a wheelchair) is inaccessible to some of the people who most need it. As micromobility has evolved, cities have begun to understand the need to allocate and manage the space allocated to them. Sometimes that allocation is done in mobility hubs – whether on the tiniest scale by allocating a car parking space or two to e-scooter or bike parking, possibly with planters and/or a seat to create an attractive 'parklet' or on a bigger scale by collocating with bus or rail infrastructure.

In addition, mobility hubs enable car club cars to be parked in a visible area with space for branding and instructions. Information boards are important elements, giving people the chance to read about and understand shared mobility – knowledge is the first step in changing behaviour. The profile that mobility hubs have can be leveraged as new elements are added, for instance with the introduction of electric vehicle charging to help increase confidence in electric vehicles and ensure that people driving electric vehicles can locate and navigate charging easily.

The vehicles within mobility hubs have become more diverse – many car clubs have added vans to their offer and shared e-cargo bike companies have increased their footprint. The profile of the mobility hub makes it easier for people to see the options that are open to them and try out modes they might not have considered otherwise (such as e-bikes or scooters). The integration of these facilities encourages the switch to electrification whether through EV car club or personal vehicles.

Much as railway stations have traditionally attracted services – from cafes to dry cleaners – into or adjacent to their buildings, mobility hubs are also about creating focal points for social elements and services. Seating, planters, and lighting help to create a secure and well-used area, cafes are encouraged and lockers for parcel deliveries are regularly provided so people can pick up parcels as they arrive or change mode.

Signage is crucial – both wayfinding signage so people understand where they are and how to get to places and also the options on offer. Mobility hubs are an opportunity to encourage people to share more – and provide links to services that enable that. The links with MaaS are obvious, as the platform offers both route planning and the modes available. Having those modes colocated in an attractive space only reinforces the message and the opportunity.

The indications are that mobility hubs will be increasingly adopted. In Europe, the European Commission is considering requiring more than 400 urban areas (known as 'urban nodes') to develop 'multi-modal passenger hubs' to improve first- and last-mile connections and to enhance long-distance connectivity. On a global level, mobility hubs featured in the UITP's knowledge brief on rural mobility, 'The Rural Mobility Challenge for Public Transport: How combined Mobility Can Help' published in 2022. Whilst much of the document featured the modes that improve rural mobility (including, among other things, on-demand transport, peer-to-peer car sharing, and better active travel infrastructure), it also looked at the role of mobility hubs and how they increase the reach of public transport into rural areas. For instance, the public transport network of the Provinces of Groningen-Drenthe in the Netherlands has built a network of mobility hubs with on-demand/shared services to extend the reach of the 'backbone' of high-quality public transport services delivered via train and bus rapid transit. These rural mobility hubs have led to better integration of new modes such as demand-responsive transport and carpooling with the cycling infrastructure.

Similarly in Flanders, where mobility hubs and MaaS are already being co-developed this corresponds with a move to 'demand-driven public transport'. In rural areas, where service frequency is low, there is a move to provide different forms of public transport depending on the area, including 'buses, taxis, small buses, shared cars, and shared bikes, etc.' The transfer between one mode and another will be seamless with mobility hubs to ensure end-to-end transport and MaaS to facilitate booking and planning [8].

This is further reinforced in the 2023 UITP policy brief on mobility hubs. The document, 'Mobility hubs: Steering the shift towards integrated sustainable mobility', calls for decision makers to work with public transport operators and service providers to develop integrated mobility hubs that work for everyone.

> By focusing on making active, shared and public travel modes the simplest choice for people, mobility hubs can deliver wider benefits and unite different stakeholders, including residents, in the transition to sustainable mobility.

The UITP calls for a **regulatory framework** to balance what it deems as the 'volatile' shared mobility market – possibly referring to the tendency for new entrants to rise rapidly with venture capital and, in some instances, collapse spectacularly. The tension between the consistency and reliability that public transport aims for and the less stable offers that have emerged in shared transport is clear. The UITP calls for clarity about responsibilities between stakeholders and multilateral commitments to create stable, long-term offers – which must underlie integrating hubs into other urban development plans.

It calls on public authorities and public transport operators to take strong leadership, especially in the branding, design, and communication and a coordination role to build and orchestrate a network of mobility hubs, possibly including privately run hubs from other sectors such as parking, energy, and housing.

A final matter to consider on the technological side, which is still early in its development and in its interface with MaaS, is the deployment of digital twins in transport infrastructure. A digital twin is essentially a digital representation of a physical form, that can reflect its physical behaviour by applying models, monitoring data, and performing analysis in real-time.

To date, the consideration of digital twins in transport infrastructure (streets, railways, stations, etc.) has been in understanding the behaviour of vehicles and people using them [9]. This moves from existing models that are primarily predictive to three-dimensional representations of infrastructure where performance can be monitored in real-time [10]. Should digital twins become common elements of understanding the performance of transport infrastructure and services, interfaces with MaaS platforms seeking this performance information may be feasible.

15.4 Summary of Mobility Hubs and MaaS

Whilst the formal concept of Mobility Hubs is a relatively recent development, the key components of Mobility Hubs have long been delivered as part of transport networks to varying degrees. What elements are present in individual hubs will vary according to the circumstances of each hub, but this often includes providing physical infrastructure for a variety of transport modes, a convenient location in which to interchange between such modes, and a variety of technologies like digital signage to provide a quality customer experience for those interchanging. They can also include other operational elements, like layover facilities.

Their full integration with MaaS, consequently, depends on their ability to enable interchange between different modes of transport and consequently support MaaS in providing a competitive alternative to the private car. Consequently, Mobility Hubs can be considered a complimentary service to MaaS in this regard.

References

[1] Final Report Analysis of the Impacts of Car-Sharing in Bremen, Germany [Internet]. Available from: https://share-north.eu/wp-content/uploads/2018/08/Analysis-of-the-Impact-of-Car-Sharing-in-Bremen-2018_Team-Red_Final-Report_English_compressed.pdf & Mobility Hubs: The Problem-Solving Approach to Congestion and Parking CITY OF BREMEN CASE STUDY [Internet]. 2021 [cited 14 Aug 2023]. Available from: https://uploads-ssl.webflow.com/6102564995f71c83fba14d54/618d2af52a57df5c3f4c6ab6_CoMoUK%20Mobility%20hubs_Breman%20case%20study_Jan%202021.pdf

[2] Mobility Hubs: Steering the Shift Towards Integrated Sustainable Mobility [Internet]. UITP. Available from: https://www.uitp.org/publications/mobility-hubs-steering-the-shift-towards-integrated-sustainable-mobility/

[3] Kubitz, B *et al.* (2018) *Mobility as a Service: The Second Annual Survey of MaaS in the UK*, Landor LINKS, ISBN: 978-1-899650-85-9

[4] eHUBS – Smart Shared Green Mobility Hubs [Internet]. www.nweurope.eu. Available from: https://www.nweurope.eu/projects/project-search/ehubs-smart-shared-green-mobility-hubs/

[5] eHUBS – Smart Shared Green Mobility Hubs [Internet]. vb.nweurope.eu. 2023 [cited 14 Aug 2023]. Available from: https://www.nweurope.eu/projects/project-search/ehubs-smart-shared-green-mobility-hubs/#tab-6

[6] Ancoats Mobility Hub | Ancoats | Manchester City Council [Internet]. www.manchester.gov.uk. Available from: https://www.manchester.gov.uk/info/200079/regeneration/8462/ancoats/2

[7] Mobility Hubs: Steering the Shift Towards Integrated Sustainable Mobility [Internet]. UITP. Available from: https://www.uitp.org/publications/mobility-hubs-steering-the-shift-towards-integrated-sustainable-mobility/

[8] The Rural Mobility Challenge for Public Transport: How Combined Mobility Can Help [Internet]. UITP. 2022 [cited 14 Aug 2023]. Available from: https://cms.uitp.org/wp/wp-content/uploads/2022/02/Knowledge-Brief-Rural-Mobility_FEB2022-web.pdf

[9] Jafari M, Kavousi-fard A, Chen T, and Karimi M. A Review on Digital Twin Technology in Smart Grid, Transportation System and Smart City: Challenges and Future. *IEEE Access*. 2023. p. 1–1.

[10] Gao Y, Qian S, Li Z, Wang P, Wang F, and He Q. Digital Twin and Its Application in Transportation Infrastructure [Internet]. *IEEE Xplore*. 2021. p. 298–301. Available from: https://ieeexplore.ieee.org/abstract/document/9540108

Chapter 16
Policy making for MaaS

16.1 Introduction

Mobility-as-a-Service interfaces with a number of regulatory and policy areas outside of just transport. The complexity of MaaS regulation, and the subsequent challenges around ensuring a degree of regulatory consistency – something that has long been a policy goal of MaaS advocates – reflects the varying scales, structures, and attitudes towards governance in different countries and regions. These regulations and policies are not just set by the public sector, but also by industry representative bodies and in some cases the private sector, all of which have had a significant impact on the ability of MaaS to become established.

Regardless of the regulatory environment, governments have a variety of regulatory tools that they can apply at a variety of stages of maturity to develop MaaS. The UK Policy Lab refers to this as 'Government-as-a-system'. The ability of policymakers to effectively utilise different policy measures to develop and establish MaaS, or in some cases to not utilise policy measures at all and instead leave to the market to deliver and develop solutions, depends on the capacity and capability of different organisations to wield them (Figure 16.1).

In the early establishment of MaaS, government appetite towards risk-taking on a concept largely unproven determined the location of early adopters of MaaS. The most notable initial advances that embraced MaaS as a different concept to more traditional operator ticketing and journey planning were in Scandinavian countries such as Sweden and Finland.

16.2 MaaS as public policy

The transport policy environment of a nation is significantly influenced by its decision-making structures, notably the degree to which transport powers are devolved away from national government to regional and city government. Whilst transport policy is traditionally considered in terms of strategy documents and funding projects, the reality is that transport policy is a complex mix of functions delivered by different partners to achieve transport outcomes [2].

The result of this is a complex mix of policy initiatives and regulatory powers of transport services, infrastructure investment, and protection of consumers that makes scaling Mobility-as-a-Service style services between countries a significant

Figure 16.1 Government as a system [1]

challenge. A notable exception to this is within the European Union through its Sustainable and Smart Mobility Strategy, which contains a commitment 'making connected and automated multimodal mobility a reality'.

The emergence of MaaS in the policy space has mainly focused on two key functional areas of policy making. The first is in transport policy making, and the second is within business policy. Regardless of whether MaaS is considered in either area, the public sector policy outcomes of MaaS usually link to wider policy goals such as the UN Sustainable Development Goals or socio-economic outcomes, reflecting the priorities of the government at the time.

It should be noted that, between transport and business policy, tension has often been observed in the policy discourse. A common critique of business policy has been the focus on technological utopianism in the rhetoric surrounding the business opportunities that can be secured through MaaS [3].

16.3 Mobility-as-a-Service as transport policy

Within the context of transport policy, delivering MaaS is rarely seen as an end in itself or a policy goal in itself. But it is seen as a mechanism by which a number of wider transport policy goals can be achieved, and then typically as a part of wider digital initiatives to achieve transport policy goals.

Goals in transport policy have shifted in recent years from outcomes relating specifically to transport, to wider policy goals. Such goals have been linked to the United Nations' 17 Sustainable Development Goals. Common themes across transport policies include tackling climate change, tackling social inequity, improving access to key services and employment, and contributing to overall community well-being (Figure 16.2).

Transport policy itself has increasingly sought to identify the outcomes within its direct control with these wider objectives [5]. Such areas within control include building, maintenance, and operation of transport infrastructure, the specification and operation of transport services, and supporting regulatory functions relating to matters like safety, environmental impact, and customer service standards.

Within this context, MaaS is often considered as part of a wider digitalisation of transport which seeks to achieve travel behaviour change, improve the efficient operation of the transport system, and deliver an enhanced customer experience. The degree to which each is important depends on the policy-making institution in question [6].

In the West Midlands in the United Kingdom, the West Midlands Strategic Transport Plan sets out the main regional policy relating to transport. Its primary objectives relate to Economic Growth and Inclusion, Population Growth and Housing Development, the Environment, Public Health, and Social Wellbeing [7]. Contributing to all of this is the widespread adoption of smart technology to make the transport system operate more efficiently, and to improve the experience of customers using the system, to the point where the strategy has its own 'Smart Mobility Tier'. The strategy document states that:

Figure 16.2 The United Nations sustainable development goals [4]

> The West Midlands will have an effective and well used intelligent mobility solution which supports integrated travel across all means of transport. People and businesses will be enabled and incentivised to make cost effective, informed and sustainable travel choices using 'live' travel information and seamless payment systems which span multiple modes.
>
> We will work with others to ensure that developments in technology and innovation are encouraged and harnessed effectively to ensure the best practicable level of service can be provided.
>
> We will have a coordinated approach to responding to the challenges of air quality targets through effective management of road traffic, innovation in vehicle and road infrastructure technology that supports efficient mobility.

A key action was to develop a Personal Mobility Platform that supports an integrated journey planning, navigation, and payment system across all modes. Whilst the app Whim had been operational in the West Midlands, as part of the Future Transport Zone the West Midlands is developing its own MaaS app. The aims of this project are: [8]

- Build trust in our regional mobility system by improving user experience
- Broaden choice through integration, centralised information, and simple ticketing
- Reduce emissions and improve health by incentivising shared and active travel
- Enhance the economic sustainability of our network by reducing costs and encouraging innovation

In its Land Transport Masterplan 2040 [9], the Singapore Land Transport Authority states that technology is an enabler in several of its key initiatives. These initiatives focus on ensuring that Singapore is a 45 min city with 20 min towns, that it provides transport for all, and that it contributes to healthy lives and safer journeys. This links into wider aims and objectives that seek to deliver:

- a transport network that is convenient, well-connected, and fast;
- a transport ecosystem characterised by gracious behaviour and inclusive infrastructure; and
- a transport environment that supports healthy lives and enables safer journeys.

Specifically in relation to technology as an enabler of journeys, the Masterplan states:

> Today, many people rely on their smartphones to plan their journeys. Expectations about transport data have changed fundamentally since the last [Land Transport Masterplan] in 2013. Hence, we will make more real-time information available to app developers. We are also working with the Singapore Land Authority (SLA) on a trial to map barrier-free routes in the vicinity of the Enabling Village in Redhill by end-2019. If successful, we will explore mapping barrier-free route locations across our towns and city. To help people better plan their routes, we will also explore providing more real-time information about lift and escalator maintenance works at our MRT stations and pedestrian overhead bridges. Together, this data will help developers create apps that can make journeys easier for all. Already, the ongoing trial of the assistive passenger information software known as Mobility Assistance for the Visually Impaired and Special Users (MAVIS) has made bus journeys easier for some people with special needs.

Whilst Singapore has done much work in providing a 'sandbox' environment to test new mobility services, MaaS included, this master plan seeks to provide a more definitive direction for the policy goals that MaaS should seek to achieve. Whilst this does not prevent MaaS solutions from being delivered that are primarily commercial in nature, and therefore may not seek to achieve public policy goals as their stated objective, this document provides a clear direction for any operators who seek to partner with the Land Transport Authority (LTA) as to what their solution should help to enable.

16.4 Mobility-as-a-Service as economic policy

Much of the interest of policymakers in the last decade has been to use advances in transport technologies as an economic advantage and capability. This is often with the twin purposes of capturing a significant proportion of stated economic opportunities associated with such technologies (usually stated as a future global market value), or to encourage domestic firms to develop their capabilities to export products and services in the future.

Within the United Kingdom, the future of transport is stated as one of the Grand Challenges as part of the UK's Industrial Strategy [10]. This has been built on in the more recent Plan for Growth [11], which highlights that for the potential economic benefits of MaaS (and other transport technologies) to be realised, regulatory reform is needed to establish the United Kingdom as a testbed for such innovations in transport. This Future of Transport programme, being led by the UK's Centre for Connected and Autonomous Vehicles (a joint unit of the Department for Business, Energy, and Industrial Strategy and the Department for Transport, but hosted by the latter) has already considered a number of regulatory areas which it has reviewed to maximise the economic potential of MaaS.

The most directly applicable is a consultation on the Future of Transport regulatory review: call for evidence on micromobility vehicles, flexible bus services, and Mobility-as-a-Service [12]. This specifically sought views from local government and industry on what the opportunities and risks of MaaS platforms might be, and what role central and local governments should play in their development. A key outcome of this consultation in relation to the economic opportunities posed by MaaS was concern over competition issues. The outcomes report [13] noted that

> There was a range of competition issues presented that may not be addressed through existing regulation with the dominant themes concerning private sector monopolies, data sharing, integration of transport operators onto a MaaS platform and due to the collaborative nature of MaaS to agree commercial agreements, the risk of organisations violating existing competition laws.

An associated consultation sought feedback on the establishment of 'regulatory sandboxes', where new regulations and standards could be trialled with the intention of stimulating competition and market activity within future transport industries, MaaS included [14]. The outcomes of this consultation were yet to be published by mid-2023.

When initially enacted, the changes to Finland's Transport Code initially envisaged a new approach to transport that not only sought to deliver a more data-centric industry and open up services to competition but also to establish Finland as a centre of excellence in such services to export.

Critical to the approach of the Finnish government in providing an enabling economic policy was a close collaboration between a number of different ministries to establish mutually supportive policy programmes [15]. Whilst the Transport Code reforms have provided an enabling framework for digital solutions in transport, supportive policy measures were necessary from a number of government departments to enable MaaS to be established as an economic opportunity [16]. The most notable partnership emerged between the Ministry of Transport and Communications and the Finnish Innovation Agency, Tekes.

In 2015 and 2016, Tekes funded 31 MaaS projects with a total of 5.5 million euros in funding [17], reflecting the agency's joint work with the Ministry for Transport and Communications on Smart Transport as a thematic area. When undertaking the work, Tekes estimated that such programmes would provide

2,100 million euros in additional export opportunities and create 3,500 new jobs within Finland. In addition to funding trials, Tekes funded trade missions and encouraged overseas companies to invest in Finland.

This approach has not been sustained, however. In recent years, the approach of Business Finland (the successor to Tekes) has shifted its focus towards more significant grand challenges. Notably carbon neutrality in energy systems, zero waste and a circular economy, and merging the physical and virtual worlds [18]. Meanwhile, the Ministry of Transport and Communications has sought to further roll out the remaining elements of the Transport Code.

The overall impact of this economic policy has not been determined, and no assessment has been published. This reflects a wider issue concerning economic policy in relation to MaaS, where success metrics have often been loosely defined, and success has been difficult to determine. This is with the exception of levels of activity, where data can be found on the number of MaaS projects and the value of public investment in supporting companies developing MaaS solutions.

16.5 Enabling key capabilities to support MaaS

The delivery of MaaS requires the establishment of a variety of new skills that various transport industries have not had the necessary requirements to possess, or at least are not required as part of their core functionality. Transport industries have traditionally been focused on engineering and other STEM-related capabilities to operate transport systems and networks, although, with the advent of Intelligent Transport Systems (ITS) and various transport modelling packages, some specialist technological capability is present in the transport sector.

In a wider exploration of the skills gap in what it terms 'intelligent mobility', the Transport Systems Catapult identified that the future of transport – MaaS included, would require a wide variety of non-STEM skills to realise. Within its skills matrix, it identifies that a variety of skills in technology, social sciences, and human sciences are needed to develop the systems and business models required to realise the potential of future transport technology (Figure 16.3).

This poses a significant policy challenge. Such high-value digital and social science skills are in short supply, with the United Kingdom estimating that there is a current shortfall of 281,000 people with such skills [19]. A skills gap that is only likely to grow over the coming years.

For transport policymakers seeking to develop MaaS in their areas, this is a significant policy challenge. The development of 'base' digital skills is often seen as, and dealt with, by departments for education at a regional and national level as part of a wider initiative to develop digital skills. Whilst transport policymakers may have relationships with universities and their equivalents, and may offer research funding and competitions to develop MaaS offerings where the digital skills of such institutions are applied, this does not establish a core digital skills capability.

Numerous technical roadmaps on the application of MaaS and other transport technologies identify skills as a significant challenge [20]. For instance, techUK

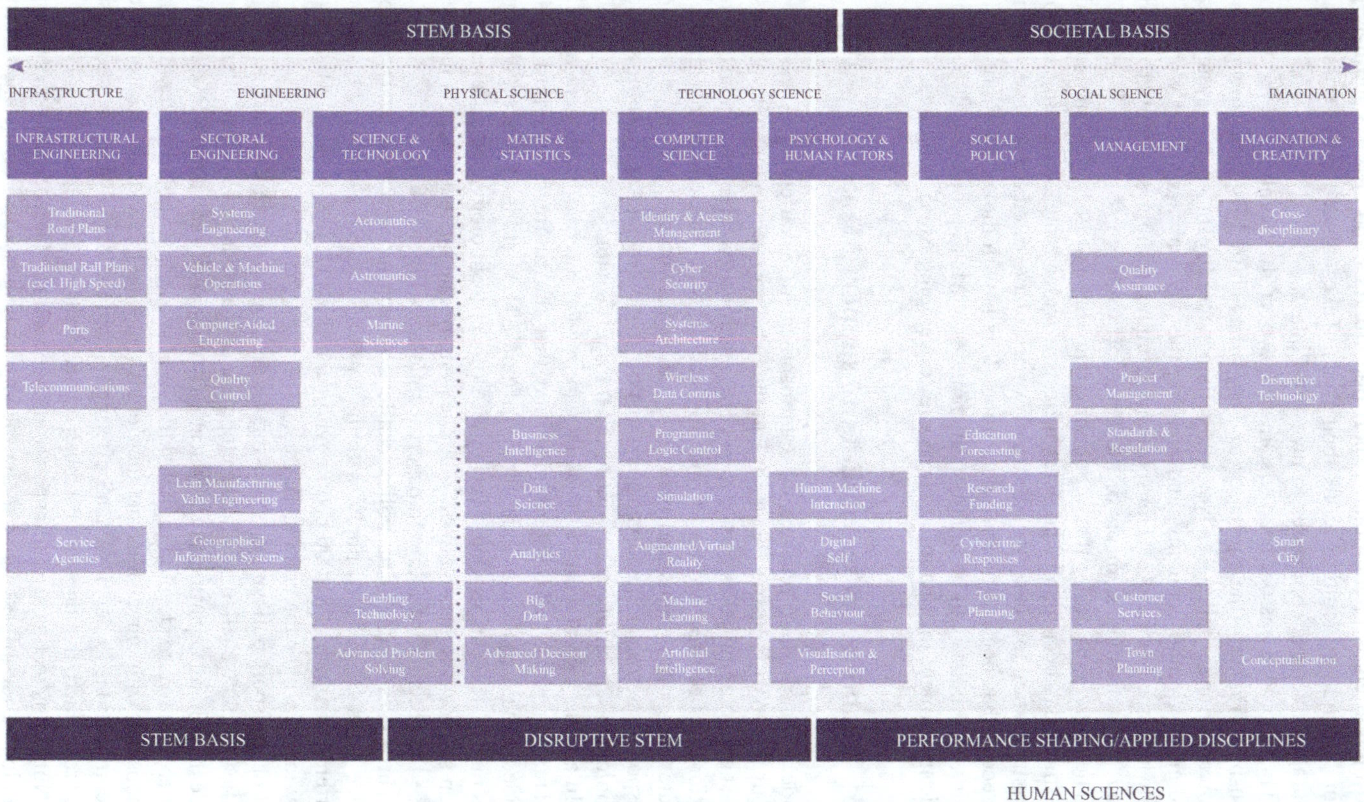

Figure 16.3 Intelligent mobility skills matrix [19]

identifies several actions that need to be taken to tackle an ongoing transport skills gap. Most notably the following [21]:

1. The Department for Transport should publish a policy roadmap for addressing issues identified within its transport labour market and skills consultation.
2. The Apprenticeship Levy needs to be reformed to transform its impact across transport and the wider economy.
3. The creation of a Digital Skills & Productivity Tax Credit to drive business investment in training and productivity-enhancing technology.
4. Government should lead the creation of new digital skills for the transport task force.
5. Government must work with the industry to create a Transport Digital Champion Advisory Council.

To date, however, there has been no significant and consistent policy effort to develop and establish these skills across the transport industry. There has been work to establish delivery capability at city and some regional levels, for example in the establishment of a Future Mobility and Innovation Team at Transport for the West Midlands.

The development of skills is not the only means by which policymakers can have a meaningful impact in establishing core capabilities to enable MaaS to be delivered. One of these is data standards, which is discussed more in the following section of this book. But another common area is establishing the 'rules of engagement' between operators and MaaS providers in a manner that enables MaaS operators to offer a truly multi-modal solution, but also respects issues such as commercial confidentiality and even privacy.

A key policy consideration in this regard is enabling competition through the sharing of data. The nature of competition within the transport industry is that it is primarily between modes of transport, for example, private car club companies potentially competing for trips with established public transport, the latter of which is typically heavily regulated but with exceptions, such as the UK Bus industry.

Even if the business incentive is strong but relationships with the public sector are not adequate, then there is a risk that one of the many business models of MaaS could be developed in a way that bypasses the public transport system, creating greater competition with private industry, and further encouraging less sustainable means of transportation [22]. A number of new mobility companies establish as part of their business plans that their solution is complementary to other solutions in tackling common transport policy issues. For example, in 2018 Lyft mentioned a number of initiatives that it had undertaken to be complementary to public transport. Notably the following [23]:

- A New Approach to Suburban Mobility and Dial-a-Ride Service | City of Monrovia, California**: **We have partnered with the City of Monrovia to launch a new model for suburban mobility, enabling residents and visitors in this community to access a Lyft ride anywhere in the GoMonrovia service area for just $0.50, and to easily connect to the LA Metro Gold Line. This innovative public-private partnership provides the community with an enhanced and affordable transportation option.

- Transforming Lives with On-Demand Paratransit | Regional Transportation Commission of Southern Nevada, Nevada: Lyft is working with the Regional Transportation Commission of Southern Nevada (RTC) to provide on-demand transportation to a subset of their paratransit community for the first time. Customers in the pilot programme have noted the life-transforming impact that RTC and Lyft are providing by enabling same-day, on-demand travel when most passengers could previously only book rides days out from the date they needed them.
- Providing free first-last mile access to core transit centres for bus riders | Pierce County Transit, Washington: With support from the Federal Transit Administration's Mobility on Demand (MOD) Sandbox grant programme, Limited Access Connections, we have partnered with Pierce County Transit to bridge first-last mile gaps to bus and rail stations throughout Pierce County, introducing added mobility and new multi-modal options.
- Mitigating regional congestion with multi-modal programmes | Transportation Authority of Marin, California: We are working with the Transportation Authority of Marin (TAM) to provide a discount for passengers using the Lyft Line to get to and from the region's new SMART rail line. Working with TAM we are expanding residents and employees access to and from this important new commuter rail.
- Providing greater connectivity to new rail systems | Charlotte Area Transit System, North Carolina: We have partnered with the Charlotte Area Transit System (CATS) to provide discounted first-last mile trips to Charlotte's new light rail extension. These are just some of the many ways we are working with public transportation agencies to reshape the way communities move – making it easier for communities across the country to access transit solutions.

Evidence from across the world gives an indication as to the degree of competition between modes that can be enabled through a MaaS platform. Within the city of Cairo, the modal shift towards ridesharing services was from traditional taxis (30% of modal shifted estimated), private cars (30%), and public transport (24%) [24]. A study of data from the National Household Travel Survey in the United States states that the degree of substitution between mobility on demand and public transport is highly content-dependent, with other factors such as age, work status, employment location, and availability of a car playing a more dominant role [25].

This potential loss of passengers to competing modes, and even competing companies with the mode, makes organisations unwilling to share data that may give them a commercial advantage [26]. Policymakers have two primary tools by which data sharing can be mandated among mobility service providers. The most significant is the passing of laws and regulations that make it a requirement for any organisation providing transport services to share data.

This approach has most notably been trialled as part of the Finnish Transport Code. The Transport Code places an obligation on incumbents as well as new entrants to the transportation market to provide their operational data as well as their single tickets for third-party resale and use [15]. The Data Regulation that forms part of this code sets a number of data-related requirements on a variety of bodies, including the following [27]:

> The Finnish Transport Agency shall open the data it has received as well as statistics and studies compiled and conducted by it through an open interface in a machine-readable format to make them freely available. The data that are opened must be in a format that makes it impossible to identify individual users and does not compromise business or professional secrets.
>
> Regardless of the mode of transport, a provider of passenger mobility services shall ensure that essential, up-to-date data on its services is freely available from an information system (open interface) in a standard, easy to edit, and computer-readable format. At minimum, this essential data shall include information on routes, stops, timetables, prices, availability and accessibility.
>
> The Finnish Transport Safety Agency shall be notified of the web addresses of an interface referred to in subsection 1 above and web addresses of any additional information needed to use the interface as well as any address updates before operation is initiated or, with regard to updates, as soon as a new address is known.
>
> The Finnish Transport Agency shall offer a technical service through which the data referred to in subsection 1 may alternatively be provided.
>
> Providers of road and rail passenger transport services, providers of brokering and dispatch services, or actors managing a ticket or payment system on behalf of these shall give mobility service providers and providers of integrated mobility services access to the sales interface of their ticket and payment systems, through which it is possible to: (1056/2017)
>
> 1) purchase a ticket product at a basic price that, at minimum, entitles the passenger to a single trip; the travel right based on this ticket shall be easily verifiable using generally applied technology; or
>
> 2) reserve a single trip or a transportation, the exact price of which is unknown when the service begins or which for some other reason will be paid by mutual agreement after the service has been provided.
>
> Mobility or integrated mobility service providers or actors managing a ticket or payment system on behalf of these shall give mobility or integrated mobility service providers access to the sales interface of their ticket and payment systems, or, if necessary, provide access to the systems via another electronic transaction channel and allow the mobility or integrated mobility service providers with the right of access to purchase ticket products at the service user's request on his or her behalf or other products entitling them to use the mobility services using the identification and user information of the service user's existing user account.

16.6 The establishment of data standards

An area where the public sector has been especially active is the establishment of common data standards underpinning MaaS offerings. With MaaS operators providing an integration function, and the associated possibilities of greater data sharing among all MaaS ecosystem organisations, there has been much activity in the public sector on this enabler of MaaS.

Public sector organisations have been particularly active in two areas of data standards:

- The development of the standards themselves, especially in the early stages of the development, and often in close collaboration with sector partners;
- Enabling the role out of these standards, through the establishment of codes of practice, through licencing or purchasing requirements, or through primary legislation.

Public sector authorities play both of these roles as a means of reducing barriers to entry for potential MaaS providers. If data is provided in a consistent manner by all operators, this is not only a benefit in terms of technical integration (e.g., through APIs), but the associated ruleset with data standards can reduce the administrative burden on the public and private sectors as all parties are signed up to common rules [28].

An example that has been increasingly adopted globally is the Mobility Data Specification (MDS) (Section 12.5.1). Pioneered by the Los Angeles Department of Transportation (LADOT), the City of Santa Monica, the City of Austin, the San Francisco Metropolitan Transit Authority, the Seattle Department of Transportation, Harvard Kennedy School, Bird, Spin, and Lime, it is a digital tool that helps cities to better manage transportation in the public right of way. MDS standardises communication and data-sharing between cities and private mobility providers, such as e-scooter and bike-share companies. This allows cities to share and validate policy digitally, enabling vehicle management and better outcomes for residents.

The core of data challenges in micromobility are due to variances in interpretation. Some of them are rather cut and dry, allowing for a simple, programmatic solution. But sometimes operators use fields differently than anticipated.

For example, the cost is supposed to be a positive value, but sometimes operators will indicate a refunded ride with a negative number. It may make sense to the operator, but it does not fit the specification. The types of data collected can also vary by vendor. Some operators omit certain events. Some operators will decide not to serve unavailable events, which makes it more difficult to present accurate information. For example, if 300 vehicles are available at the onset, but there are no unavailable events, we have no way of knowing if there are actually only 200 vehicles available.

Authentication and rate limiting, which every operator does differently, can also cause issues with data. Rate limiting is the number of requests we can make to an API in a given time frame. If exceeded, you get an error message. Sometimes there will be limits on data that can be obtained after a period of time, for example, 30 days, reducing the data to which cities or operators have access.

In effect, although there are data standards and regulations, the industry is not standardised to the extent that it is fully transparent.

16.7 Policies at different scales

Globally, transport policies can be adopted at a variety of geographical scales. This is often closely linked with a public sector body with some degree of statutory or policy-making power, with the commitments contained within those plans representing their degree of authority. There is significant variance in policy-making responsibilities between different countries and regions, although these usually manifest themselves within a plan or strategy.

At the international scale, there are very few examples of MaaS being an explicit policy commitment, primarily as international bodies have limited policy-making authority. The notable exceptions to this rule are the United Nations, which is guided by the Sustainable Development Goals that contain no specific reference to MaaS [29], and the European Union.

The United Nations itself has only published a single publication on MaaS, through the United Nations Economic Commission for Europe in 2020. This document contained a series of policy recommendations for local authorities [30], which are summarised below, but as this is a technical report it is not formal UN policy on MaaS.

Recommendations of the UNECE Report on Transport Trends and Economics 2018–19 on Mobility-as-a-Service:

MaaS deployment can be accelerated with supportive actions and policies. It is recommended to local authorities to:

- consider appropriate methods of sharing data in a way that is fair to consumers and businesses to increase opportunities of integrating into a MaaS platform;
- set out a clear vision and strategy on how to integrate car and bike sharing as well as other mobility services into a MaaS platform in a way that is cost-competitive and reduces private car ownership;
- support business and industry in elaborating interoperability standards for facilitating communication between mobility systems;
- support – in collaboration with national authorities – innovation and MaaS development by incorporating the notion that no transportation title is anonymous;
- address the physical infrastructure challenge by working with the car and bike sharing providers and other mobility service operators/providers to deliver adequate infrastructure for car sharing, bike sharing, and adequate cycling lanes for safe bike riding;
- support and become a member of MaaS partnership model and help to establish a level-playing field for the mobility service operators/providers;

- orchestrate MaaS, as a member of the MaaS partnership towards high-level objectives such as decreasing the impact of transport and mobility on the environment and human health;
- support transport infrastructure providers (TIPs) to apply integrated network management and assist the coordination between TIPs for different modes and between them and other partners such as MaaS providers and mobility service operators/providers; and
- provide – in collaboration with national authorities – platforms for knowledge exchange for integrated network management.

In a similar manner, there are a number of international bodies that, whilst they do not have a formal policy-making role, provide guidance to policymakers in their roles as forums or advisory bodies. For instance, the International Transport Forum has published a number of technical studies related to MaaS that have contained a number of policy recommendations, which are summarised in the below table. However, whilst such documents may provide useful guidance to policymakers, they are not formal policy (Table 16.1).

The European Union itself does not have a formal MaaS policy. But the work undertaken by the EU itself has sought to enable MaaS through a number of practical policies and initiatives. This is often with the common theme of European-wide integration of systems. For instance, in Roadmap to a Single European Transport Area – Towards a Competitive and Resource-efficient Transport System [31], a number of policy commitments are made that have a strong MaaS theme to them. Most specifically the development of a multi-modal transport information management and payment system by 2020.

To date, much of the work of the EU has been in supporting the development of MaaS solutions through research and development funding. Funding programmes such as Frameworks 4 to 7, Horizon 2020, and the Horizon Programme have funded a significant number of projects of which the development or testing of a number of MaaS-style solutions has been a key part.

MaaS is also an element of the European Union's Mobility Strategy and Action Plan, underneath a more general umbrella of Smart Mobility [32]. Under this Smart Mobility policy, there are two flagship policy priorities making connected and automated multi-modal mobility a reality, and Innovation, data and Artificial Intelligence for smarter mobility. Under both of these policy priorities, the European Commission realises that there are still significant barriers at an international level facing the delivery of the capabilities necessary to deliver MaaS.

> The EU needs to transform its legal framework to support multimodal travel information, booking and ticketing services, while at the same time looking at the rights and obligations of online intermediaries and multi-modal digital service providers selling ticketing and/or mobility services. The Commission will examine whether data sharing, including on fares, in road and rail passenger transport services, and selling arrangements are fit for purpose. Meanwhile, smart and interoperable payment services and

Table 16.1 Recommendations of ITF reports relevant to Mobility-as-a-Service

Document title	Publisher	Published	Key policy recommendations
Ticketing in Mobility -as-a-Service	UITP	July 2022	• Establish cooperation between mobility actors on an organisational basis • Arrange technical interoperability after organisational interoperability has been set • Arrange MaaS according to governance needs • Align MaaS and the business model of each mobility partner • Make MaaS a success whilst attracting transport users to choose MaaS • Create an ecosystem for data sharing accepted by all parties involved • Make agreements based on needs in sharing data • Make use of standard procedures and standard protocols whilst sharing data
Innovations for Better Rural Mobility	International Transport Forum	December 2021	• Formulate a countrywide accessibility policy and implement Sustainable Regional Mobility Plans (SRMP) • Adopt a whole-of-government approach for rural public services and the local economy • Make regulations more flexible to allow for the development of innovative, cost-effective mobility solutions • Combine public mobility budgets to achieve cost savings • Fund pilot schemes to test innovative mobility concepts • Prioritise financial support for innovative services according to higher impact levels rather than the use of high-tech • Use innovative financing approaches to increase funding pools and the viability of individual transport services • Increase central government funding for shared and active travel in rural areas • Provide technical assistance for rural mobility at the national or regional level • Promote mobility hubs to connect local services to the core network • Support the development of national or regional Mobility-as-a-Service (MaaS)

(Continues)

Table 16.1 (*Continued*)

Document title	Publisher	Published	Key policy recommendations
Using Business Models for Better Integrated Mobility	UITP	December 2021	• Many pilots are designed to gain a better understanding of the impact of a new service and finalise a business case to justify public spending if needed • Individual business models normally concentrate on their own efficiency and not on the efficiency of the integrated offer or the wider policy goal of sustainability • Use a Business Model Canvas to identify and overcome critical challenges in the ecosystem and achieve a more integrated multi-modal system with attractive and sustainable mobility services
Adapting Mobility-as-a-Service for Developing Cities: A Context-Sensitive Approach	World Bank	December 2021	• Regulations of MaaS should focus on desired outcomes and principles, rewarding positive impacts and taxing negative externalities • Any MaaS governance and regulatory framework in developing countries need to address information, payment, and service integration, but also equity and climate considerations • Data is the new regulatory currency in the MaaS governance framework • A specialised agency will likely be necessary for MaaS to thrive
Integrating Public Transport into Mobility as a Service	International Transport Forum	October 2021	• Adopt a light and flexible regulatory approach that allows Mobility-as-a-Service to evolve • Integrate the governance of Mobility-as-a-Service into broader sustainable mobility policies • Allow public transport operators the freedom to negotiate with Mobility-as-a-Service providers • Create data-sharing frameworks that are as open as possible, as constrained as necessary • Define common building blocks for sharing data within a Mobility-as-a-Service ecosystem
Customer Attraction Playbook	UITP	October 2021	• Become more deliberate and commercially minded about attracting customers back to public transport • Get the basics right – real-time info (arrivals, crowding, etc.), hygiene, safety • Engage in dynamic feedback with customers – digital tools/app interactions • Data is critical • End-to-end journey: partners together delivering seamless mobility

The Innovative Mobility Landscape: The Case of Mobility as a Service	International Transport Forum	July 2021	• Anchor the governance of Mobility-as-a-Service in a strategic vision, applied to the whole functional urban area and informed by effective digital monitoring • Seek a greater understanding of how Mobility-as-a-Service can add value for the user • Guide Mobility-as-a-Service where necessary to achieve agreed societal outcomes • Adopt a flexible and light-handed regulatory approach towards Mobility-as-a-Service platforms • Adopt a predictable regulatory approach and allow for evolution • Enhance public transport authorities' and operators' ability to negotiate terms of sale and reuse of tickets with Mobility-as-a-Service providers • Base data-sharing frameworks on the principle of 'as open as possible, as closed as necessary' • Build data portability into the MaaS ecosystem by default • Consider common building blocks for sharing data • Establish data-reporting requirements that are proportionate and targeted to outcomes • Adopt complementary policies in other areas to ensure that the Mobility-as-a-Service ecosystem contributes to desired policy outcomes • Invest in the built environment and interchange facilities • Skill sets will need to evolve to improve the public authority's capacity to regulate and assess the digital market
Piecing Together the Puzzle: Mobility as a Service from the User and Service Design Perspectives	International Transport Forum	January 2021	• Taking a systems approach to understanding MaaS developments can aid public authorities and actors to better understand which enablers and barriers exist • A lot is being asked of potential MaaS users, as integrated mobility services can be difficult to grasp, especially whilst they remain largely unobservable and untestable • There is limited evidence to either support or contradict the many hopes projected onto MaaS
Mobility as a Service: A New Ambition for Public Transport Authorities	International Transport Forum	October 2020	• MaaS can help with setting up pricing policies that penalise the private car and more generally the modes of transport that are most harmful to the environment • The distribution of the road network between the different modes of transport and traffic regulation (including pricing) must be on the same hand as the definition of the public transport offer

(Continues)

Table 16.1 (*Continued*)

Document title	Publisher	Published	Key policy recommendations
			• Public authorities must acquire new skills and present themselves as 'trusted third parties' capable of offering all stakeholders the integration of data in a public database ensuring equal rights and duties for all
How Mobility as a Service Impacts Public Transport Business Models	International Transport Forum	October 2020	• Business model discussions are very much in their infancy, with key discussions and future research needing to be directed at understanding better the links between business models and organisational forms and governance • What the transition between MaaS as niche and MaaS being pervasive might be
The Future of Mobility Post-COVID	Arthur D. Little / UITP	July 2020	• Think and act at the system level: vision, regulation, execution planning, and mobility funding equation • Foster innovation through public-private collaborations on technology & innovative business models • A Unified Mobility Management Model beyond MaaS, enabling real-time optimisation of flows and assets
Regulating App-Based Mobility Services	International Transport Forum	August 2019	• Maintain a permissive regulatory environment for new app-based mobility services • Treat incumbent mobility providers and new market entrants equally • Revise outdated and fragmented regulatory frameworks for mobility services • Focus regulation on addressing clearly identified market failures • Take the broader urban policy environment into account when designing regulations • Consider subsidies for app-based mobility services where appropriate and invest in supporting infrastructure
Ready for MaaS?	UITP	May 2019	• Authorities need to take action now as global platforms are building solutions that do not necessarily have a non-discriminatory approach • If deployed in the right framework centred on high-capacity public transport and active mobility it can be a tool to change citizen's travel behaviour towards more sustainable modes and optimise urban mobility systems • MaaS is the business model of the future

Optimising New Mobility Services	International Transport Forum	December 2018	• Design and pilot regulations informed by city goals • Understand the costs of new mobility services • Evaluate system performance and not just those of mobility operators • Develop new mobility ecosystems
Mobility-as-a-Service: Implications for Urban and Regional Transport	POLIS	September 2017	• Where new mobility services do develop, the policy environment (at EU, national, or local level) should ensure these contribute to sustainable mobility goals, for example by being complementary to public transport provision and the encouragement of active modes (walking and cycling), i.e., truly intermodal trips • Whether it makes sense to encourage the development of third-party private MaaS platforms will depend on local circumstances and, in particular, on the level of integration of existing services • City and regional authorities need to be involved in the development of policy around MaaS at the EU and national levels, through new models of governance and with public sector leadership, to avoid environmental, economic, and social dysfunctions • MaaS should not be regarded as a distinct player from policy; it can only achieve its goals if integrated with other measures such as low emission zones, pedestrianised areas, on-street parking policies, personal/workplace mobility management, etc.

tickets require further development; making them a standard requirement in any relevant public procurement contract would promote greater use.

Central to this will be ensuring that all aspects of transport are ticketless and interoperable. The Commission commits to updating technical specifications for interoperability, including the form of electronic consignment notes, easy and affordable use of cross-border car rentals, contactless payments for parking and tolls, and better information about areas in which cities or local authorities restrict car use to tackle congestion or improve air quality would all contribute to a smoother driving experience.

The strategy also places a focus on developing key enabling technologies and skills. A key area of focus is in creating the Common Mobility Data Space. The strategy states the following:

> It will take into consideration the horizontal governance set out in the data strategy and the Data Act and the principle of technology neutrality. The aim is to collect, connect and make data available to meet EU objectives, from sustainability to multimodality. This Mobility Data Space should function in synergy with other key systems, including energy, satellite navigation and telecommunications, while being cyber safe and compatible with Union data protection standards. At the same time, a level playing field for data in the value chain must be preserved so that innovation can thrive and new business models emerge. The Commission will consider different regulatory options to give operators a safe and trustworthy space to share their data within and across sectors, without distorting competition and while respecting privacy and the Union's international obligations.

16.8 Summary of policy making for MaaS

MaaS is often found as part of various policies at local, national, and international level. However, its presence is generally supportive of wider goals. For instance, it is expected to support sustainable development, social policies (as an element of access to services and opportunities), and economic policies (through transport to work).

MaaS sits on the interface between public and private services, and there are policy questions over whether it should be public or privately delivered. The answers depend very much on the context, attitudes to public and private service delivery more generally or within transport in particular.

Degrees of devolution at the national level also affect the development of MaaS policy, with highly devolved areas often aspiring to MaaS solutions as policy.

A further element of the relationship between MaaS and policy is the extent to which other policies indirectly affect MaaS, for instance, open data policies, digitalisation, environmental, and transport policies will all affect the development of MaaS.

References

[1] Introducing a "Government as a System" Toolkit – Policy Lab [Internet]. openpolicy.blog.gov.uk. Available from: https://openpolicy.blog.gov.uk/2020/03/06/introducing-a-government-as-a-system-toolkit/

[2] Future of Mobility: Governance of UK Transport Infrastructures [Internet]. GOV.UK. Available from: https://www.gov.uk/government/publications/future-of-mobility-governance-of-uk-transport-infrastructures

[3] Pangbourne K, Mladenović MN, Stead D, and Milakis D. Questioning Mobility as a Service: Unanticipated Implications for Society and Governance. *Transportation Research Part A: Policy and Practice*. 2020;131:35–49.

[4] United Nations. The 17 Sustainable Development Goals [Internet]. United Nations. 2015. Available from: https://sdgs.un.org/goals

[5] Ponti M, Boitani A, and Ramella F. The European Transport Policy: Its Main Issues. *Case Studies on Transport Policy*. 2013;1(1–2):53–62.

[6] Smith G, and Hensher DA. Towards a Framework for Mobility-as-a-Service Policies. *Transport Policy*. 2020;89:54–65.

[7] Movement for Growth: The West Midlands Strategic Transport Plan [Internet]. Transport for West Midlands (TfWM.org.uk); 2016 [cited 14 Aug 2023]. Available from: https://corporate.tfwm.org.uk/media/1099/movement-for-growth.pdf

[8] Mobility as a Service [Internet]. Transport for West Midlands (www.tfwm.org.uk). Available from: https://www.tfwm.org.uk/who-we-are/our-strategy/innovation-and-future-transport/mobility-as-a-service/

[9] Land Transport Master Plan 2040 [Internet]. Land Transport Authority (Singapore); 2019 [cited 14 Aug 2023]. Available from: https://www.lta.gov.sg/content/dam/ltagov/who_we_are/our_work/land_transport_master_plan_2040/pdf/LTA%20LTMP%202040%20eReport.pdf

[10] UK Government. The UK's Industrial Strategy – GOV.UK [Internet]. www.gov.uk. 2019. Available from: https://www.gov.uk/government/topical-events/the-uks-industrial-strategy

[11] Build Back Better: Our Plan for Growth (HTML) [Internet]. GOV.UK. 2021. Available from: https://www.gov.uk/government/publications/build-back-better-our-plan-for-growth/build-back-better-our-plan-for-growth-html

[12] Future of Transport Regulatory Review: Call for Evidence on Micromobility Vehicles, Flexible Bus Services and Mobility-as-a-Service [Internet]. GOV.UK. Available from: https://www.gov.uk/government/consultations/future-of-transport-regulatory-review-call-for-evidence-on-micromobility-vehicles-flexible-bus-services-and-mobility-as-a-service

[13] Future of Transport Regulatory Review Summary of Responses Moving Britain Ahead [Internet]. 2020. Available from: https://assets.publishing.service.gov.uk/government/uploads/system/uploads/attachment_data/file/989583/Future-of-Transport-Regulatory-Review-Summary-of-Responses.pdf

[14] Future of Transport Regulatory Review: Regulatory Sandboxes [Internet]. GOV.UK. Available from: https://www.gov.uk/government/consultations/future-of-transport-regulatory-review-regulatory-sandboxes/future-of-transport-regulatory-review-regulatory-sandboxes

[15] Smith G, Sochor J, and Sarasini S. Mobility as a Service: Comparing Developments in Sweden and Finland. *Research in Transportation Business & Management*. 2018;27:36–45.

[16] Innovative Mobility Services in Finland Key Messages [Internet]. [cited 14 Aug 2023]. Available from: https://www.oecd.org/climate-action/ipac/webbooks-practices/dynamic/ipac-case-studies/bc4ce864/pdf/innovative-mobility-services-in-finland.pdf

[17] Korkiakoski M. Mobility as a Service and Intelligent Mobility by Tekes and Finpro [Internet]. MaaS – Tekes. Tekes and Finpro; 2016 [cited 14 Aug 2023]. Available from: https://vdocuments.mx/maas-tekes.html?page=1

[18] Business Finland [Internet]. [cited 14 Aug 2023]. Available from: https://www.businessfinland.fi/48fc14/globalassets/finnish-customers/about-us/bf-temaattiset-painopisteet-en.pdf

[19] Intelligent Mobility Skills Strategy Growing New Markets in Smarter Transport [Internet]. 2016 [cited 14 Aug 2023]. Available from: https://cpcatapult.wpenginepowered.com/wp-content/uploads/2021/07/IM_Skills_Strategy.pdf

[20] Crozet Y, and Coldefy J. Mobility as a Service (MaaS): A Digital Roadmap for Public Transport Authorities [Internet]. CERRE Centre on Regulation in Europe; 2021 [cited 14 Aug 2023]. Available from: https://cerre.eu/wp-content/uploads/2021/01/CERRE_EN_Mobility-as-a-Service-MaaS_A-digital-roadmap-for-organisational-authorities_January2021.pdf

[21] admin. techUK Launches New Transport Skills Report [Internet]. transportandenergy. 2023 [cited 14 Aug 2023]. Available from: https://transportandenergy.com/2023/03/29/techuk-launches-new-transport-skills-report/

[22] Butler L, Yigitcanlar T, and Paz A. Barriers and Risks of Mobility-as-a-Service (MaaS) Adoption in Cities: A Systematic Review of the Literature. *Cities*. 2020;103036.

[23] Lyft [Internet]. Lyft.com. 2021. Available from: https://www.lyft.com/blog/posts/how-lyft-works-with-public-transit-agencies-across-the-country-to-eliminate-transportation-barriers

[24] Mostofi H. The Frequency Use and the Modal Shift to ICT-Based Mobility Services. *Resources, Environment and Sustainability*. 2022;9:100076.

[25] Shamshiripour A, Rahimi E, Shabanpour R, and Mohammadian A (Kouros). Dynamics of Travelers' Modality Style in the Presence of Mobility-on-Demand Services. *Transportation Research Part C: Emerging Technologies*. 2020;117:102668.

[26] Turoń K. Open Innovation Business Model as an Opportunity to Enhance the Development of Sustainable Shared Mobility Industry. *Journal of Open Innovation: Technology, Market, and Complexity*. 2022;8(1):37.

[27] Act on Transport Services, Ministry of Transport and Communications (Finland) [Internet]. [cited 14 Aug 2023]. Available from: https://www.finlex.fi/en/laki/kaannokset/2017/en20170320.pdf

[28] GBFS and Shared Mobility Data Policy for European Cities [Internet]. MobilityData. 2021 [cited 14 Aug 2023]. Available from: https://mobilitydata.org/gbfs-and-shared-mobility-data-policy-in-europe/

[29] United Nations. The 17 Sustainable Development Goals [Internet]. United Nations. 2015. Available from: https://sdgs.un.org/goals

[30] Mobility as a Service Mobility as a Service Inland Transport Committee [Internet]. Available from: https://unece.org/DAM/trans/main/wp5/publications/Mobility_as_a_Service_Transport_Trends_and_Economics_2018-2019.pdf

[31] EUR-Lex – tr0054 – EN – EUR-Lex [Internet]. eur-lex.europa.eu. [cited 14 Aug 2023]. Available from: https://eur-lex.europa.eu/EN/legal-content/summary/roadmap-to-a-single-european-transport-area-towards-a-competitive-and-resource-efficient-transport-system.html

[32] Putting European Transport on Track for the Future #MobilityStrategy [Internet]. Available from: https://transport.ec.europa.eu/system/files/2021-04/2021-mobility-strategy-and-action-plan.pdf

Chapter 17
MaaS regulation

17.1 Introduction

Policymakers globally have, for a number of years, sought to understand what changes in regulations are required to enable MaaS to become more widely adopted within their jurisdictions. The regulatory frameworks and procedures for adopting laws vary significantly between states and even between different regions within states, depending on the degree of devolution of transport and other powers from national governments.

Whilst governments have a number of options to enact policy to enable the adoption of MaaS (see the previous chapter), the changing and adoption of regulation is unique among the policy tools. Regulations and laws establish enforceable common rules for markets and common standards to which mobility providers are expected to adhere. They are the most prescriptive policy tools and consequently can be seen as a significant barrier to the emergence of new MaaS solutions [1]. However, legislation can also be written in a manner that defines end goals and objectives, potentially facilitating the emergence of MaaS solutions.

At the core is the regulatory debate is balancing the need for competition, and the need for collaboration between the different mobility partners to achieve a wider public policy objective. This in turn affects the governance structures and roles of public authorities and mobility actors within it. The Union Internationale des Transports Publics (UITP) [2] has identified three broad governance methods, whereby public authorities and operators take different roles:

- In a **commercial integrator model**, MaaS operators sign bilateral commercial agreements with a variety of operators to provide services through its platform. The role of the public authority is minimal, save for policing competition and establishing common, but voluntary, frameworks for cooperation on matters like data sharing and fares. This approach was used in the West Midlands, where MaaS operator Whim signed commercial agreements with rail operators, taxi and car hire companies, and became part of the West Midlands Combined Authority joint ticketing scheme for local buses.
- In the **open back-end platform model**, the public authority sets up a platform in which data from the various mobility services is provided (ticketing, timetables, route calculation, booking, etc.) and is provided in an integrated fashion. MaaS operators then use this platform to build their solutions. In Vienna,

for example, two public sector organisations established a subsidiary company to provide this platform, which is used by local MaaS operators.
- In the **transport as an integrator model**, the public transport operator develops this service and invites other mobility operators to be part of the MaaS solution. This corresponds to the example of Hanover, where Mobilitätsshop is led by the main urban transport operator (Üstra).

The relative benefits of each governance model have yet to be fully established, and evidence of the impact of each is currently lacking. There is a perception that the commercial integrator model will give commercial organisations the opportunity to innovate and provide more customer-focused solutions that are both widely applicable to a general audience and serve niche markets. This approach is also emerging in sectors where there is a significant degree of private sector activity, for example through Uber providing bicycles and e-scooters through its app. However, the pivot by mobility operators to B2G models (see Section 5.10) indicates that this is not a clear-cut issue.

There is also a perception that the transport as an integrator model may lead to the achievement of wider policy objectives. In many instances, local public transport operations are a role of the city or regional authority, meaning that any MaaS solutions could be designed with wider public policy objectives in mind.

17.2 Data protection and security

The delivery of MaaS is sometimes considered to be predicated on the sharing of personal data, primarily about travel habits but also socio-economic data (e.g., age) to provide add-on services, and in some instances to provide identity verification. The operation of transport systems and the delivery of personalised MaaS solutions relies on the availability of good quality data, including [3]:

- Public data – typically held by national, regional, and local public sector authorities and agencies, and in some cases by local communities;
- Private data – typically held by private organisations, is often not publicly available or only available via licencing, and may include sensitive or proprietary information such as bank details, mobile phone usage statistics, and historical location tracking data;
- Data Exhaust – also known as passive data that is collected by agencies but is often non-core. Whilst on its own it is not valuable, it can add value to other datasets. For example, web search histories are combined with observed travel behaviours;
- Community data – typically unstructured and primarily relating to social interactions, especially social media data;
- Self-quantification – when individuals collect data on themselves, often to track progress against a specific goal. For example tracking runs, cycle rides, and health data.

A significant amount of data that is potentially useful for MaaS is collected every day. By 2025, it is estimated that 463 exabytes of data will be collected every day, and the average connected car will collect 2 terabytes of data every day [4]. Added to this, the regulatory landscape for the collection, handling, and processing of data as well as privacy laws is highly fragmented between different countries. For example, a company may be headquartered in the United States, have data processing centres based in the European Union, serving customers in Japan. Within the context of MaaS, this example may be even more complex when multiple actors are working in collaboration to deliver a transport service, each of which will have their own data storage and processing arrangements.

Within the context of the European Union, the application of the General Data Protection Regulation (GDPR) is particularly relevant, particularly as it applies to all companies that process the data of EU Citizens regardless of the geographical location of their headquarters. The key changes introduced by GDPR include [5]:

- The GDPR increases the territorial scope of applicability, as it applies to all companies that process personal data of EU citizens, regardless of where the company is located.
- Requirements for obtaining consent have also been strengthened and place more onus on companies to provide consent language that is clear and easily accessible to the user. It must also be as easy to withdraw as to provide consent.
- The requirement to consider Privacy by Design has been codified into the regulation, which indicates that: 'The controller shall . . . implement appropriate technical and organisational measures . . . in an effective way . . . in order to meet the requirements of this Regulation and protect the rights of data subjects'.
- Those data controllers and processors whose core activities consist of processing operations requiring large-scale, regular, and systematic monitoring of data subjects or monitoring of special categories of data will be required to appoint a Data Protection Officer (DPO).
- Individuals are provided with the 'right to be forgotten', i.e., the user may request that the responsible data controller remove his or her personal data and cease any further dissemination of the data. This is also aligned with the increased rights of the individual to obtain information on whether his or her personal data are being processed and, if so, where and for what purposes.

Within the MaaS ecosystem, MaaS operators can act as both data controllers and data processors. This is complicated by the fact that different organisations within the MaaS ecosystem can act in the role of both data controller and data processor and as the sharing of data about travellers is considered to be a necessary pre-requisite for MaaS to work, understanding these roles and responsibilities is essential. Views on this vary significantly. For example, urban mobility data platform provider, Vianova, states that within their preferred data licensing regime [6]:

- cities are data controllers, for they collect and use mobility data for their own defined purposes;

Table 17.1 Data present in mobility data specification and its applicability to GDPR [7]

Types of data	Present in MDS	Subject to GDPR
Personal information about riders (name, address, payment, etc.)	No	Yes
Vehicle location data connected to anonymous rider's trip, trip data, vehicle IDs (and associated data)	Yes	Yes, in most circumstances
Aggregated trip data, fleet information (without vehicle IDs), and location data were not connected to any rider's trip	Yes	No, in most circumstances
City policy and geographic definitions	Yes	No

- mobility operators are data sources, from which such aforementioned data is collected;
- third-party platforms and technical service providers such as we, Vianova, are data processors.

Cotterill, in contrast, states that there is still significant scope for evaluation of the most appropriate practice. Cotterill recommends that the following key practices be codified in any MaaS agreement.

- Development of consistent consent language that is clear and understandable, and adequately represents the underlying practices of relevant parties;
- Incorporation of Privacy by Design principles in the system architecture, particularly in instances where the sharing of data between processors is necessary for service provision or utilisation;
- Determine ways to minimise collection of data where possible;
- Establish consistent practices for users to request the removal of data and methods of compliance across involved partners;
- Appoint a Data Protection Officer

The Open Mobility Foundation, which is responsible for the development of the Mobility Data Standard, has come to the view that it is lawful to collect and process MDS data even in a non-aggregated form, including data that would be considered personal under GDPR. They have summarised the different types of data and whether they are covered under GDPR in Table 17.1.

17.3 Competition against collaboration

Initial regulatory efforts associated with Mobility-as-a-Service were focused on enabling a competitive market in the sphere of MaaS providers through technical integration enabled by legislation. Common examples include enforcing transport service providers to open up data sources on their operations and their pricing, with the public sector playing the role of both enforcers of the legislation and enabling compliance with transport operators.

This notion of opening up the MaaS operator space to competition varies significantly from the concept of MaaS which requires close collaboration between organisations at all levels. Specifically within the realms of data sharing. The ability of MaaS to provide an open marketplace for different mobility options based (partly) on the knowledge of personal preferences and trip choices whilst requiring them to collaborate in a manner beneficial to the system as a whole is an unresolved challenge of MaaS; and one where the solutions are likely to be based as much on existing practice as it is on legislation.

The evidence concerning the impact of MaaS projects on competition and collaboration is so far limited. Few regulatory authorities have undertaken impact studies to date in terms of the impact of regulatory changes on competition and collaboration. Regardless of this lack of evidence, it is important to consider a basic tension in the transport industry (and congruently in MaaS) between the benefits of competition and the benefits of collaboration.

Where Mobility-as-a-Service is being offered in a competitive market setting, MaaS operators, service providers, and regulators need to ensure that MaaS and its associated sharing of data do not result in market collusion, whereby dominant organisations distort the market in a manner that inhibits competition. Evidence from the operation of passenger transport markets indicates that markets tend towards concentration of a few providers at the route, local, and regional levels [8]. There is a significant risk that dominant operators in a market could share data in a way that further entrenches their market position, to the detriment of open competition.

Furthermore, collusion and monopolies in a MaaS setting are not just horizontal (i.e., at the MaaS operator or service provider level), but throughout the vertical supply chain. Platforms and operators can design the system architecture to exclusively obtain data from various sources – especially private data sources – and accordingly, create a silo or 'walled garden' which becomes the industry standard purely through its scale [9].

However, there is also the potential for collaboration to achieve public policy goals, and also benefit businesses. One way to reimagine this tension is to consider collaboration as establishing a system of standardised rules, processes, and procedures that create a homogenous and open playing field [10].

A further challenge to the effective delivery of a collaborative environment to develop MaaS is regulatory procedures in other, related fields, that restrict the ability of collaborative approaches to be delivered. For example, traditional procurement processes (including those of procurement of transport services) seek to internalise the value added from a procurement within the service itself, as opposed to a more collaborative approach whereby value is generated outside the immediate service offering and is accordingly considered as part of the procurement [11].

17.4 Consumer protection

There have been limited attempts currently to apply specific consumer protection legislation to MaaS. Governments have instead followed the approach to adapt existing consumer protections in the realms of transport service operations, privacy, digital economy, and e-commerce to MaaS. For instance, the requirements to produce a Data Privacy policy have their origins in the General Data Protection Regulation from the European Union.

Within the realm of consumer protection, there are several aspects to consumer protection:

- Data protection and security, as outlined previously in this chapter;
- Minimum expected service levels, especially minimum customer service standards that may be industry practice or set down in legislation that are relevant to MaaS – for example a requirement to respond to an issue raised with a transport service within 28 days;
- The right to redress in the event of disruption and alteration/cancellation of travel plans, including rights associated with making alternative travel arrangements and compensation for delays.

Current legal frameworks have been developed from legacy transport laws and e-commerce legislation often first passed when e-commerce began to emerge as a significant element of the economy. For instance, the European Union has recently begun to update its rules concerning digital contracts and a revised Payment Service Directive [12], replacing rules that have been in place since the early 2000s.

The result of this is that not only are the levels of protection for consumers highly variable across different transport modes and technological intermediaries but there is often little by way of strict definition as to what good consumer protection entails [13]. For example, for a MaaS user who transfers from a bus to a train during their trip, there are still questions over whether they are afforded the level of consumer protection of a bus passenger, a rail passenger, or the user of a technological platform.

This is a current significant legal grey area, and the lack of relevant definitions has resulted in courts interpreting who is responsible for consumer protection in the light of existing laws. For example, the Court of Justice in the European Union in Case C-434/15 *Asociación Profesional Elite Taxi v Uber Systems Spain SL* was required to judge whether Uber was a transport service or an information service. It judged that [14]:

> Article 2(a) of Directive 2000/31/EC of the European Parliament and of the Council of 8 June 2000 on certain legal aspects of information society services, in particular electronic commerce, in the Internal Market ('Directive on electronic commerce') refers, must be interpreted as meaning that an intermediation service such as that at issue in the main proceedings, the purpose of which is to connect, by means of a smartphone

> application and for remuneration, non-professional drivers using their own vehicle with persons who wish to make urban journeys, must be regarded as being inherently linked to a transport service and, accordingly, must be classified as 'a service in the field of transport' within the meaning of Article 58(1) TFEU.

This essentially means that Uber is classified as a transport service, for which the technology (the app) is simply a linked part of that service. Accordingly, consumer (and wider market) protections should be applied in a manner that reflects Uber being a transport service.

But such judgements do not provide clarity on how transport services and information intermediary services (e.g. MaaS operators) are defined, as the appropriate legal framework is either uncertain or missing. Several recommendations have been made in academic study [15], notably the following:

- Clarify the qualification of MaaS operators as an Information Society Service or a Transport Service provider, with a strong recommendation being stated that it should be defined as the service is presented to the user;
- Develop common standards and regulatory frameworks governing the conduct of digital intermediaries;
- Develop an appropriate legal framework for multi-modal contracts, especially the creation of 'omnimodal' passenger rights;
- Effectively regulate conflicts of interest, especially where there is a strong degree of vertical integration between digital intermediaries and services (e.g., the MaaS operator);
- Improve the effectiveness of extrajudicial redress procedures, for example through having a binding Alternative Dispute Resolution Procedure.

The approach adopted for the regulation of consumer rights so far has been primarily to utilise existing rights within the context of MaaS. There has also been a significant movement towards self-regulation of rapidly emerging mobility markets, including MaaS and ride-hailing [16], so that the potential of such options is understood prior to regulating in a wider consumer and social interest.

17.5 Pricing

The pricing of transport has, for many years, been used as a mechanism by which demand for transport, especially during the traditional transport peak hours, has been regulated. In addition to this function of regulating demand through peak pricing, the use of price discrimination has also been utilised as a means of financing transport systems at least in part. In the wake of the COVID-19 pandemic, public transport revenues were particularly hard hit as a result of restrictions and ongoing recovery. For instance, in the Netherlands, the impact on farebox revenue of the COVID-19 pandemic was estimated to be EUR 1 billion in 2020 [17].

The importance of regulating travel by way of pricing is articulated by Horcher and Graham [18]. In that, the price of travelling by a set mode, much like pricing elsewhere in the economy, should ideally reflect the wider marginal social cost of travelling. Failure to do so means that some of the trips made on transport networks are wasteful and their social cost exceeds the social benefits, or some of the useful trips are not realised.

In the majority of places where MaaS has so far been deployed, one of two pricing mechanisms has been deployed. The first is a 'pay-as-you-go' offer that can vary according to the time of day or the distance travelled, which is often sold through a single or return ticket. The other is what is broadly referred to as a 'non-uniform tariff structure', but in reality is a collection of different types of tickets such as zonal fares and block tickets where a discount is secured by paying for travel in a block. These types of tickets typically give heavy discounts for regular use, sometimes tantamount to unlimited travel within a set zone. Many areas charge a premium for travel during the traditional hours of peak demand, especially during the morning peak hours from (typically) 7 a.m. to 9 a.m.

MaaS operators have typically sought to replicate these different pricing regimes through their standard service offerings. However, the degree of direct replication varies according to the local context and the degree of independence of the MaaS operator. For MaaS operators who operate with a degree of independence from public authorities and transport operators, and who are free to establish their own business deals, a variety of service bundles are often offered that are highly targeted towards the target audience. In terms of the business deals of such MaaS operators, a common pattern is to sign up to the commercial terms of existing transport providers, with MaaS operators then seeking to generate profit through users not fully utilising the trips available to them. For example, the Citymapper Pass [19] in London offers the use of the local public transport network (the Super Pass), the Santander Cycle Hire Scheme, e-scooters, taxis, and car clubs as part (the Super Duper Pass). The Super Pass is available at less than the price of a weekly Oystercard, because whilst Citymapper has signed the standard commercial agreement for reselling tickets with Transport for London, its route to profitability is for people to take fewer trips than the equivalent Oyster ticket (see Section 5.4).

In addition to the intention to recover the operating costs of transport services, specific fares may be offered for equity reasons. For example, research has indicated that, if crowding externalities are not considered, the provision of travel passes is a progressive policy that favour lower income groups [20], groups that typically use local public transport (especially buses) more [21]. Free travel can also be offered to specific groups, such as schoolchildren and the elderly. To date, there is little evidence that the welfare impacts of specific pricing mechanisms or regulations play a role in pricing MaaS.

Subscription-based service offerings seem to be a popular method of delivering MaaS operations where they have been deployed globally.

In classic economics, the relationship between price, supply, and demand is supposed to arrive at an optimal value to consumers and fit demand to supply (i.e., be high priced at peak times to avoid overcrowding) and produce the best income for the transport network. Flat fare models distort this relationship as once a pass has been paid for it can be used intensively (beyond optimal for the transport system if this is a factor in crowding) without additional costs.

The psychological draw of simple flat fares is evidenced by the increase in revenue experienced by Transport for London on the introduction of the Oyster pass. This key event is cited by those advocating subscription models for MaaS.

Whilst transport economists see dynamic pricing as a way to manage transport networks [22], there is a tension with the simple pass or subscription which reduces the cognitive load of making multiple small decisions about transport every day and equates the payment with a car finance payment.

Despite this, academic research is becoming more interested in the potential role of MaaS as a price regulation tool to implement marginal cost pricing in urban transport. Estimates from the surge pricing mechanism adopted by Uber in North America show this potential, where a consumer surplus of $1.60 for every $1 spent on Uber rides [23]. The implementation of the surge pricing dynamic within the Uber app indicates that pricing for the marginal costs of transport is technically feasible. Whether or not it is desirable is another matter.

17.6 Interfaces with transport mode regulation

Existing deployments of Mobility-as-a-Service have taken place in a regulatory framework that has been developed over many decades. Accordingly, the MaaS solutions deployed have had to be mindful of a number of regulatory issues across a wide variety of modes of transport when seeking to deploy a multi-modal solution.

Within most nations, there is often no single governance of the wider transport system, but a multitude of governance networks closely aligned with single modes or with specific transport industries. This makes integrated outcomes, crucial for the achievement of MaaS in many areas, difficult to achieve.

Part of this complexity is due to the organisations which are responsible for the governance of different modes of transport. Arrangements vary significantly between countries, but some areas of commonality are the presence of a national government department with overall responsibility for the transport system, and local or regional transport and highway authorities delivering transport responsibilities at a local level. Both types of organisation may establish Executive Agencies with the necessary authority to undertake the statutory duties (Table 17.2).

A similar complexity is a question of which organisations are directly responsible for specific aspects of regulation in a manner that could affect MaaS operations. For instance, for the bus market in the United Kingdom, the primary regulatory role is that of the Area Traffic Commissioner. The Traffic Commissioner is the registration body for bus services and is also responsible for enforcement action. Local Transport Authorities typically play a more limited role other than tendering supported bus services and home-to-school transport services. The exception to this is London, for whom Transport for London lets contracts for all local bus services in the city. Additionally, local transport authorities have a

Table 17.2 Overview of responsibilities for different aspects of mobility regulations

	International	National	Regional or local
Highways	Trans-Continental Networks	National Department for Transport National Government Departments (other) National Infrastructure Authorities Road Traffic Safety Driving and Vehicle Standards	Local Transport Authorities Local Highway Authorities
Buses		National Department for Transport National Government Departments (other) Competition authorities	Regional Government Regional Transit Agencies Local Transport Authorities
Rail	European Commission	National Department for Transport National Government Departments (other) National Infrastructure Authorities Train Operating Companies Safety Authority	Regional Government Regional Transit Agencies Local Planning Authorities Local Transport Authorities
Ferries	International Maritime Organisation European Maritime Safety Agency	National Department for Transport National Government Departments (other) Coastguard (Civil or Military) Maritime Safety Agencies Inland Waterway Agencies Marine Conservation Agencies Lighthouse and Navigation Authorities	Port Authorities and Harbour Masters Local Planning Authorities Local Transport Authorities

statutory duty under the Bus Services Act 2017 to ensure that data on local buses is published openly.

Within each transport mode can also be found a wide variety of types of regulation, for which again there are varying standards between modes. Common areas of transport mode regulation include:

- **Safety standards** pertaining to the operation, vehicles, infrastructure, liability, and management of transport services that can be integrated into MaaS offerings;

- **Licencing of vehicles and services**, including the organisations responsible, minimum standards for licenced vehicles and services, and appropriate notice periods for the commencement or cessation of new services;
- **Governance of competition**, including whether competition between services is permissible, the rules governing competition, regulation of transport monopolies, and issues associated with collusion;
- **Procurement rules for transport services**, including when it is permissible to procure transport services, minimum requirements for contracts of different values, the need to ensure a competitive process when tendering, and rules on openness and transparency.

The implications for MaaS operators within this context are significant. This is not only because MaaS operators have been required to navigate complex regulatory requirements governing each transport mode, but they also need to demonstrate how their service offering will ensure that existing operators do not become in breach of regulations themselves.

A notable example from the United Kingdom is the requirements facing the bus industry concerning joint ticketing – a key offering of MaaS. Under competition law, it is normally illegal for commercial bus operators to come together and agree on prices or agree to share customers and a market. Accordingly, new ticketing schemes including a MaaS offering need to demonstrate that it is lawful under competition law. In the United Kingdom, this is most often achieved through the demonstration of (or building on an existing) block exemption, exempting a scheme from competition rules, where the criteria include [24]:

- It must have an open door policy and not prevent an operator from joining the scheme without good reason.
- It must not restrict how operators run their own operations (e.g., choice of route, ticket types, individual pricing, and fare structure).
- It must not restrict how operators make their own commercial decisions (e.g., vehicle numbers, timetable, areas of operation).
- It must not result in operators exchanging commercially sensitive information, such as future pricing intentions or strategy, unless indispensable for the scheme to operate.

There has been a significant effort to overcome such issues through regulatory harmonisation – applying the same basic set of rules across markets. This has primarily been focused on different modes of transport, particularly on matters such as common technical standards and competition rules. This has been particularly the case in Europe, which has gone further and is seeking to establish common standards in the digital space to enable a competitive digital marketplace, including for MaaS.

There are two notable pieces of legislation passed by the European Union that play a role in establishing this digital marketplace for MaaS. The first is the Digital Markets Act [25]. What this Act does is provide a regulatory framework for digital companies, and sets down clear rules for companies – including MaaS providers – to stop them from imposing unfair rules and conditions (i.e., to prevent them from abusing their dominant market position) and to force interoperability in relation to messaging and data.

The second is the Digital Services Act [26]. This act concerns itself with the interaction between service users and digital providers. This includes new rules concerning illegal content, giving users a greater say over the content and services that are made available to them, and ensuring that products and services sold online are safe to use. In relation to MaaS, this would give consumers a greater say over the mobility options made available to them and provide minimum standards for the mobility services that are sold through digital means.

The regulation of digital services is a new frontier for MaaS. MaaS providers might fall under the different scopes of EU legislation based on their character. For example, MaaS providers might act in EU legislation as a transport service provider or digital service provider, and based on this there are different rules for service providers playing in the same market. In this sense, the proposed Data Act and Digital Services Act (DSA) should ensure that the legislative framework for MaaS is clear and that the same data-sharing rules apply to similar services [27]. The challenges posed by such common regulatory standards are likely to be better understood over the coming years as the regulations become embedded.

17.7 Regulatory change for the benefit of MaaS

'Even from early modelling work, it is already clear, therefore, that no amount of smart technology (or big data) will overcome the need for good policy, planning, and governance of the smart transition' [28,29].

Few nations have undertaken regulatory change purely with the aim of stimulating growth in MaaS or creating a marketplace for MaaS. The notable exception is the Transport Code of Finland, which was adopted in 2017. The longer-term impacts of this change in legislation (being mindful that the delivery of the legislation and understanding its impact was disrupted by the COVID-19 pandemic) have been assessed by the Finnish Government [30].

The conclusion of this analysis has shown that, whilst the Code has been successful in placing Finland at the forefront of the development and delivery of new mobility services, it was not possible to determine the systemic changes achieved as a result of the Code being delivered. An example is the emergence of new transport service operators and generating competition in road and passenger transport. However, the technical assessment has identified a number of challenges to delivering the code, including:

- The impacts of the COVID-19 pandemic on total demand for travel, and consequently the development of the new transport market envisioned by the Code;
- Significant challenges in implementing associated data regulations, for example, the use of ticket and payment interfaces and the development of technical interface solutions;

- Varying degrees of success, at least in the short term, of opening up competition in taxi operations, and no significant effects on road freight transport or the use of buses;
- Conflicts with wider regulatory issues for specific sectors meaning that conformity with regulations is still largely done on a sector-by-sector basis, especially where international regulations play a significant role (e.g., GDPR);
- Whilst the opening of interfaces and the interoperability of data has progressed, notably in Helsinki, the adoption has been slower than expected, primarily due to challenges associated with business models and the COVID-19 pandemic.

The Finnish Transport Code highlighted that regulatory change for the benefit of MaaS is not simply a case of passing a new law, but there is a significant degree of working through associated legislation and changes to both primary and secondary legislation needed to enable MaaS. Consequently, the approach increasingly being taken by governments is that either single, large, all-encompassing pieces of legislation covering all aspects of the changes needed, or smaller, more incremental changes along a regulatory path to eventually create a regulatory environment favourable for MaaS.

An example of the former is the loi d'orientation des mobilités (LOM), or the French Mobility Orientation Law [31]. This is a significant new law, covering four key objectives:

1. Get out of car addiction,
2. Accelerate the growth of new transport modes,
3. Achieve the ecological transition, and
4. Schedule investments in transport infrastructure.

The result is that the law enacted to enable the delivery of new mobility services also covers initiatives such as rolling out safe cycle infrastructure, the delivery of electric vehicle charging points, a ban on the sale of fossil fuel cars by 2040, and achieving net zero carbon emissions from transport by 2050.

More specifically for MaaS is the empowering of the Organising Authorities for Mobilities (AOMs) to include new mobility services under their jurisdiction. This includes car sharing, carpooling, and self-driving vehicles. The AOMs also are required to speed up the opening up of all data sources for services under their jurisdiction. They are charged with coordinating the opening up of data and the transport regulatory authority (ART) with monitoring and settling disputes. Data must be disseminated either statically (a file) or dynamically via an API (programming interface) on the National Access Point (NAP), transport.data.gouv.fr. This NAP collects the data and shares it with reusers.

Further to this, are also required to provide a digital interface accessible to parties offering MaaS. Also new in the LOM is the definition of two legal categories of digital services for information, reservations, and selling mobility services:

- the 'contact platform', which simply allows parties to deliver their own fare products;

- the 'distributor', which can, if the AO agrees, set the price a) for selling its own fare products and b) for reselling those of the transport operator.

The Finnish and French examples are the only two examples that can be found globally of legislation passed for the specific purpose of enabling MaaS or establishing a MaaS marketplace. Other legislation passed nominally for the purposes of future transport has tended to focus on potential service offerings that could be part of MaaS in the future, notably connected and autonomous vehicle legislation in Japan [32] and Germany [33], or has plugged regulatory gaps.

A key challenge when it comes to legislation on MaaS is the ability to strictly define what MaaS is for the purposes of legislation that is enforceable, whilst ensuring a degree of flexibility to meet broader standards and definitions of MaaS that may be applied globally. This matter is not assisted by the lack of consensus on what constitutes MaaS.

Additionally, there is a general unwillingness among policymakers to adopt a strong, regulatory approach to a market that is still emerging and being defined. An approach recommended in the United Kingdom, for example, has been to develop a voluntary code of practice, which the UK government consulted on in February 2022 [34]. Any new legislation is recommended to focus on gaps in the regulatory framework that apply to MaaS, with existing laws instead updated to reflect any future vision of MaaS [35]. This reflects both that MaaS is still hard to define, as well as the practical considerations around ensuring there is sufficient time in the parliamentary schedule to consider new laws.

17.8 Summary of MaaS regulation

Few places so far have specifically implemented regulations to promote the adoption of MaaS – the key exceptions are Finland and France. The laws which particularly enable MaaS are strong open data laws, ensuring that transport providers are transparent about their service data, fares and timetables, and legal entitlements to public transport services. Both these have helped promote transport innovation and digitalisation to fit the supply of transport with the needs of people.

There are elements of data laws that are important to inspire confidence in MaaS – such as the GDPR which protects personal data. However, whilst personal data should rightly be protected, the learning possible from aggregate data could benefit societies and improve future transport provision. The key role of governance is to regulate so that societal benefits accrue whilst individuals are protected.

Fragmented transport regulation and applying competition regulation to transport has consequences for treating public and shared transport as a single system, particularly in the United Kingdom. There is a real tension between truly open market approaches and the benefits of ensuring access to mobility within the constraints inherent to transport networks and infrastructure.

Whilst some modes are price regulated, and price transparency is often valued in regulation, there is a tension with dynamic pricing in on-demand systems and MaaS that includes them.

References

[1] Smith G, Sochor J, and Karlsson ICM. Public–private innovation: barriers in the case of mobility as a service in West Sweden. *Public Management Review*. 2019c;21(1): 116–137.

[2] UITP (2019) Mobility as a Service. UITP.

[3] George G, Haas MR, and Pentland A. Big Data and management. *Academy of Management Journal*. 2014;57(2):321–6.

[4] Desjardins J. How much data is generated each day? [Internet]. *World Economic Forum*. 2019. Available from: https://www.weforum.org/agenda/2019/04/how-much-data-is-generated-each-day-cf4bddf29f/

[5] Cottrill CD. MaaS surveillance: privacy considerations in mobility as a service. *Transportation Research Part A: Policy and Practice*. 2020;131:50–7.

[6] F T. Practical Guide to Mobility Data Sharing & Personal Privacy under GDPR Ruling [Internet]. Vianova. 2022 [cited 14 Aug 2023]. Available from: https://medium.com/vianova/practical-guide-to-mobility-data-sharing-personal-privacy-under-gdpr-ruling-1ea73950c21d

[7] Using MDS Under GDPR | Open Mobility Foundation [Internet]. www.openmobilityfoundation.org. 2021 [cited 13 Aug 2023]. Available from: https://www.openmobilityfoundation.org/using-mds-under-gdpr/

[8] UK Government Web Archive [Internet]. webarchive.nationalarchives.gov.uk. [cited 14 Aug 2023]. Available from: https://webarchive.nationalarchives.gov.uk/ukgwa/20140402161606/http:/oft.gov.uk/OFTwork/markets-work/buses#named2

[9] Björn Lundqvist, and Murati E. Collaborative platforms and data pools for smart urban societies and Mobility as a Service (MaaS) from a competition law perspective. *MPI Studies on Intellectual Property and Competition Law*. 2020;191–226.

[10] Merkert R, Bushell J, and Beck MJ. Collaboration as a service (CaaS) to fully integrate public transportation – lessons from long distance travel to reimagine mobility as a service. *Transportation Research Part A: Policy and Practice*. 2020;131:267–82.

[11] Smith G, Sochor J, and Karlsson ICM. Public–private innovation: barriers in the case of mobility as a service in West Sweden. *Public Management Review* [Internet]. 2018 [cited 31 Oct 2019];21(1):116–37. Available from: https://research.chalmers.se/publication/506505/file/506505_Fulltext.pdf

[12] e-Commerce Rules in the EU | Shaping Europe's Digital Future [Internet]. digital-strategy.ec.europa.eu. Available from: https://digital-strategy.ec.europa.eu/en/policies/e-commerce-rules-eu

[13] Faculteit Rechten Transport Services in the Digital Era: Consumer Protection and the Concept of "Mobility-as-a-Service" [Internet]. Available from: https://repository.uantwerpen.be/docman/irua/fba228/188853.pdf

[14] CURIA – Documents [Internet]. curia.europa.eu. Available from: https://curia.europa.eu/juris/document/document.jsf

[15] Pavlovski-Dikker B. Transport Services in the Digital Era: Consumer Protection and the Concept of "Mobility-as-a-Service" [Internet]. lib.ugent.be. 2022 [cited 13 Aug 2023]. Available from: https://lib.ugent.be/nl/catalog/rug01:003053918

[16] Napalang MSG, and Regidor JRF. Innovation versus regulation: an assessment of the Metro Manila experience in emerging ridesourcing transport services. *Journal of the Eastern Asia Society for Transportation Studies [Internet]*. 2017;12:343–55. Available from: https://www.jstage.jst.go.jp/article/easts/12/0/12_343/_article/-char/ja/

[17] UITP projects €40bn hit for European public transport in 2020 [Internet]. *International Railway Journal*. 2020. Available from: https://www.railjournal.com/financial/uitp-projects-e40bn-hit-for-european-public-transport-in-2020/

[18] Hilary. Pricing and Efficient Public Transport Supply in a Mobility as a Service Context [Internet]. ITF. 2020 [cited 13 Aug 2023]. Available from: https://www.itf-oecd.org/pricing-and-efficient-public-transport-supply-mobility-service-context

[19] Citymapper Pass [Internet]. Citymapper.com. 2019. Available from: https://citymapper.com/pass

[20] Jara-Díaz S, Cruz D, and Casanova C. Optimal pricing for travelcards under income and car ownership inequities. *Transportation Research Part A: Policy and Practice*. 2016;94:470–82.

[21] Cui B, Boisjoly G, Miranda-Moreno L, and El-Geneidy A. Accessibility matters: exploring the determinants of public transport mode share across income groups in Canadian cities. *Transportation Research Part D: Transport and Environment*. 2020;80:102276.

[22] Hörcher D, and Graham D. Corporate Partnership Board CPB Pricing and Efficient Public Transport Supply in a Mobility as a Service Context Discussion Paper [Internet]. [cited 22 Nov 2022]. Available from: https://www.itf-oecd.org/sites/default/files/docs/pricing-efficient-public-transport-maas.pdf

[23] Cohen P, Hahn R, Hall J, Levitt S, and Metcalfe R. Using Big Data to Estimate Consumer Surplus: The Case of Uber [Internet]. National Bureau of Economic Research Working Paper Series. 2016. Available from: https://www.nber.org/papers/w22627

[24] Competition and Markets Authority. Transport Ticketing Schemes: What's Allowed under Competition Law [Internet]. 2016. GOV.UK. [cited 13 Aug 2023]. Available from: https://www.gov.uk/government/publications/public-transport-ticketing-schemes-block-exemption-guidance-cma53/transport-ticketing-schemes-whats-allowed-under-competition-law

[25] The Digital Markets Act: Ensuring Fair and Open Digital Markets [Internet]. commission.europa.eu. Available from: https://commission.europa.eu/strategy-and-policy/priorities-2019-2024/europe-fit-digital-age/digital-markets-act-ensuring-fair-and-open-digital-markets_en

[26] European Commission. The Digital Services Act: Ensuring a Safe and Accountable Online Environment [Internet]. commission.europa.eu. 2022. Available from: https://commission.europa.eu/strategy-and-policy/priorities-2019-2024/europe-fit-digital-age/digital-services-act-ensuring-safe-and-accountable-online-environment_en

[27] Karjalainen P. MaaS Alliance Position Paper Concerning the Proposed Digital Services Act Package by the European Commission [Internet]. [cited 13 Aug 2023]. Available from: https://maas-alliance.eu/wp-content/uploads/sites/7/2021/02/MaaS_Alliance_DSA_Final070920.pdf

[28] Marsden G, and Reardon L. *Governance of the Smart Mobility Transition*. Leeds: Emerald Group Publishing, 2018.

[29] New Governance Challenges in the Era of Smart Mobility, Iain Docherty, Governance of the Smart Mobility Transition, Greg Marsden Louise Reardon, 2017.

[30] Ala-Heikkilä E. Lausunto: Liikennepalvelulain pidemmän aikavälin vaikutukset [Internet]. Taksiliiton kotisivut. [cited 20 Aug 2023]. Available from: https://www.taksiliitto.fi/lausunto-liikennepalvelulain-pidemman-aikavalin-vaikutukset/

[31] LOI n° 2019-1428 du 24 décembre 2019 d'orientation des mobilités (1) – Légifrance [Internet]. Gouv.fr. 2019 [cited 12 Aug 2023]. Available from: https://www.legifrance.gouv.fr/loda/id/JORFTEXT000039666574

[32] Author N. Japan's Traffic Law Revised to ADD RULES for Next-Gen Vehicles [Internet]. *The Japan Times*. 2022. Available from: https://www.japantimes.co.jp/news/2022/04/19/national/crime-legal/next-generation-mobility-law/

[33] BMDV – Germany Will Be the World Leader in Autonomous Driving [Internet]. bmdv.bund.de. Available from: https://bmdv.bund.de/SharedDocs/EN/Articles/DG/act-on-autonomous-driving.html

[34] Mobility as a Service: Code of Practice [Internet]. GOV.UK. Available from: https://www.gov.uk/government/consultations/mobility-as-a-service-code-of-practice

[35] Mobility as a Service – Report Overview – Transport Committee [Internet]. Parliament.uk. 2018. Available from: https://publications.parliament.uk/pa/cm201719/cmselect/cmtrans/590/report-overview.html

Chapter 18
The future of MaaS

The preceding chapters of this book have looked at the evolution and many versions of MaaS developed by 2023. The work on MaaS to date has laid the foundations of what the future will hold for MaaS, however, there has also been extensive work undertaken to understand what the future of MaaS may hold. The key themes of which are summarised in this chapter.

18.1 Lessons of the early adopters of MaaS

The ability of MaaS to realise its potential is to understand who its likely early adopters are, and its subsequent pathway to a broader mass market appeal. This is challenging in circumstances where there may be well-established ticketing systems or other solutions that integrate elements of MaaS already and have been established for a number of years. Wider research on travel behaviour change indicates that once travel habits are set, they are difficult to break [1]. Whilst unsustainable habits make change difficult, their features (frequency, automatic, resistance to change) can also make them desirable sustainable behaviours to obtain [2].

Research into the early adopters of MaaS is primarily dictated by identifying the potential markets for MaaS technology products based on largely conceptual models of changes in behaviour and stated preference surveys. Research from the Netherlands [3], for instance, surveyed 1,547 people via a series of 25 statements and questions to develop a MaaS Potential Index (MPI). A Lasso regression analysis was then used to link to the respondents' personal characteristics. The MPI indicators were:

1. Tech-savvy – use of and interest in innovative digital technologies
2. Renting or sharing – use of and interest in renting or sharing mobility
3. Multi-modal – a need to take multi-modal trips
4. Travel information – a need for more detailed and up-to-date travel information

The key characteristics of the most promising groups were:

1. Young people (not specifically defined, but broadly referred to as those aged under 30 years old)
2. People who frequently use public transport
3. Frequent flyers

4. Highly educated people
5. People more deeply concerned about the environment
6. People who frequently take leisure trips (day trips and visits to restaurants/cafes)

An international Delphi study of experts also revealed some perspectives on who the early adopters of MaaS would be [4]. This research indicated that the anticipated early adopters of MaaS are likely to be younger people (referred to as 'Gen Z') and frequent users of public transport.

Additional factors have been identified as potentially being indicators of early adopters of MaaS. These include a high degree of engagement and skills with digital technology [5], higher socio-economic status [6], and younger adults being more willing to adopt such solutions than older adults [7,8].

However, analysis of the early adopters of MaaS through the early trials has indicated something slightly more nuanced. One of the overarching conclusions is that merely introducing an app or multi-modal information or integrated payment solutions will not be enough to adopt MaaS, but 'services of the service' will need development too [8]. This includes a process of onboarding, support functions, user responsibilities, and instructions.

A study of the early adopters of the UbiGo service in Sweden [9] focussed on the technical aspects of the trial, but contained some insight into those willing to participate in the trial of new mobility services. Approximately half were men and half were women with a mean age of 38.4 years. The majority of the households consisted of two (or more) adults, with or without children. According to initial data collection, a large majority of participants used the internet and apps on computers, tablets, and smartphones on a daily basis.

Particularly relevant is the introduction of policies to restrain car use, specifically in urban areas. Experience of understanding the first adopters of MaaS for the new neighbourhood of Viva in Gothenburg, Sweden, respondents revealed that the lack of dedicated parking spaces caused them to evaluate their own travel behaviours and seek out alternative options, of which the MaaS solution was one [9].

Research from adjacent fields indicates that, whilst young people do seem to be more likely to adopt products similar in nature to MaaS, their household situation has a significant bearing on the attractiveness of these service offerings. For instance, the adoption of electric car sharing in Germany is prevalent among young people who live as a couple but without cars, or are starting a family and are using car sharing as a supplement to an existing car [10].

Other research around car club adoption indicates that whilst relatively affluent and highly educated people are prevalent among the first adopters, there is also a cohort of adopters who do so because they are experiencing pressure on their finances and lifestyle which are alleviated by car club membership [11].

18.2 Achieving social goals

As covered in the regulation and policy chapters of this book, MaaS has traditionally been viewed (and implemented) as a transportation solution addressing policy challenges, with the attainment of broader policy objectives seen as an additional benefit. Some areas have also been considered and sought to accelerate MaaS as a business opportunity. However, increasingly the positioning of MaaS in achieving wider social goals is of increasing interest to practitioners.

A significant intervention in this debate was from a paper by Pangbourne *et al.* [12] on questioning mobility as a service. Whilst a significantly theoretical paper, it identified a number of potential societal, environmental, and health implications of the delivery of MaaS as envisioned at the time. Such implications include the following:

- Having in mind a potential for 'technological gentrification' of transport, there is little discussion to date on how MaaS can work effectively to tackle urban transport problems if there are urban residents outside the system;
- MaaS' reliance on registration and digitalisation might further exclude social groups experiencing difficulties in handling new technologies or having access to banking;
- Current MaaS designs may not address transport poverty, defined as unaffordability, lack of accessible mobility options, and disproportionate exposure to transport externalities;
- The absence of active travel in many packaged MaaS products of the time may result in a lack of evidence to support the claims that bundling access to transport via MaaS packages improves transport sustainability;
- There is evidence that public transport use is being reduced by the business models of ride-hailing brokers like Uber or Lyft that make impulsive door-to-door service exceptionally convenient.

Of particular interest has been the relationship between MaaS and wider social goals concerning equity and inclusion. Much work to date has identified a number of different social factors that have affected the adoption of MaaS. Zhang and Kamargianni [13] summarise many of these social factors. There is an inverse relationship between age and the propensity to utilise MaaS packages. A study focusing on respondents from the Netherlands shows that younger people between 18 and 34 years old tend to be more willing to adopt MaaS schemes. Similar conclusions were also drawn from a study in Australia, which indicates that full-time employed young individuals are the potential MaaS users, whilst older retired individuals lack interest in MaaS. Meanwhile, there are mixed results on the effects of gender.

There has been an increasing consideration of gender equity in MaaS research in recent years. Up until 2022, there had been relatively little discussion about the gender equity implications of MaaS, and the evidence of the impact on different genders has been mixed. Though some thought has been given to this matter in the academic literature (Table 18.1).

Some assessments have also been undertaken to understand the relationship between MaaS adoption and age. Of particular interest has been the impact of MaaS on age groups who often find car ownership especially challenging, notably younger (typically people aged 21 years old or less) and older (typically people aged 65 years old or over) populations.

Table 18.1 The strengths, weaknesses, opportunities, and threats of MaaS in assisting gender equality [14]

	Positive for gender equity	**MaaS unlikely to help**
Inherent to MaaS	**Strengths** • MaaS already supports women's current travel behaviours: ○ Use public transport more. ○ More likely to combine modes. ○ Less likely to commute 'traditionally'. • Aligns with women's higher environmental values. • Aligns with women's lower value placed on car ownership.	**Weaknesses** • MaaS unlikely to counter entrenched division of family care and stereotyped societal roles. • MaaS cannot take away all the difficulties associated with being 'encumbered'. • MaaS is not about bike lanes (or other infrastructure). • MaaS unlikely to change biases in desire to engage with vehicle technologies.
Potential for the Future	**Opportunities** • Capitalise on strengths and facilitate access to mobility modes and behaviours already favoured by women. • Differences in MaaS package preferences imply that subscription models could be designed to reduce inequity. • MaaS could help with perceptions of safety by including shared and on-demand services. • In-app information on security and safety (safety from attack, safety from collisions) can help address mobility security issues.	**Threats** • Getting bundle design wrong could increase inequity. • A focus on micromobility could favour men. • MaaS could push women into less sustainable travel. • MaaS uptake highly influenced by existing habits. • Inappropriate technology (app) design could represent an additional barrier to women. • Going for low-hanging fruit (for mode shift) could be at the cost of equity.

Young people are often considered to be an integral part of the future of MaaS, with assumptions made about their willingness to adopt new technologies, lack of willingness to own a car, and propensity to live in urban areas. Some of the early research into MaaS adoption has identified young professionals as key early adopters of MaaS. Ethnographic research in smart cities indicates that technology has a key role to play for children and young people in shaping how they experience urban areas.

An example of this is the Smarter Together Project in the Simmering district of Vienna [15]. The focus of this project was the promotion of renewable energies, the refurbishment of buildings to improve energy efficiency, and ecological change in mobility and transport. Citizen participation played an important role in all subprojects. For this purpose, information campaigns, festive events, games, and interactive workshops took place during the different phases of implementation.

During the refurbishment of a school complex in the district centre, the SC project coordinators closed a street next to the school to put up a car-free zone, which could be used for temporary classroom containers and as an open space during breaks. Furthermore, they installed mobility stations to foster sustainable means of transport. They supported rentable cargo bikes, offered e-car sharing, e-bike sharing, e-charging stations, and bicycle parking boxes with a bicycle pump and an information screen. One major mobility station was installed close to the metro stop with the goal of making the ride to and from the metro line easier for people living in the district.

Despite the low awareness of overarching goals or a common framework formulated by the project, several narratives resonated with some of the main concerns and interests of young people in Simmering, such as environmental protection, better futures, and strengthening the ties between various city actors. Making urban mobility more sustainable was, for example, a priority for some participants. Notably, participants consistently expressed concern about the crisis discourses on climate change and their own observations about the lack of attention to the environment in their neighbourhood. Interestingly, the interactions with technology, whilst useful, were not of primary concern.

Meanwhile, the concerns of young people in relation to MaaS also relate to the modes that underpin MaaS offerings. Work on understanding this has identified that young people have a variety of complex needs that are not often considered in MaaS projects. In qualitative research concerning the perspectives of young individuals on the possibilities of Mobility-as-a-Service (MaaS), it is evident that their primary apprehensions revolve less around the concept of MaaS itself and more around fundamental concerns related to using public transportation for any form of travel. In line with other transport work, choices are not made on a cost-benefit rational basis but on perceptions, affective desires, and the experience of practice, including perceptions around reliability and frequency of services. In this way, choices are similar to those of adults, but other choices are specific to those of young people, such as the acute anxieties expressed around personal safety particularly when travelling alone on public transit. It was notable not only that young people felt threatened but also that they felt they were less likely to be helped [16]. There were also minor misgivings about the notion of MaaS, particularly around articulating the potential of shared modes. As with much other shared travel research, ultimately even young people value privacy and space.

Young people also perceive advantages in terms of flexibility, particularly during disruption. They have concerns about shared vehicles, a degree of continued desire to own and use a private vehicle [17] and have ambivalence towards sharing data with transport providers. They also root their understanding and interpretation of MaaS within their own travel experience, but also their experience of other digital services [18].

With regard to the elderly, it is acknowledged that there is a significant gap in the research on the impact of new transport innovations, including MaaS, on the elderly. Specific opportunities for innovation that have been identified across the development of new transport technologies include [19]:

- New vehicles, including Autonomous Vehicles, need to be designed to be accessible for people with limited mobility.
- Addressing the door-to-vehicle portion of the journey, which can be the biggest barrier to using public transport for people with health conditions.
- Designing for rural provision where there are higher percentages of older people and a greater reliance on private vehicles than in urban centres.
- Developing non-digital ways for customers to interface with MaaS and other transport information/ticketing/planning services, such as in-person or through voice-activated options.
- Build in consideration of the accessibility needs of users (including both the accessibility of vehicles and the interchanges between modes) in MaaS and other transport planning services.
- Development of equitable models of ownership and cost (including exploring community, shared, and public ownership).

Much of the academic literature focuses on the future roles that the government could play in ensuring that the delivery of MaaS-style services considers the significant variety of needs of the elderly population. In terms of government-sponsored organisational innovation, ageing populations especially require proper organisation and adaptability to best meet their needs. Some organisational innovations can only be completed by way of authorised, legal action. A key limitation to any legally enforced organisational innovation is its prohibitive cost. Moreover, the role of government to provide mobility as a service (MaaS), includes carpooling and car-sharing schemes easily usable by the elderly [20]. The second part of this, if not initially crafted with the users' needs in mind, might isolate the very users it aims to cater to.

Similar to young people, a significant future challenge for the elderly related to MaaS is in relation to MaaS's relationship with transport services. Studies into this have revealed that technological adoption and proficiency across elderly groups is not a consistent thing, contrary to perceptions of poor technological literacy. Whilst technology adoption rates are higher in younger age groups they are also rising in elderly age groups. For example, in the United Kingdom the proportion of those aged 75 years and over who are recent internet users nearly doubled since 2013, from 29% to 54% in 2020 [21].

Lessons can be learned from innovations in similar fields and a necessary view of the future is that as long as the elderly are included in future MaaS design, they can become regular users. An example offering such insight was the Breng flex, a public shared on-demand transport service that began its experimental operation on December 18, 2016, operating in the city of Nijmegen, the Netherlands until 2019. Interviews with experts and the users themselves found that in the context of the ageing population, the new innovative Demand Responsive Transport (DRT) is a prospective mobility solution for the elderly. With the support of technology, mobile internet, and smartphones, travellers can plan and reserve DRT services from their smartphone. Digital enhancements alone do not adequately address the preferences of the elderly. The inclusion of other features, such as short walking distance to embarkation points and the possibility to reserve service by telephone, is also essential in making the service elderly-friendly [22].

A matter of increasing interest is also the interface between transport and data ethics. Existing big datasets that are often utilised to develop new online services contain biases in terms of the representativeness of different groups in the population compared to each other.

It should be noted that much effort is being made to tackle the biases in numerous datasets, so that computational factors in online services are tackled, and such work is necessary for tackling such bias. Yet, human and systemic institutional and societal factors are significant sources of bias as well and are currently overlooked. Successfully meeting this challenge will require taking all forms of bias into account. This means expanding perspectives beyond the data pipeline to recognise and investigate how this technology is both created within and impacts our society [23].

Such debates are relatively unexplored in the context of MaaS specifically, but an increasing number of academics are exploring the implications of the interface between mobility justice and data justice. For instance, the work of Behrendt and Sheller [24] proposes a Mobility Data Justice Framework with a focus on distributive, procedural, and epistemic elements:

- A distributive mobility data justice perspective considers (a) how mobility and data are accessed, accumulated and distributed in (un)equal ways, (b) how data and mobility infrastructures lead to (in)equitable risks and benefits distributions, and (c) how moral subjects are constituted with relation to mobility and data, i.e., who is considered deserving or worthy of access to various mobility data or data mobilities, and who is targeted by such data collection, surveillance, and control.
- A procedural perspective on mobility data justice explores who participates in the decision making and the design of infrastructure that is relevant to the intersection of mobility and data, as well as its use. This involves considering the wide range of infrastructures relevant to mobility data (in)justices, but also the design, economic, and policy processes around them, and who gets to participate or not.
- An epistemic perspective of mobility data justice explores what counts as ‘data’ and knowledge in relation to mobility, and how it is applied to knowing the world. This includes considerations of the ways data systems are implicated in the production of intersectional (im)mobilities, including age, gender, race, sexuality, class, and disability.

It should be noted that much of this work to date has been largely theoretical in its approach, with any study of social inclusion in relation to MaaS focussing on characteristics of the population and their propensity to adopt MaaS solutions that are trialled. The results of such studies indicate that whilst age is heavily considered and monitored as part of numerous trials and studies, gender is considered less, and factors such as income and household composition are considered even less so (Table 18.2).

Table 18.2 Number of MaaS studies where socio-economic characteristics were considered [25]

	Age	Gender	Education levels	Limitations (physical)	Household	Income	Employment status
Number of studies	15	11	7	0	6	6	5

18.3 MaaS and climate change

Across the transport sector globally, the issue of reducing carbon emissions and tackling climate change is of significant concern to policymakers and practitioners. In 2016, transport was responsible for 16% of global carbon emissions [26], primarily through the use of the internal combustion engine in private cars.

It is anticipated that the ability of MaaS to enable transport systems across the world depends on its ability to influence the travel behaviours of its users. This can be summarised in how MaaS can achieve three things:

- Reduce the total amount of trips that are taking place, especially using higher carbon emission modes such as single occupancy private cars. Achieving this is essential for meeting climate change targets [27].
- Enabling a shift in the trips that are made to modes that are significantly less polluting modes. The least polluting modes are walking, cycling, the use of public transport, and car sharing [28].
- Improving the efficiency of the use of vehicles when they are used, so that carbon emissions per person and overall are reduced. The most notable is the use of public transport, but this also involves car sharing and car clubs.

A study was undertaken based on hypothetical adoption scenarios of a MaaS solution in Amsterdam (Table 18.3). The results were then linked to an emission model and the impacts on emission levels were compared for various scenarios with different MaaS bundles. Whilst the study urged caution on MaaS being seen as a solution to a significant number of urban emission issues, the results revealed that MaaS could play a role in reducing urban emissions.

Whilst all three MaaS scenarios generally show a reduced reliance on private cars and an increased preference for public transport, the optimistic scenario stands out as the one with the most significant reduction in private car usage, ultimately resulting in public transport becoming the predominant mode of transportation in terms of modal share. For this scenario, however, the share of active mode trips decreased substantially. The impacts of MaaS were then assessed for the emissions. In the conservative scenario, the total emissions decreased by around 3%–4%, whilst in the balanced scenario, the decrease amounts to 14%–17%. The most substantial decline in emissions (as expected) is associated with the optimistic scenario in which the total emissions dropped 43%–54% for various pollutants.

Table 18.3 Emissions impacts of hypothetical adoption scenarios of a MaaS solution in Amsterdam, the Netherlands [29]

Characteristic	C – Conservative	B – Balanced	O – Optimistic
Monthly fee	€240	€210	€150
Pricing scheme			
Public Transport	Pay per ride	Unlimited rides in one zone	Unlimited rides
Electric Bike Share	Pay per ride	One hour free per day	Pay per ride, 50% discount
Electric Car Share	Pay per use, 20% discount	300 min included	120 min included
Taxi	50 km included, then pay per ride	Pay per ride, 40% discount	30 km included
Electric Car Rental	Two days included	Pay per ride	Four days included
Ridesharing	Pay per ride	100 km included	Unlimited rides
On-Demand Bus	Unlimited rides in one zone	Pay per use, 20% discount	Pay per ride

Some initial research has also been undertaken to study the potential for the deployment of MaaS within the context of a carbon charge, similar to either a congestion charge or other forms of environmental taxation within the context of Changchun, China. In practice, this would present an additional charge for more polluting modes of transport through the MaaS app. The findings of this study show that, when compared to a scenario without the charging scheme, the utilisation rate of non-clean energy-powered shared cars decreases from 39.27% to 18.90%, whilst shared mobility services and public transport both experience an increase of 8.23% and 6.46%, respectively. The total carbon emissions decreased by 19.24%. Travellers tend to choose shared cars or carpools rather than public transport [30].

This field is characterised by ongoing research, with current models and analyses relying on behaviour change assumptions that need to be tested through real-world trials to confirm their accuracy. Preliminary research suggests that achieving carbon reduction goals through MaaS will hinge on the extent to which MaaS can bring about behavioural changes in how people utilise various modes of transportation.

18.4 Future perspectives on MaaS

The future of MaaS is currently clouded in uncertainty to a significant degree. Now that early trials and deployments of MaaS have been undertaken, a variety of different futures are possible. The predominant view among practitioners is that whilst there may be market growth potential for specific MaaS applications, the application of MaaS as a service offering is uncertain, and in some cases treated with scepticism.

This scepticism is somewhat rooted in the fact that a number of policy and regulatory matters related to MaaS still need development [31]. A specific issue identified by researchers is the role of government in sponsoring the enabling infrastructure for MaaS (e.g., data standards) as well as procuring the services themselves. There is also considerable uncertainty as to how the impacts of MaaS on different modes of transport can be understood, and potentially mitigated against. A notable example is the impact on public transport use, something often stated to be a key pillar of any MaaS offering.

Some scenario planning work has been undertaken to identify potential futures for the development of MaaS. In an exploration of MaaS market futures, Enoch and Potter [32] identify four distinct future scenarios.

Scenario 1 – Highly consolidated market, Highly consolidated services (local/regional scale)

This scenario represents the model seemingly most expected by many local and regional governments, particularly in Europe, whereby each public transport authority would establish its own MaaS platform monopoly across its operating region. Qualifying transport service providers, potentially a mix of commercial, public-sector and/or community operators of both traditional public transport and shared transport services, would then join each platform and could apply for subsidies to provide non-commercial but socially necessary journeys. Current examples of such schemes are being piloted in Helsinki, Finland and the West Midlands in the UK.

Scenario 2 – Highly consolidated market, and highly fragmented services (global scale)

This global oligopolistic future sees multi-national commercial organisations each entering the MaaS marketplace to capture the business in their particular field of operation. Transport operators such as Stagecoach and Deutsche Bahn would seek to dominate the provision of traditional public transport; car rental companies like Avis and Hertz would manage most of the car-sharing operations; Uber and Lyft would lead the provision of taxi-like lift-sharing options; and big car park operators (such as NCP in the UK) would provide the car parking spaces. Crucially almost everyone would connect with each other through a fully open and accessible Digital Service Provider (DSP) (perhaps provided by a global IT provider such as Google) and pay through credit card payment systems.

Scenario 3 – Highly fragmented market. Highly consolidated services (National scale)

A range of service providers (large and small) that individually offer different components form partnerships and/or merge with others. A hypothetical example might occur where an Amazon-Arriva-HSBC partnership would compete with MTR-Mastercard-Yahoo-Toyota, and Uber-Diners Club-SNCF to deliver seamless end-to-end journey solutions by the full range of modes to transport users in a monopolistic environment. Users making trips that are recognised as being ‘socially desirable’ are directly subsidised with there being no supply-side subsidies. All but the most rural areas would have at least two or three providers, and users would choose the package that suits them best. Perhaps the closest parallel to how such a market might look is the telecommunications market in nation-states across the European Union. Here, on average there are 3–4 mobile network operators per member state which compete by offering a range of subscription and pay-as-you-go 'packages' comprising telephone, text, data, and other ancillary services to consumers who then choose the offer that best meets their needs.

Scenario 4 – Highly fragmented market. Highly fragmented services (Multi-scale, from neighbourhood to global)

This model envisages a highly fragmented, multi-scale marketplace, almost from the neighbourhood level to the global level. This is highly competitive and comprises many thousands of transport service providers (ranging from individual 'little vehicle' owners renting out a scooter for a limited time period or car owners offering lifts for one-off trips, to local bus companies and multi-national multi-modal operators) promoting their service offerings to a multitude of MaaS DSPs. Users pay via their own favoured banking service thanks to the adoption of standardised data and payment API structures. Fare levels change in real-time (within bounds), according to the service attributes required by the user, the level of demand for the service, and the quality of service delivered. Such an environment is already beginning to evolve in London, whereby cashless contact credit/bank cards are set to replace the dedicated Oyster transport payment card and a whole range of new mobility products are now being trialled – e.g., Via, Uber Pool, Smart Ride and Chariot (bus-taxi hybrid schemes); Urbo, Mobike, Ofo (shared bike schemes); Drivy, DriveNow, Zipcar (car club schemes); GoCarShare, BlaBlaCar, liftshare.com (lift sharing schemes).

A critical uncertainty that affects the future of MaaS is the role to be played by the public and private sectors in the development and provision of such services. Within the literature, there is no clear consensus as to what is the most desirable method of operation of MaaS in the future. Meanwhile, current trials have delivered a variety of approaches varying by the involvement of the public sector and based on the characteristics of the local area. However, much of the future-facing literature proposes several alternatives.

There is a case for a government-contracted model where the government directly procures a mobility broker through a competitive tender, with the opportunity to negotiate contract renewal (under actionable benchmarking) at subsequent rounds once the market has matured. The move to outcome-based contracts will mean more complex performance measures (e.g., related to customer accessibility rather than vehicle kilometres) but developments in technology are fast making this possible. Under the terms of the contract, the government sets accessibility standards, which might be defined as delivering X percentage of people services within Y minutes, for a given period using any mode of the broker's choosing. One of the most promising opportunities under a contracted MaaS model is the ability of government to regulate network efficiency by incorporating a road user charge as an input into the package price.

Smith *et al.* [33] state that the degree of engagement of the public sector will be an especially critical question for the involvement of public transport in any MaaS mix. Three predictive scenarios were identified by the authors – market-driven, public-controlled, and public-private – and the implications for future public transport, in terms of the scope, usage, access, business model, competence structure, and brand value, were identified. The conclusions were that regardless of market development for MaaS, there is a regulatory requirement on behalf of public authorities to maintain a core public transport service.

Zhao *et al.* [34] take a different approach in terms of seeking to understand the value of impact and the outcomes of MaaS in the future, which may consequently indicate the value of public sector intervention. They note that policy and regulatory actions in relation to MaaS are one of the key strategic uncertainties facing the MaaS market, so policy and regulations are considered an uncertainty, first because there is limited understanding of MaaS regulatory schemes, and second, because of the possibility that such policies will be unsuccessful. Their analysis shows that MaaS can lead to sustainable reductions in mobility and CO_2 emissions from the private car sector, ranging from 2.5% to 50% if the vehicle capacity can be optimised and technologies enable the use of green fuel.

Several people have explored the development of Mobility-as-a-Service within the context of a changing world for transport and cities, often using scenario planning and systems thinking as a basis for their analysis. Utilising the Avoid-Shift-Improve Framework, Mårtensson *et al.* [35] identified a series of causal loops where MaaS could play a variety of roles across the Avoid-Shift-Improve Framework. They conclude that depending on the policy measures introduced and the stated intentions of policy (e.g., increasing or decreasing vehicle kilometres), the deployment of MaaS can have a number of positive and negative implications for achieving these goals. Examples of these implications include:

- Mobility and accessibility services potential to contribute to local lifestyles, which in turn can contribute to longer journeys to a certain type of activity due to urban sprawl;
- When aiming to realise the potential contribution to environmental sustainability, the complexity of the socio-technical context in combination with the degree of interconnectedness between mechanisms may prove difficult to navigate;
- In relation to mobility and accessibility more specifically, problems such as congestion, air pollution, noise, and greenhouse gas emissions are cumulative effects of individual activities, which can stand in conflict with higher-level aims such as a more efficient transport system.

When exploring the future of mobility in the city of Jakarta, four scenarios were developed for the future based on literature reviews and stakeholder engagement [36]. Within that, MaaS only played a key role in one scenario, referred to as One Seamless Ecosystem. It comments on the importance of MaaS within such a future as the government strives to have an integrated transportation system and sufficient infrastructure that can provide high-quality service and seamless experience to all its users, supported by MaaS technology. The use of MaaS applications includes the shared mobility services currently emerging in Indonesia.

Should ride-hailing be fully utilised, it can help increase mobility quality since it can increase the usage of vehicles. It also facilitates those who do not have access to personal mobility, such as non-owners and those who cannot drive, thus increasing the average mobility miles caused by the rebound effect.

MaaS, in the minds of policymakers and other practitioners, is considered likely to be part of the future of transport, especially in cities. But instead of being a central focus of the future of transport, MaaS is being seen in terms of achieving wider social, economic, and environmental goals.

In consideration of what the future may hold with regard to MaaS, practice from scenario planning work to date has largely been qualitative in nature. Current evidence of the impact of MaaS from existing trials is still emerging, and therefore difficult to quantify in a meaningful manner to form a qualitative measure of the impact of MaaS in different places in the future. Consequently, many of the perspectives on the future of MaaS are based on subjective or collective views of transport and MaaS experts interviewed in the development of these scenarios.

At this stage, it can be concluded that in the minds of practitioners, the future of MaaS is subject to a significant degree of uncertainty. Key questions around policy, regulation, the role of the private sector, the viability of different business models and interfaces with different modes remain subject to significant uncertainty. Across the transport industry, the collective view of the role of MaaS in the future as well as what MaaS will look like in the future is uncertain and subject to debate.

In 2020, the management consulting firm Gartner placed MaaS in the 'trough of disillusionment' in its Connected Vehicles and Smart Mobility Hype Cycle (Figure 18.1). Like many other technologies, MaaS was strongly hyped when it emerged and did not immediately live up to the hype. Although disillusionment is commonplace, it has followed the pattern that more iteration and development have established more realistic use cases.

The trough of disillusionment is where new technologies are made or broken. Many of these failures centre on inappropriate uses of technology. Because the

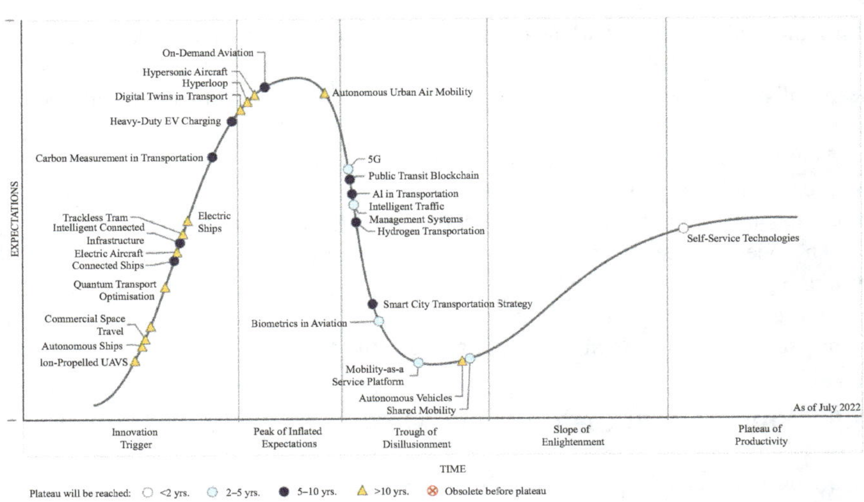

Figure 18.1 Gartner Hype Cycle for transportation and Smart Mobility [37]

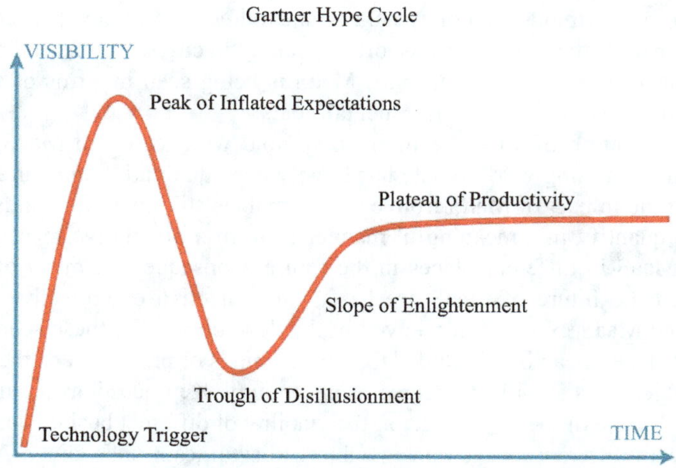

Figure 18.2 Example Gartner Hype Cycle

technology does not live up to the early overinflated expectations of enterprises and the media, it is rapidly discredited. There is not always an actual drop in the overall adoption numbers as an innovation slides into the trough. Instead, the anticipated rapid growth in adoption may simply be delayed. What suppliers and investors anticipated as a 'hockey stick' uptake remains a depressingly slow growth path. The length of the trough is one of the most variable parts of the Hype Cycle (Figure 18.2).

MaaS is a complex example as it relies on several different possible markets. Whilst most technology depends mainly on individual use, adoption, and enthusiasm, the trials of MaaS and the development of business models have increasingly revealed that it also depends on the enthusiasm of the public sector for MaaS as a means of delivering transport policy or other social, environmental, and societal benefits.

Whilst the technology itself is increasingly mature, its adoption depends on complex relationships between cities or regions and public and private operators, sometimes helped and sometimes hindered by regulatory mechanisms.

The need for better transport systems and the adoption of digitisation in all corners of people's lives creates a fertile seed for the wide adoption of MaaS. But the timescales are subject to significant uncertainty particularly as there is always the potential for unforeseen disruptions to change the context.

18.5 Summary of the future of MaaS

The relentless pace of digitisation means that MaaS is used where conditions enable it, however, the history so far implies that its widespread adoption will be a result of government intervention.

The exciting elements of MaaS that tailor transport services to tricky areas, which have the potential to make the biggest difference to individuals or areas, are also the most dependent on local and community engagement and the most resistant to scaling.

Likewise, whilst early adopters among users were younger and more affluent, the people for whom it could provide the most benefits are less likely to adopt it without significant engagement.

The many dimensions and factors make predictions difficult however the future of MaaS is unlikely to be homogenous. There are many potential scenarios, and some are more credible for some countries or regions than others.

References

[1] Marsden G, Mullen C, Bache I, Bartle I, and Flinders M. Carbon reduction and travel behaviour: Discourses, disputes and contradictions in governance. *Transport Policy*. 2014;35:71–8.

[2] Verplanken B, and Whitmarsh L. Habit and climate change. *Current Opinion in Behavioral Sciences*. 2021;42:42–6.

[3] Waterstaat M van I en. Promising groups for Mobility-as-a-Service in the Netherlands - Document (research publication) – Netherlands Institute for Transport Policy Analysis [Internet]. english.kimnet.nl. 2019 [cited 2023 Aug 13]. Available from: https://english.kimnet.nl/publications/documents-research-publications/2019/08/15/promising-groups-for-mobility-as-a-service-in-the-netherlands

[4] Jittrapirom P, Marchau V, van der Heijden R, and Meurs H. Future implementation of mobility as a service (MaaS): Results of an international Delphi study. *Travel Behaviour and Society*. 2020;21:281–94.

[5] Hilary. Piecing Together the Puzzle: Mobility as a Service from the User and Service Design Perspectives [Internet]. ITF. 2020 [cited 2023 Aug 12]. Available from: https://www.itf-oecd.org/piecing-together-puzzle-mobility-service-user-and-service-design-perspectives

[6] Zijlstra T, Durand A, Hoogendoorn-Lanser S, and Harms L. Early adopters of mobility-as-a-service in the Netherlands. *Transport Policy*. 2020;97:197–209.

[7] Alonso-González MJ, Hoogendoorn-Lanser S, van Oort N, Cats O, and Hoogendoorn S. Drivers and barriers in adopting Mobility as a Service (MaaS) – A latent class cluster analysis of attitudes. *Transportation Research Part A: Policy and Practice*. 2020;132:378–401.

[8] Karlsson ICM, Sochor J, and Strömberg H. Developing the "Service" in Mobility as a Service: Experiences from a Field Trial of an Innovative Travel Brokerage. *Transportation Research Procedia* [Internet]. 2016 [cited 2019 May 29];14:3265–73. Available from: https://www.sciencedirect.com/science/article/pii/S2352146516302794

[9] Smith G, Sochor J, and Karlsson ICM. Adopting Mobility-as-a-Service: An empirical analysis of end-users' experiences. *Travel Behaviour and Society*

[Internet]. 2022 [cited 2022 Jun 3];28:237–48. Available from: https://www.sciencedirect.com/science/article/pii/S2214367X22000345

[10] Burghard U, and Dütschke E. Who wants shared mobility? Lessons from early adopters and mainstream drivers on electric carsharing in Germany. *Transportation Research Part D: Transport and Environment*. 2019;71:96–109.

[11] Commission on Travel Demand: Shared Mobility Inquiry Response from CoMoUK [Internet]. [cited 2023 Aug 11]. Available from: https://www.creds.ac.uk/wp-content/uploads/05-CoMoUK.pdf

[12] Pangbourne K, Mladenović MN, Stead D, and Milakis D. Questioning mobility as a service: Unanticipated implications for society and governance. *Transportation Research Part A: Policy and Practice*. 2020;131:35–49.

[13] Zhang Y, and Kamargianni M. A review on the factors influencing the adoption of new mobility technologies and services: autonomous vehicle, drone, micromobility and mobility as a service. *Transport Reviews*. 2022;43:1–23.

[14] McIlroy RC. Mobility as a service and gender: A review with a view. *Travel Behaviour and Society*. 2023;32:100596.

[15] Ghafoor-Zadeh D. Moving through, interacting with, and caring for the city. Children's and young people's everyday experiences in smart cities. *Digital Geography and Society*. 2023;4:100051.

[16] Children, Young people and Mobility as a Service: Opportunities and barriers for future mobility. Transportation Research Interdisciplinary Perspectives [Internet]. 2020 Mar 1;4:100107. Available from: https://www.sciencedirect.com/science/article/pii/S259019822030018X

[17] Storme T, De Vos J, De Paepe L, and Witlox F. Limitations to the car-substitution effect of MaaS. Findings from a Belgian pilot study. *Transportation Research Part A: Policy and Practice*. 2020;131:196–205.

[18] Schikofsky J, Dannewald T, and Kowald M. Exploring motivational mechanisms behind the intention to adopt mobility as a service (MaaS): Insights from Germany. *Transportation Research Part A: Policy and Practice*. 2020;131:296–312.

[19] Ageing and Mobility: A Grand Challenge [Internet]. 2019. Available from: https://ageing-better.org.uk/sites/default/files/2019-09/Ageing-and-mobility-grand-challenge.pdf

[20] Cirella GT, Bąk M, Kozlak A, Pawłowska B, and Borkowski P. Transport innovations for elderly people. *Research in Transportation Business & Management* [Internet]. 2019 [cited 2023 Aug 12];30:100381. Available from: https://www.academia.edu/90780748/Transport_innovations_for_elderly_people

[21] Ons.gov.uk. Internet users, UK – Office for National Statistics [Internet]. www.ons.gov.uk. 2021. Available from: https://www.ons.gov.uk/businessindustryandtrade/itandinternetindustry/bulletins/internetusers/2020

[22] Jittrapirom P, van Neerven W, Martens K, Trampe D, and Meurs H. The Dutch elderly's preferences toward a smart demand-responsive transport service. *Research in Transportation Business & Management*.

2019;30:100383. https://www.sciencedirect.com/science/article/abs/pii/S2210539519301002

[23] Schwartz R, Vassilev A, Greene KK, Perine L, Burt A, and Hall P. Towards a standard for identifying and managing Bias in Artificial Intelligence. NIST [Internet]. 2022. Available from: https://www.nist.gov/publications/towards-standard-identifying-and-managing-bias-artificial-intelligence.

[24] Mobility data justice. Mobilities [Internet]. 2023 [cited 2023 Aug 12]. Available from: https://www.sciencedirect.com/org/science/article/abs/pii/S1745010123000954

[25] Kriswardhana W, and Esztergár-Kiss D. A systematic literature review of Mobility as a Service: Examining the socio-technical factors in MaaS adoption and bundling packages. *Travel Behaviour and Society*. 2023;31:232–43.

[26] Ritchie H, and Roser M. Emissions by Sector [Internet]. Our World in Data. 2020. Available from: https://ourworldindata.org/emissions-by-sector

[27] Climate Change Committee. 2022 Progress Report to Parliament [Internet]. Climate Change Committee. 2022. Available from: https://www.theccc.org.uk/publication/2022-progress-report-to-parliament/

[28] Ritchie H. Which Form of Transport Has the Smallest Carbon Footprint? [Internet]. Our World in Data. 2020. Available from: https://ourworldindata.org/travel-carbon-footprint

[29] Labee P, Rasouli S, and Liao F. The implications of Mobility as a Service for urban emissions. *Transportation Research Part D: Transport and Environment* [Internet]. 2022 [cited 2022 Mar 24];102:103128. Available from: https://www.sciencedirect.com/science/article/pii/S1361920921004235

[30] Zong F, Li YX, and Zeng M. Developing a carbon emission charging scheme considering mobility as a service. *Energy* [Internet]. 2023 [cited 2023 Aug 11];267:126495. Available from: https://www.sciencedirect.com/science/article/abs/pii/S0360544222033813

[31] Habib MA, and Lynn R. Planning for connected, autonomous and shared mobility: A synopsis of practitioners' perspectives. *Procedia Computer Science*. 2020;170:419–26.

[32] Enoch M, and Potter S. MaaS (Mobility as a Service) market futures explored. *Transport Policy*. 2023;134:31–40. https://www.sciencedirect.com/science/article/pii/S0967070X2300032X

[33] Smith G, Sochor J, and Karlsson ICM. Mobility as a Service: Development scenarios and implications for public transport. *Research in Transportation Economics*. 2018;69:592–9.

[34] Zhao X, Andruetto C, Vaddadi B, and Pernestål A. Potential values of maas impacts in future scenarios. *Journal of Urban Mobility*. 2021;1:100005.

[35] Mårtensson HB, Larsen K, and Höjer M. Investigating potential effects of mobility and accessibility services using the avoid-shift-improve framework. *Sustainable Cities and Society* [Internet]. 2023 [cited 2023 Aug 11];96:104676. Available from: https://www.sciencedirect.com/science/article/pii/S2210670723002871

[36] Sunitiyoso Y, Wicaksono A, Pambudi NF, *et al.* Future of mobility in Jakarta Metropolitan Area: A multi-stakeholder scenario planning. *Transportation Research Interdisciplinary Perspectives* [Internet]. 2023 [cited 2023 Jun 1];19:100810. Available from: https://www.sciencedirect.com/science/article/pii/S25901 9822300057X#s0050

[37] Summary Translation: Hype Cycle for Transportation and Smart Mobility, 2022 [Internet]. Gartner. [cited 2023 Aug 20]. Available from: https://www.gartner.com/en/documents/4018051

Chapter 19
Conclusions

The history of MaaS reveals a broad diversity of digitalised transport services that go by the name. Services that identify as MaaS have sprung from many origins – cities desiring next-level transport that enables them to function better, transport operators (both public and private) seeking new customers or market share, technology innovators seeking markets for their innovations, and academics validating their research and modelling. MaaS has originated in public institutions and from private capital.

However, the convergence that has been charted has been towards MaaS as an instrument of local, regional, or even national communities. The rising trend involves technology innovators who are increasingly targeting public transport authorities, whilst cities and regions are exploring technological solutions to better engage and involve citizens in achieving policy and climate objectives.

Whilst these larger trends have shaped the last 20 years, there are still many granular developments that work within the larger frameworks and general direction of travel. There will always be diverse layers and operations that create tailored solutions for individuals or businesses. However, the basic fact about transport is that it is based on a set of constrained resources – road space, railways, vehicles, and time – within a larger constrained environment (space for homes, businesses, agriculture, and the broader natural environment on which all life depends).

Managing the apportionment of space between these competing needs requires a degree of organisation that is difficult to conceptualise without functioning governance. In an increasingly digital age, MaaS is a tool to help cities and regions communicate with citizens and visitors about the transport that is available to them. It can clarify choices and promote better choices. It is also a means for people to take charge of their travel, and evaluate the cost and convenience of their choice or adapt it to their particular needs.

MaaS is a concept that has captured significant interest from a number of parties. From policymakers seeking new solutions to long-standing transport problems to technology entrepreneurs seeking to disrupt the transport sector, there has been significant debate over the concept, numerous trials of it, and much in the way of academic research.

It is difficult to draw any definitive conclusions from the current state of MaaS because beyond the technical feasibility, and the fact that the technology for implementing MaaS exists even if it is unequal, assessing whether MaaS is truly effective is a complex task. Success depends on a number of complex factors outside of the direct control of the MaaS solutions themselves.

The future of MaaS is likely to be diverse and reflect the environment in which it is deployed. Transport systems have generations of investment embodied in them. There will be a certain amount of inertia reflecting local history wherever last mile (generally lightweight) modes and long distance (generally expensive and long term) infrastructure are combined.

As combined digitalised transport services become more ubiquitous, the term itself may well fade and it is unlikely to be called MaaS.

MaaS is possibly just the way that transport is organised in the future and it will look very different in different places.

What we have sought to articulate in this book is this complexity. Transport systems and the trips that are undertaken on them are inherently complex, and consequently, the types of services that will be provided will reflect this complexity.

Regardless of your own views on whether MaaS, however you define it, to date has been a success or the role you see it playing in the future of transport and our wider society, it is a concept that has become established, and consequently, it will be part of the future of transport over the coming years, and maybe decades.

To paraphrase the science fiction writer William Gibson, the future of transport – including MaaS – is here. It is just not evenly distributed.

Index

www.ingramcontent.com/pod-product-compliance
Lightning Source LLC
LaVergne TN
LVHW011845200926
848484LV00002B/402

* 9 7 8 1 8 3 9 5 3 1 4 3 9 *